WirelessMAN®

Inside the IEEE 802.16™ Standard for Wireless Metropolitan Area Networks

Carl Eklund
Roger B. Marks
Subbu Ponnuswamy
Kenneth L. Stanwood
Nico J.M. van Waes

Published by
Standards Information Network
IEEE Press

Trademarks and Disclaimers

IEEE believes the information in this publication is accurate as of its publication date; such information is subject to change without notice. IEEE is not responsible for any inadvertent errors.

Library of Congress Cataloging-in-Publication Data

WirelessMAN : Inside the IEEE 802.16 standard for wireless metropolitan area networks / Carl Eklund ... [et al.].

    p. cm. -- (IEEE standards wireless networks series)

Includes bibliographical references and index.

ISBN 0-7381-4842-3

1. Metropolitan area networks (Computer networks). 2. IEEE 802.16 (Standard) 3. Wireless communication systems. I. Eklund, Carl, 1970- II. Series.

TK5105.85.W57 2006
621.384--dc22

                2006041723

12906107

IEEE
3 Park Avenue, New York, NY 10016-5997, USA

IEEE, 802, and WirelessMAN are registered trademarks in the U.S. Patent & Trademark Office, owned by the Institute of Electrical and Electronics Engineers, Incorporated (www.ieee.org/).

IEEE Standards designations are trademarks of the IEEE (www.ieee.org/).

DOCSIS is a registered trademark of Cable Labs (http://www.cablelabs.org).

Wi-Fi is a registered trademark of the Wi-Fi Alliance (http://www.wi-fi.org/).

WiMAX Forum is a registered trademark of the WiMAX Forum (http://www.wimaxforum.org).

The following figures in this book are the property of Doceotech, Incorporated, and are reprinted with permission: 1-1, 3-5, 6-3 through 6-7, 6-9 through 6-11, 6-21, 6-22, and 9-4.

Jennifer McClain Longman, Managing Editor
Linda Sibilia, Cover Designer

Review Policy

Standards Information Network/IEEE Press publications are not consensus documents. Information contained in this and other works has been obtained from sources believed to be reliable and reviewed by credible members of IEEE Technical Societies, Standards Committees, and/or Working Groups, and/or relevant technical organizations. Neither the IEEE nor its authors guarantee the accuracy or completeness of any information published herein, and neither the IEEE nor its authors shall be responsible for any errors, omissions, or damages arising out of the use of this information.

Likewise, while the authors and publisher believe that the information and guidance given in this work serve as an enhancement to users, all parties must rely upon their own skill and judgment when making use of this information and guidance. Neither the authors nor the publisher assumes any liability to anyone for any loss or damage caused by any error or omission in the work, whether such error or omission is the result of negligence or any other cause. Any and all such liability is disclaimed.

This work is published with the understanding that the IEEE and its authors are supplying information through this publication, not attempting to render engineering or other professional services. If such services are required, the assistance of an appropriate professional should be sought. The IEEE is not responsible for the statements and opinions advanced in the publication.

The information contained in Standards Information Network/IEEE Press publications is reviewed and evaluated by peer reviewers of relevant IEEE Technical Societies, Standards Committees and/or Working Groups, and/or relevant technical organizations. The authors addressed all of the reviewers' comments to the satisfaction of both the Standards Information Network/IEEE Press and those who served as peer reviewers for this document.

The quality of the presentation of information contained in this publication reflects not only the obvious efforts of the authors, but also the work of these peer reviewers. The IEEE Press acknowledges with appreciation their dedication and contribution of time and effort on behalf of the IEEE.

To order IEEE Press Publications, call 1-800-678-IEEE.

Print: ISBN 0-7381-4842-3   SP1146

See other IEEE standards and standards-related product listings at http://standards.ieee.org/

Dedication

The authors dedicate this effort to the many members of, and participants in, the IEEE 802.16 Working Group on Broadband Wireless Access. Their tireless efforts since 1998 have led to the completion of a dozen standards with potential to bring the world closer together. At the very least, the project has already brought a world of participants closer together.

Acknowledgments

The authors acknowledge the detailed reviews provided by Dr. David R. Smith and Dr. Claude Weil of the (U.S.) National Institute of Standards and Technology (NIST), as well as the assistance of several anonymous reviewers.

Portions of this work were contributed by NIST and are not subject to copyright protection in the United States.

The authors further acknowledge Doceotech, Incorporated, which provided some of the figures in this book. These figures, reprinted with permission, are drawn from one of the company's courses providing training for engineers on wireless technologies.

Authors

Carl Eklund received his M.S. in engineering physics from Helsinki University of Technology in 1996. He joined the Communication Systems Laboratory of Nokia Research Center in 1998, working mainly on radio protocol design and standardization. In the IEEE 802.16 effort, he chaired the MAC Task Group that developed the IEEE 802.16 medium access control layer (MAC) protocol for IEEE Std 802.16-2001. He also served as the technical editor for the protocol implementation conformance statement (PICS) and test suite structure and test purposes (TSS&TP) specifications for IEEE Std 802.16-2001. Eklund currently is a principal engineer in the Radio Communications Laboratory of Nokia Research Center, Helsinki, Finland. Since October 2005, he has been heading the research and standardization program for WiMAX and IEEE 802.16 in Nokia.

Roger B. Marks initiated, in 1998, the effort leading to the formation of the IEEE 802.16 Working Group on Broadband Wireless Access, chairing it since inception and serving as Technical Editor of the group's first two standards. He also serves actively on the IEEE 802 Executive Committee and holds the position of China Liaison Official. Marks is a physicist with the (U.S.) National Institute of Standards and Technology (NIST) in Boulder, Colorado, USA. He received his A.B. in physics in 1980 from Princeton University and his Ph.D. in applied physics in 1988 from Yale University. A Fellow of the IEEE and an IEEE Distinguished Lecturer, Marks developed the IEEE Radio and Wireless Conference and chaired it from 1996 through 1999. He is the author of over 90 publications and the recipient of numerous awards, including the Individual Governmental Vision Award from the Wireless Communications Association and the IEEE Technical Field Award in measurement technology. He has received the U.S. Department of Commerce Gold, Silver (three times), and Bronze Medals.

Subbu Ponnuswamy was one of the early participants in the IEEE 802.16 Working Group and a contributor to the IEEE 802.16 and IEEE 802.11 standards. He is also a coauthor of a WiMAX course for development engineers, offered by Doceotech. He has many years of industry experience in the design and development of wireless local area network (LAN) and metropolitan area network  (MAN) products, including those based on the IEEE 802.16 and IEEE 802.11 standards. As the director of engineering at Kiwi Networks, Ponnuswamy led the design and development of interference-resilient IEEE 802.16 and IEEE 802.11 systems in the license-exempt bands for indoor and outdoor applications. He also led IEEE 802.11 MAC application-specific integrated circuit (ASIC) and software development at Vivato for smart antenna systems. During his tenure at Malibu Networks, he designed and developed a quality-of-service-centric broadband wireless MAC. He has also held various technical positions with Honeywell, Sequent Computer Systems, and Lincom Wireless. He is currently with Aruba Networks. Ponnuswamy is the author of many publications and patents in the areas of wireless communication, real-time systems, and multiprocessor communication networks. He graduated with an M.S. in computer engineering from Wayne State University and a B.E. in electronics and communication engineering from the University of Madras, India. He is a member of the Association for Computing Machinery (ACM) and the IEEE Communications Society.

Kenneth L. Stanwood is president and chief executive officer of Cygnus Communications, which makes products for wireless multimedia distribution. He was previously chief technology officer of Ensemble Communications, which produced local multipoint distribution services (LMDS) equipment and provided key technology to  IEEE 802.16 and WiMAX. As a representative of Ensemble, Stanwood was one of the founders of the WiMAX Forum® and served on its board of directors. Stanwood is vice-chair of the IEEE 802.16 Working Group and has been involved with IEEE 802.16 and the European Telecommunications Standards Institute (ETSI) Broadband Radio Access Networks (BRAN) Technical Committee for over 6 years. He was a primary designer of the

IEEE 802.16 MAC. He holds 11 patents and has numerous patent applications, all related to broadband wireless access. He received his master's degree from Stanford University.

Nico J.M. van Waes received an M.S.E.E from the Technical University Delft in the Netherlands in 1994 and a Ph.D. from New Jersey Institute of Technology in 1998. He joined the Wireless Router Division of Nokia Networks in 1999 as a systems engineer, working primarily on physical layer (PHY) and radio frequency (RF) issues as well as standardization. From 1999 till 2004, van Waes held various standards-related public positions such as chief technical editor of IEEE Std 802.16a™, IEEE P802.16.2a, and early versions of IEEE P802.16d; area coordinator and editor for ETSI BRAN HiperMAN; and chair of the OFDM Forum's fixed wireless access (FWA) working group. From 2004 till 2005, he led Nokia Research Center's efforts in IEEE P802.11n standardization. Since early 2006, van Waes has been a manager with Nokia IPR, responsible among others for the IEEE 802.16 and IEEE 802.11 portfolios. He has half a dozen patents filed and is the author of several published papers.

Foreword

Because of the explosive global growth of the Internet and its applications, this authoritative, comprehensive, and well-written insider's guide to broadband wireless standards is truly a book whose time has come. Published by the Standards Information Network/IEEE Press, a unit of the Institute of Electrical and Electronics Engineers (IEEE), *WirelessMAN* serves as a companion to IEEE Std 802.16, the WirelessMAN standard for broadband wireless metropolitan area networks. This book provides an overview of the technology, explaining the rationale behind the choices made. This overview of the complex technical specification is valuable for the tens of thousands of engineers involved in R&D and deployment of relevant technologies—as well as to less technical opinion leaders in related fields of marketing, sales, finance, government affairs, and media/public relations. *WirelessMAN* is so clearly organized and written that it can be a rewarding reading experience even for those with no formal technical training.

The authors are and have been leaders of the related standardization projects. IEEE Std 802.16 was developed by the IEEE 802.16 Working Group on Broadband Wireless Access, beginning in 1999. The book's authors include the IEEE 802.16 founding chair, Dr. Roger B. Marks; its vice chair, Kenneth L. Stanwood; and three of its most visionary and dedicated practitioners, Carl Eklund, Subbu Ponnuswamy, and Nico J.M. van Waes. Those involved in standards work can appreciate that it is a process whose success depends upon far more than just vision and technical expertise. That's only the beginning. To be as successful as the WirelessMAN standards effort has been, it's necessary also to draw on the skill sets commonly associated with master politicians and corporate strategists because the stakes can be so huge—especially in this instance, where the goal is nothing short of gradually replacing multibillion-dollar proprietary technologies with lower cost, standards-based interoperable products. If the upside is high for some players, so is the downside for others. In other words, creating a successful standard is not in everyone's interest, and so whatever method of collegial input is used may also be abused.

The IEEE 802.16 Working Group has made impressive progress in its standards effort, as measured by objective data. It has held more than 43 weeklong sessions around the world, from the group's official founding in 1999 to the publication of this book in 2006. By working cooperatively, the group has developed and advanced its standards in near-record time. The current version of its fixed wireless standard was most recently updated as IEEE Std 802.16-2004 [B20], and an amendment adding mobility support (IEEE Std 802.16e™-2005 [B24]) was published in late 2005. The IEEE 802.16 standard is at the core of the fast-growing certification effort by the WiMAX Forum®, whose expanding activities have been generating news and magazine articles at the rate of six hundred a month at times.

By attending the organizational strategy meeting in July 1998 of the predecessor group that led to the IEEE 802.16 Working Group, I had the privilege of observing the growth of this process right from the beginning despite my lack of any formal training in the relevant technology. Fortunately, I did have good advice from my mentor in this area, Dr. Weston E. Vivian, the longtime University of Michigan professor who holds the distinction of being the first person ever elected to the U.S. Congress with a doctorate in either engineering or science. His counsel was to "sit in on every technical meeting you can – and don't say much." The assumption was that over time, even I would learn something. That strategy meeting was called by Roger Marks at the Boulder, Colorado, offices of the National Institute of Standards and Technology (NIST) to create what became known as the National Wireless Electronic Systems Testbed before its assimilation the next year into the IEEE standards process. The few others at the meeting had enormous subject-matter expertise, most with doctorates in electronic engineering and affiliations with major companies. Despite the additional handicap of not being in a position to make substantive commitments because Wireless Communications Association International (WCA) is technology neutral, I attended as an observer, encouraged by knowledge that standards were regarded by our members as potentially vital, but also extremely difficult to achieve in practice.

We at WCA have been extremely impressed by the dedication of the WirelessMAN pioneers, including its founding chair. Although WCA continues to adhere to a technology-neutral standpoint in order to showcase all relevant technologies on a fair basis, it seems appropriate nevertheless to share a few personal observations at this juncture: First, the leadership of this group is particularly dedicated, well-organized, and enthusiastic about creating new networks. These standards meetings are all-day, weeklong affairs, and more than once I have called upon leaders in the middle of their night to learn of developments. Second, and more important, *WirelessMAN*, an authoritative insider's account, is certain to prove valuable as a guide to understanding the history and implications of the technology—not just the IEEE 802.16 version, but in the context of competing and complementary broadband technologies, both wireless and wireline. The book is highly recommended for designers of components, systems, and test equipment based on the WirelessMAN standards, plus a variety of other telecommunications professionals and students. This is an important and useful book for anyone in this fast-exploding field.

Andrew Kreig
President and Chief Executive Officer
Wireless Communications Association International
1333 H Street, NW; Suite 700 West
Washington, DC 20005
United States of America

Preface

Broadband access has become a critical bottleneck in the ongoing information technology revolution across the globe. The growing popularity of multimedia applications has made the requirements on broadband access even more demanding. In order to serve this growing market need, the reach of existing broadband networks needs to be extended, and new broadband infrastructures need to be deployed in developing nations where no wired broadband infrastructures exist.

The advent of standards-based and interoperable wireless local area networks (WLANs) and wireless personal area networks (WPANs) has resulted in widespread adoption of wireless networks in homes as well as in consumer and enterprise markets. The productivity gains and increased flexibility that comes with wireless have transformed the way many businesses operate.

Broadband wireless access (BWA) is the next logical step in extending fiber optic backbone networks to enhance broadband proliferation and to provide broadband access in areas where no other methods are available. BWA is also attractive as a network infrastructure that can be deployed much more quickly than its wired counterparts. Moreover, broadband wireless has the potential to offer capacity on demand.

Proprietary BWA solutions have existed for some time, with limited market penetration in licensed and license-exempt bands. However, the lack of a global standard and lack of multivendor interoperable equipment have hampered widespread adoption. Given the infrastructure-intensive nature of broadband wireless deployments, widespread adoption requires interest from a wide range of service providers. All of the potential providers, large or small, are seeking standardized rather than proprietary solutions. With nomadic and mobile systems, roaming operations force questions of cross-vendor interoperability to the forefront. However, even for non-nomadic fixed broadband wireless networks, in which roaming is virtually nonexistent, operators are hesitant to accept unnecessary risk in the current economic climate, and the use of proprietary equipment from a small number of vendors is an unnecessary risk. Good standards activities, such as those that are well established in IEEE 802®, not only bring operators reasonably priced equipment but also put them on a curve of declining costs and growing performance. The IEEE 802.16 standard for wireless metropolitan area networks (MANs) has been developed by the IEEE 802.16 Working Group to address this BWA market. The activity has been in progress since 1999; it has been highly productive, and the standard continues to evolve.

As you will find in this book, IEEE 802.16 is quite different from other IEEE 802 standards. The technology is more oriented toward carrier-class services whose infrastructure development is designed and deployed by professionals affiliated with experienced telecommunications operators. Also, compared to some other IEEE 802 standards, IEEE 802.16 leaves more of the implementation decisions to the implementor, particularly at the base station (BS) side. This provides the system developer with the opportunity to develop a simple, low-cost system or to apply extra effort to create a substantially better performing system that may fetch a higher price. The goal is a wide deployment of a broad range of IEEE 802.16-based systems that are interoperable but highly differentiated. This should help to keep the market dynamic and innovative and provide many attractive options to the carriers.

Objective

The objective of this book is to provide a detailed overview of the IEEE 802.16 standard for fixed BWA, including detailed descriptions of the IEEE 802.16 medium access control layer (MAC) and physical layer (PHY) operation, with emphasis on the design and the technology behind the standard. This book explains why certain design choices were made and how recent technological developments, real-world experience, and lessons learned from previous projects were used throughout the IEEE 802.16 development process to make critical tradeoffs.

The book also reviews the chronological development of the IEEE 802.16 MAC and PHY and the rationale behind the development decisions. It provides a summary of ongoing projects, related standards, and future extensions to IEEE 802.16. The optional and mandatory parts of the

IEEE 802.16 standard are clearly identified and explained throughout the book, wherever applicable.

Organization

Chapter 1 provides an introduction to BWA and identifies the key market segments and basic requirements. Chapter 2 overviews the IEEE 802.16 Working Group, including the organizations under which it works and the procedures it uses, summarizing the history of the group and its past and current projects. Chapter 3 defines the basic MAC, PHY, radio frequency (RF) and other protocol concepts necessary for understanding the IEEE 802.16 standard. The final introductory chapter, Chapter 4, describes the overall architecture of IEEE 802.16 and introduces various components and key features of IEEE 802.16 in detail.

The rest of the book describes the components of the IEEE 802.16 architecture in a top-down fashion. Chapter 5 describes the asynchronous transfer mode (ATM) convergence sublayer (CS) and packet convergence sublayer (PCS) of IEEE Std 802.16. In Chapter 6, the basic concepts of the IEEE 802.16 MAC, supporting all of the IEEE 802.16 PHY alternatives, are covered. Chapter 7 provides details of the IEEE 802.16 MAC operation, including network entry, initialization, PHY support, automatic repeat request (ARQ) and quality of service (QoS). The security sublayer, including encryption methods and key exchange mechanisms, is described in Chapter 8. Chapter 9 describes the mesh extensions to the IEEE 802.16 MAC and the additional scheduling methods defined to support the mesh option.

The next three chapters describe three major PHY alternatives specified in IEEE 802.16. Chapter 10 addresses the single-carrier (WirelessMAN-SC) PHY specified for use above 10 GHz. Chapter 11 describes the orthogonal frequency division multiplexing (WirelessMAN-OFDM) PHY for frequencies below 11 GHz, and Chapter 12 describes the orthogonal frequency division multiple access (WirelessMAN-OFDMA) PHY for the same frequencies.

Chapter 13 explores the IEEE 802.16 standard's support for multiple antenna systems, including adaptive antenna systems (AAS), as an extension of the IEEE 802.16 point-to-multipoint (PMP) topology. Chapter 14 addresses the performance of the MAC and various PHYs. A summary of the IEEE 802.16 conformance standards and an introduction to the interoperability work being done by the WiMAX Forum are given in Chapter 15. Chapter 16 provides a summary of related standards and standardization activities.

Appendix A provides a list of IEEE 802.16 headers, subheaders, and MAC management messages for ready reference.

Reading and style

This book is intended for anyone who wishes to understand the IEEE 802.16 standard and operation, including designers, engineers, students, and deployment professionals. This book is self-contained, and no specific knowledge of any other wireless protocols is assumed. However, a basic knowledge of communication protocols and concepts is necessary for understanding the key MAC and PHY chapters of this book. An extensive bibliography is given at the end for anyone who would like to read more on specific topics that are not covered here in detail.

While the introductory chapters provide basic information for all types of readers, the MAC- and PHY-specific chapters can be treated independently, with the exception of the MAC-PHY interface, by designers who are interested in designing only the MAC or PHY portion of a IEEE 802.16 system. The mesh and AAS chapters require the understanding of the MAC architecture and PHY interface defined in earlier chapters. This book can be used for self-study of the IEEE 802.16 standard, as a reference, or as a designer's handbook.

Advanced readers should look at the book as a companion to IEEE Std 802.16. Readers should note that IEEE 802 standards are available for free download through the Get IEEE 802® program <http://standards.ieee.org/getieee802>. Readers may also wish to consult the web pages of the IEEE 802.16 Working Group on Broadband Wireless Access <http://WirelessMAN.org>, which include thousands of working group and contributed documents upon which the standard, including its amendments, are based.

Table of Contents

List of Figures. xxvii

List of Tables. xxxi

Acronyms and abbreviations. xxxiii

Chapter 1 Broadband wireless access (BWA):
Applicable market segments and requirements. 1
    Commercial fixed broadband wireless: fiber extension. 6
    Residential fixed broadband wireless: digital subscriber line (DSL) and
        cable modem alternative 8
    Quality of service (QoS) 11
    Throughput requirements. 11

Chapter 2 IEEE 802.16 standards:
The working group and documents. 13
    Background 13
    IEEE Standards Association (IEEE-SA) 13
    IEEE 802® LAN/MAN Standards Committee (LMSC). 15
    Standards development in IEEE 802 16
        Study group stage. 16
        Working group development of draft. 16
    IEEE 802.16 Working Group: Overview. 19
    IEEE 802.16 Working Group: History 19
    Technical progress in IEEE 802.16 Working Group. 21
        Air interface: IEEE Std 802.16 23
        Conformance: IEEE 802.16/Conformance0X 26
    Coexistence: IEEE Std 802.16.2™ 27

Chapter 3 Basic concepts and definitions:
Wireless protocol and communication concepts. 29
    Frequency bands 30
        Channels. 30
        Licensed and license-exempt spectrum 30
        Spectrum and standardization 32
        Coexistence 32
    Types of wireless networks 33

Fixed and mobile networks 33
Nomadic and portable networks 34
Wireless network topologies 34
RF propagation 36
   LOS and NLOS 36
   Multipath 38
   Fading 39
Antennas 40
   Antenna parameters 41
   Directional and sectorized antennas 42
   Diversity 42
      Temporal diversity 43
      Frequency diversity 43
      Spatial diversity 43
      Polarization diversity 44
      Angle diversity 45
   Multiple antenna systems 45
      Adaptive antenna systems (AAS) 45
      Multiple-input, multiple-output (MIMO) 46
   Impact of antenna technologies on protocol design 47
   Antenna design for fixed and mobile devices 47
Physical layer (PHY) 48
   Forward error correction (FEC) 48
   Single carrier and multicarrier 49
Duplexing, multiplexing, and multiple access 49
   Duplexing 50
   Multiplexing 52
   Centralized and distributed multiple access schemes 52
      Time division multiple access (TDMA) and
         frequency division multiple access (FDMA) 53
      Orthogonal frequency division multiple access (OFDMA) 53
      Code division multiple access (CDMA) 55
Data units 56
Quality of service (QoS) 57
   Per-flow and per-class QoS 60
   Wireless QoS 61
Medium access control layer (MAC) 62
   Fragmentation and packing 62
   Automatic repeat request (ARQ) 64

Chapter 4 IEEE 802.16 architecture:

Overview and key features 67
    Reference model 68
    Base station (BS) and subscriber station (SS) 71
    Convergence sublayer (CS) architecture 73
    Framing and duplexing 74
        Physical slots (PS) and mini-slots 76
        TDD framing 76
        FDD framing 78
        Time relevance 81
    Subscriber-level adaptive PHY 84
        Fixed TDD vs ATDD 84
    Framed PHY 85
    MAC efficiency 85
    Mesh 87
    Directed mesh 88
    Quality of service (QoS) 88
    Security sublayer 89
    Automatic repeat request (ARQ) 90
        Hybrid automatic repeat request (HARQ) 92
    Physical layer (PHY) 93
        Multipath mitigation 94
    Mandatory and optional components 94
    Bit ordering 98

Chapter 5 Convergence sublayers (CSs):

Support for multiple protocol transport 99
    ATM CS 99
    Packet convergence sublayer (PCS) 101
        Classification 102
        Payload header suppression (PHS) 104

Chapter 6 MAC basics:

Concepts, connections, formats, and headers 107
    Connections and addressing 107
        Service flows and service flow identifiers (SFIDs) 109
        CID allocation 109
    MAC headers and subheaders 113
        Stand-alone MAC headers 113
            BW request header 114
        MPDU header 115

Generic MAC header. 115
MAC header demultiplexing. 118
MAC subheaders. 119
   Fragmentation subheader (FSH) 120
   Packing subheader (PSH) 121
   Grant management subheader (GMSH) 122
   Mesh subheader (MSH). 123
   Fast-feedback allocation subheader (FFSH) 124
ARQ feedback. 125
Data and management PDU construction 125
Simple MPDU. 125
Subheader ordering 126
ARQ blocks. 127
Fragmentation 129
   Fragmentation without ARQ. 130
   Fragmentation with ARQ 130
Packing 131
   Packing of fixed-length SDUs. 131
   Packing of variable-length SDUs 132
   Packing with fragmentation. 132
   Packing and ARQ 134
   Packing of ARQ payload. 134
Concatenation 135
MPDU encryption and CRC 135
MAC management. 136
ARQ 137
ARQ block-based retransmissions 137
ARQ Feedback information element (IE) 139
ARQ feedback payload 143
Hybrid automatic repeat request (HARQ) 143
Compact MAP IE 144
   HARQ Control IE 144
Construction of HARQ packets. 145
Reduced connection identifier (RCID) 145
HARQ acknowledgments 147

Chapter 7 MAC operation:
Radio control, QoS, and ARQ 149
Network entry and initialization 149
Scanning and synchronization to the DL. 149
Initial ranging 149

SS basic capability negotiation 150
Authorization, security association (SA) establishment, and
   key exchange 152
Registration 153
Establish IP connectivity 153
Dynamic service establishment 154
PHY maintenance 155
MAPs and channel descriptors................................ 156
Periodic ranging.. 163
Burst profile changes 167
QoS and service flows 171
Dynamic service establishment and deletion 171
QoS model 172
QoS and traffic parameters................................. 173
   SFID and CID.. 173
   Service class name 173
   QoS parameter set type.............................. 174
   Traffic priority 176
   Maximum sustained traffic rate 176
   Maximum traffic burst 177
   Minimum reserved traffic rate 177
   Minimum tolerable traffic rate 177
   Vendor-specific QoS parameters 178
   Service flow scheduling type 178
   Request/transmission policy.......................... 179
   Tolerated jitter 179
   Maximum latency.................................... 179
   Fixed-length vs variable-length SDU indicator 180
   SDU size... 180
Interactions between QoS, CAC, and adaptive PHY............. 181
   Determining available bandwidth...................... 181
   Bandwidth on demand setting the basis for CAC 182
   Adaptive CAC philosophies for adverse conditions 187
Multicast connections...................................... 189
BW request/grant 190
Scheduling 193
Unicast polling... 197
Broadcast polling 199
Multicast polling groups................................... 200
Clock comparison 201
ARQ operation.. 203
ARQ parameters and timers................................. 203

ARQ protocol messages 206
    ARQ Feedback management message. 206
    ARQ Discard management message 206
    ARQ Reset management message. 206
BSN comparison 207
ARQ transmitter 207
    ARQ feedback processing 210
ARQ receiver. 210
    ARQ feedback generation 213
ARQ state machine reset and resynchronization. 213
Interaction with scheduler 215
HARQ operation 216
HARQ parameters 216
Subpacket transmission and acknowledgment generation 216
Performance and QoS implications. 217

Chapter 8 Security:
PKM protocol and cryptographic methods 219
Security associations (SAs) and cryptographic suites. 219
    Encrypted MPDUs. 220
    Data encryption with DES in CBC mode 220
    AES-CCM 221
Key management. 224
    AK management 224
    TEK management 226

Chapter 9 Mesh:
MAC and PHY extensions for mesh. 229
Introduction. 229
Logical mesh 232
    Mesh connections and addressing 234
    Network configuration. 234
    Network entry 236
        Scanning for active networks and coarse synchronization 237
        Obtaining network parameters. 237
        Opening sponsor channel. 238
    Distributed scheduling. 239
    Centralized scheduling 241
Directed mesh and point-to-point (PtP) 244

Chapter 10 PHY: WirelessMAN-SC:
Single-carrier PHY for 10–66 GHz. 247

    Frame structure 248

    DL channel encoding 252

    UL channel encoding 256

    Control mechanisms. 259

    WirelessMAN-SCa 260

Chapter 11 PHY: WirelessMAN-OFDM:
Multicarrier PHY for frequencies below 11 GHz 261

    Waveform construction 261

        Selection of the OFDM waveform. 261

        Selection of FFT size 262

        The OFDM waveform 265

        Subchannelization 266

        Preambles 270

    Frame structure 272

        Point-to-multipoint (PMP). 272

        Mesh. 275

    Channel encoding. 277

        Randomization 277

        Forward error correction (FEC). 278

            Reed-Solomon concatenated with convolutional coding (RS-CC) . 278

            Block turbo codes (BTCs) 279

            Convolutional turbo codes (CTCs). 280

        Interleaving. 281

        Modulation 281

    Control mechanisms. 282

        Ranging 282

        BW requests 284

        Power control 285

Chapter 12 PHY: WirelessMAN-OFDMA:
Multicarrier PHY for frequencies below 11 GHz 287

    Introduction 288

        Selection of OFDMA waveform 288

        Time-frequency mapping. 290

        Permutation examples 294

            FUSC 294

            PUSC, DL. 296

            PUSC, UL. 298

Frame structure 300
  Point-to-multipoint (PMP). 300
  Preambles 301
Channel encoding 302
  Randomization. 302
  Forward error correction (FEC). 302
    Convolutional coding (CC) 303
    BTCs and CTCs. 303
    LDPCC 303
  Interleaving 305
  Modulation 305
Control mechanisms 306
  Fast-feedback. 306
  CDMA ranging and BW requests 306
  PAPR reduction/safety zone 308
  HARQ support. 309

Chapter 13 Multiple antenna systems:

Support for advanced antennas 311
  Adaptive antenna systems (AAS) 311
    AAS support in IEEE Std 802.16 313
      DL and UL framing. 314
    MAC service and control functions. 316
      AAS MAC management messages 316
      Channel state information 317
      AAS DL synchronization and initial ranging 317
      AAS BW requests 318
      AAS support in OFDM PHY. 319
      AAS support in OFDMA PHY 320
  Open-loop transmit diversity. 321
    STC support in OFDM PHY 322
    STC, FHDC, and MIMO support in OFDMA PHY 323
  Closed-loop transmit diversity 324
    Precoding. 325
    Antenna selection 326
    Antenna grouping 326

Chapter 14 Performance analysis:

MAC and PHY performance and throughput 327
  Introduction. 327
  WirelessMAN-OFDM, fixed operation 327

Capacity analysis 327
MAC performance 329
   Full-duplex with single burst profile 331
   Mixed full-duplex/half-duplex with burst profiles case 335
WirelessMAN-OFDM, mobile operation 338
Basic PHY performance 338
Capacity analysis 340
WirelessMAN-OFDMA, mobile operation 343
Overhead and capacity 343
WirelessMAN-OFDMA vs high-speed DL packet access (HSDPA) . . 348
Basic WirelessMAN-OFDMA PHY performance 349
MIMO capacity 355

Chapter 15 Conformance and interoperability:
Conformance standards and testing 357

The WiMAX Forum 357
Conformance test standards for WirelessMAN-SC 360
The WiMAX Forum's move to lower frequencies 360
Lower frequency profiles and test specifications 361
WiMAX Forum conformance testing for fixed access 362
WiMAX Forum and mobile broadband wireless access 363

Chapter 16 Related standards:
Other wireless standards with similar applications 365

IEEE Std 802.11 365
   IEEE 802.11 MAC 366
      Medium access control layer (MAC) 366
      MAC overhead 367
      MAC summary 368
   IEEE 802.11 PHY 369
   IEEE 802.11 extensions 370
   Using IEEE Std 802.11 as a MAN 371
   Quantitative comparison of IEEE Std 802.11 and IEEE Std 802.16 . . . 371
IEEE 802.20 Working Group 373
IEEE 802.22 Working Group 374
ETSI BRAN 375
   ETSI BRAN HiperACCESS 376
   ETSI BRAN HiperMAN 377
Other regional standards activities 378
   Korean Telecommunication Technology Association (TTA) and
     WiBro 378

China Communications Standards Association (CCSA) 379
International Telecommunications Union (ITU). 379

Appendix A IEEE 802.16 headers, subheaders, and management messages **383**

Bibliography 387

Index 393

List of Figures

Figure 1–1: Typical IEEE 802.16 deployment scenarios 7
Figure 2–1: IEEE 802.16 Working Group attendance 20
Figure 2–2: IEEE 802.16 project timeline 22
Figure 3–1: Fresnel zone. 37
Figure 3–2: TDD illustration. 50
Figure 3–3: Full-duplex FDD illustration 51
Figure 3–4: H-FDD illustration. 51
Figure 3–5: OFDM and OFDMA 54
Figure 3–6: PDUs, SDUs and SAPs 57
Figure 4–1: IEEE 802.16 reference model 69
Figure 4–2: IEEE 802.16 framing. 74
Figure 4–3: PMP TDD frame structure. 77
Figure 4–4: PMP FDD frame structure. 79
Figure 4–5: H-FDD framing 80
Figure 4–6: Full-duplex FDD framing 80
Figure 4–7: Maximum time relevance of UL and DL MAPs for TDD 81
Figure 4–8: Minimum time relevance of UL and DL MAPs for TDD. 82
Figure 4–9: Maximum time relevance of DL and UL MAPs for FDD. 83
Figure 4–10: Minimum time relevance of DL and UL MAPs for FDD 83
Figure 5–1: ATM CS PDU format 100
Figure 5–2: CS PDU format for VP-switched ATM connections 100
Figure 5–3: CS PDU format for VC-switched ATM connections 100
Figure 5–4: MSDU format for PCS 101
Figure 5–5: Classification and CID mapping (SS to BS) 102
Figure 5–6: PHS operation, with masking 105
Figure 6–1: BW request header. 114
Figure 6–2: Generic MAC header. 116
Figure 6–3: FSHs 120
Figure 6–4: PSHs 122
Figure 6–5: GMSHs 123
Figure 6–6: MSH. 124
Figure 6–7: FFSH 124
Figure 6–8: MPDU structure. 125
Figure 6–9: MPDU with subheaders. 126
Figure 6–10: Ordering of MAC subheaders 128
Figure 6–11: ARQ blocks 129

Figure 6–12: MPDU with fragmentation. 130
Figure 6–13: Packing of fixed-length SDUs 131
Figure 6–14: Packing of variable-length SDUs. .132
Figure 6–15: Packed MPDU with two fragments 133
Figure 6–16: Packed MPDU with one fragment .133
Figure 6–17: ARQ feedback and packing 134
Figure 6–18: Concatenation of MPDUs 135
Figure 6–19: Format of the MAC management message 136
Figure 6–20: ARQ blocks and retransmissions. .138
Figure 6–21: ARQ Feedback IE 140
Figure 6–22: Block sequence acknowledgment MAP formats 143
Figure 6–23: Construction of HARQ packet. 145
Figure 6–24: RCID decoding. 146
Figure 7–1: SS basic capability negotiation 151
Figure 7–2: SS registration 154
Figure 7–3: DSA message flow (BS-initiated) 155
Figure 7–4: TDD frame format 157
Figure 7–5: FDD frame format 158
Figure 7–6: FDD DL subframe structure. 159
Figure 7–7: Practical TDD UL minimum-maximum relevance 161
Figure 7–8: FDD logical offset 162
Figure 7–9: Periodic ranging opportunity allocation at BS. 164
Figure 7–10: Periodic ranging receiver processing at BS 165
Figure 7–11: Periodic ranging at SS 166
Figure 7–12: CDMA periodic ranging. 168
Figure 7–13: Transition to more robust burst profile 170
Figure 7–14: Transition to less robust burst profile. 170
Figure 7–15: DSA message flow—SS initiated .171
Figure 7–16: Provisioned authorization model 174
Figure 7–17: Dynamic authorization model 175
Figure 7–18: Two terminals with same planned PHY mode. 189
Figure 7–19: Clock comparison. 202
Figure 7–20: ARQ block states at the transmitter 208
Figure 7–21: ARQ block reception 211
Figure 7–22: Transmitter-initiated reset 214
Figure 7–23: Receiver-initiated reset. 215
Figure 8–1: Encrypted payload format in AES-CCM mode 222
Figure 8–2: Initial CBC block and nonce 223

Figure 8–3: Counter blocks. 223
Figure 8–4: AK management in BS and SS 225
Figure 8–5: TEK management in BS and SS 227
Figure 9–1: Three BSs required for full coverage. 231
Figure 9–2: Two BSs required with mesh SS. 231
Figure 9–3: Neighborhood definitions 233
Figure 9–4: Mesh CIDs. 234
Figure 9–5: Mesh network entry: Opening a sponsor channel 238
Figure 9–6: Distributed scheduling. 241
Figure 9–7: Centralized scheduling. 242
Figure 9–8: Addition of PtP 245
Figure 9–9: Physical or directed mesh 246
Figure 10–1: TDD DL subframe structure 249
Figure 10–2: FDD DL subframe structure 251
Figure 10–3: DL MAP usage with shortened FEC blocks—TDM case. . . 253
Figure 10–4: Format of DL transmission CS PDU. 254
Figure 10–5: Conceptual block diagram of DL PHY 255
Figure 10–6: UL subframe structure. 257
Figure 10–7: Conceptual block diagram of UL PHY 258
Figure 11–1: OFDM symbol structure 263
Figure 11–2: OFDM frequency description 265
Figure 11–3: Subchannel subcarrier allocations 268
Figure 11–4: DL and network entry preamble structure. 271
Figure 11–5: Example OFDM structure with TDD 273
Figure 11–6: DL burst and MPDU mapping. 274
Figure 11–7: Mesh frame structure. 276
Figure 11–8: RS-CC encoding process. 279
Figure 11–9: BTC encoding block 280
Figure 11–10: CTC encoder 280
Figure 11–11: Initial ranging formats 283
Figure 12–1: Schematic indication of logical mapping elements 291
Figure 12–2: FUSC mapping example 295
Figure 12–3: PUSC DL mapping example 297
Figure 12–4: PUSC UL mapping example 299
Figure 12–5: Frame structure (TDD) 301
Figure 12–6: General shift index matrix structure 304
Figure 12–7: Single-slot initial ranging transmission for OFDMA 307
Figure 13–1: Example beam-forming network. 312

Figure 13–2: TDD framing for AAS Systems 315
Figure 13–3: FDD framing for AAS systems 315
Figure 13–4: Open-loop transmit diversity schemes 321
Figure 13–5: Four-antenna STC and SM 324
Figure 14–1: WirelessMAN-OFDM throughput per 5 MHz channel 328
Figure 14–2: WirelessMAN-OFDM outage relative to 1.5 Mbit/s 329
Figure 14–3: Average throughput delay vs offered load 330
Figure 14–4: DL, IP traffic mean transfer delay 332
Figure 14–5: UL, IP traffic mean transfer delay 332
Figure 14–6: UL, Average transmit opportunities per frame 333
Figure 14–7: UL load and utilization of transmit opportunities 334
Figure 14–8: FCH + MAP overhead 334
Figure 14–9: UL, full-duplex SSs, VoIP mean transfer delay 335
Figure 14–10: UL, half-duplex SSs, VoIP mean transfer delay 336
Figure 14–11: DL, half-duplex SSs, HTTP mean transfer delay 337
Figure 14–12: UL, half-duplex SSs, HTTP mean transfer delay 337
Figure 14–13: UL, half-duplex SSs, HTTP mean transfer delay w/ARQ . . 338
Figure 14–14: OFDM CBLERs—ITU VehA 3 km/h 339
Figure 14–15: OFDM CBLERs—ITU VehA 30 km/h 340
Figure 14–16: Channel utilization as function of offered load 341
Figure 14–17: Modulation/coding utilization as function of offered load . . 341
Figure 14–18: Call blocking (C/I < 1 dB) as function of offered load 342
Figure 14–19: DL sector capacity per cell radius—ITU VehA (60 km/h) . 344
Figure 14–20: UL sector capacity per cell radius—ITU VehA (60 km/h) . 344
Figure 14–21: DL sector capacity per cell radius (ITU outdoor-indoor) . . 345
Figure 14–22: UL sector capacity per cell radius (ITU outdoor-indoor) . . 345
Figure 14–23: Delay for VoIP using rtPS, UGS, and ertPS 347
Figure 14–24: Average cell throughput for various schemes 347
Figure 14–25: CBLER, UL PUSC, ITU VehA 3 km/h, CC, ideal 349
Figure 14–26: CBLER, UL PUSC, ITU VehA 3 km/h, CC, estimated 350
Figure 14–27: CBLER, UL PUSC, ITU VehA 120 Km/h, CC, ideal 351
Figure 14–28: CBLER, UL PUSC, ITU VehA 120 km/h, CC, estimated . . 351
Figure 14–29: CBLER, UL PUSC, ITU VehA 3 km/h, CTC, ideal 352
Figure 14–30: CBLER, DL PUSC, ITU VehA 3 km/h, CC, ideal 353
Figure 14–31: CBLER, DL PUSC, ITU VehA 3 km/h, CC, ideal 354

List of Tables

Table 2–1: IEEE 802.16 members by geography, as of January 2006 21
Table 4–1: Key IEEE 802.16 features and their status 95
Table 6–1: CID allocation and well-known CIDs. 113
Table 6–2: BW request header fields 115
Table 6–3: Generic MAC header fields 117
Table 6–4: Generic HT encodings 118
Table 6–5: MAC header demultiplexing. 119
Table 6–6: FSH fields 121
Table 6–7: Fast-feedback allocation feedback type encodings. 124
Table 6–8: Subheader and special payload ordering. 127
Table 6–9: DSA-RSP message format 137
Table 6–10: Acknowledgment types. 141
Table 6–11: RCID field interpretation 147
Table 7–1: Mapping of ATM CoS concepts to WirelessMAN. 183
Table 7–2: ARQ parameters and timers 204
Table 12–1: Subchannel allocations 289
Table 14–1: Cell-ranges DL PUSC. 354
Table 14–2: Relative performance benefit of MIMO and virtual MIMO. . 355
Table 14–3: Performance using MIMO 356
Table 15–1: WirelessMAN-SC system profiles 359
Table 16–1: Comparison of IEEE Std 802.11 and IEEE Std 802.16 372
Table 16–2: IEEE 802.20 performance targets. 373
Table A–1: MAC headers, subheaders and special payloads 383
Table A–2: Management messages. 384

Acronyms and abbreviations

The following acronyms and abbreviations are used in this book:

3DES	triple data encryption standard
AAS	adaptive antenna systems
ABR	available bit rate
ACID	ARQ channel identifier
ACK	acknowledgment
AES	advanced encryption standard
AISN	ARQ identifier sequence number
AK	authorization key
AMC	adaptive modulation and coding
AoA	angle of arrival
AoD	angle of departure
AP	access point
ARQ	automatic repeat request
ASIC	application-specific integrated circuit
ATDD	adaptive time division duplexing
ATM	asynchronous transfer mode
ATS	abstract test suite
BCC	block convolutional code
BE	best effort
BER	bit error rate
BPSK	binary phase shift keying
BRAN	Broadband Radio Access Networks (ETSI Technical Committee)
BRS	broadband radio service
BS	base station
BSN	block sequence number
BTC	block turbo code

BW	bandwidth
BWA	broadband wireless access
CAC	call admission control
CAZAC	constant amplitude zero autocorrelation
CBC	cipher block chaining
CBLER	coded block error rate
CBR	constant bit rate
CC	convolutional coding
CCM	counter with CBC-MAC
CDMA	code division multiple access
C/I	carrier to interference ratio
CID	connection identifier
CINR	carrier to interference-plus-noise ratio
CIR	channel impulse response
CLR	cell loss ratio
CP	cyclic prefix
CoS	class of service
CPE	customer premises equipment
CPS	common part sublayer
CRC	cyclic redundancy check
CS	convergence sublayer
CSCF	centralized scheduling configuration
CSCH	centralized scheduling
CSMA/CA	carrier sense multiple access/collision avoidance
CTC	convolutional turbo code
CTS	clear to send
DAMA	demand-assigned multiple access
DBPC	downlink burst profile change
DCD	downlink channel descriptor
DCF	distributed coordination function
DES	data encryption standard
DFS	dynamic frequency selection
DHCP	Dynamic Host Configuration Protocol

DIUC	downlink interval usage code
DL	downlink
DLFP	downlink frame prefix
DMZ	diversity MAP zone
DOCSIS®	data over cable service interface specification
DSA	dynamic service addition
DSC	dynamic service change
DSCH	distributed scheduling
DSCP	differentiated services code point
DSD	dynamic service deletion
DSL	digital subscriber line
DVB	digital video broadcast
EAP	Extensible Authentication Protocol
EDCA	enhanced distributed channel access
EGC	equal gain combining
EIRP	effective isotropic radiated power
EKS	encryption key sequence
FC	fragment control [field]
FCH	frame control header
FDD	frequency division duplexing
FDE	frequency-domain equalizer
FDM	frequency division multiplexing
FDMA	frequency division multiple access
FEC	forward error correction
FFSH	fast-feedback allocation subheader
FFT	fast Fourier transform
FHDC	frequency-hopped diversity coding
FHSS	frequency-hopping spread spectrum
FIFO	first-in, first-out
FSDD	frequency shift division duplexing
FSH	fragmentation subheader
FSN	fragment sequence number
FTP	File Transfer Protocol

FUSC	full usage of subchannels
GBN	go-back-n
GF	Galois field
GFR	guaranteed frame rate
GMSH	grant management subheader
GP	guard period
GPS	global positioning system
GSM	Global System for Mobile Communications
HARQ	hybrid automatic repeat request
HCCA	HCF coordinated channel access
HCF	hybrid coordination function
HCS	header check sequence
H-FDD	half-duplex frequency division duplexing
HSDPA	high-speed downlink packet access
HT	header type
HTTP	Hypertext Transfer Protocol
I	in-phase
ID	identifier
IE	information element
IFFT	inverse fast Fourier transform
IP	Internet Protocol
ISI	intersymbol interference
IUC	interval usage code
IV	initialization vector
KEK	key encryption key
LAN	local area network
LDPCC	low-density parity check coding
LLC	logical link control
LMDS	local multipoint distribution service
LMSC	LAN/MAN Standards Committee
LOS	line-of-sight
LSB	least significant bit
MAC	medium access control layer

MAN	metropolitan area network
MBS	maximum burst size
MIB	management information base
MIMO	multiple-input, multiple-output
MMDS	multichannel multipoint distribution service
MPDU	MAC protocol data unit
MPLS	multiprotocol label switching
MRC	maximum ratio combining
MSB	most significant bit
MSDU	MAC service data unit
MSH	mesh subheader
MS	mobile station
NACK	negative acknowledgment
NLOS	non-line-of-sight
nrtPS	nonreal-time polling service
nrtVBR	nonreal-time variable bit rate
OFDM	orthogonal frequency division multiplexing
OFDMA	orthogonal frequency division multiple access
O-FUSC	optional full usage of subchannels
O-PUSC	optional partial usage of subchannels
OSI	Open System Interconnection
PAN	personal area network
PAPR	peak to average power ratio
PAR	project authorization request
PCF	point coordination function
PCR	peak cell rate
PCS	packet convergence sublayer
PDU	protocol data unit
PER	packet error rate
PHS	payload header suppression
PHSF	payload header suppression field
PHSI	payload header suppression index
PHSM	payload header suppression mask

PHSS	payload header suppression size
PHSV	payload header suppression valid
PHY	physical layer
PICS	protocol implementation conformance statement
PKM	privacy key management
PM	poll-me
PMP	point-to-multipoint
PN	packet number
PoS	point of sale
PRBS	pseudo-random binary sequence
PS	physical slot
PSH	packing subheader
PtP	point-to-point
PUSC	partial usage of subchannels
Q	quadrature
QAM	quadrature amplitude modulation
QoS	quality of service
QPSK	quadrature phase-shift keying
RCID	reduced connection identifier
RCT	radio conformance test
RF	radio frequency
RLC	radio link control
RS	Reed-Solomon
RS-CC	Reed-Solomon concatenated with convolutional coding
RSSI	received signal strength indicator
RSV or Rsv	reserved
RTG	receive-transmit transition gap
rtPS	real-time polling service
RTS	request to send
rtVBR	real-time variable bit rate
SA	security association
SAID	security association identifier
SAP	service access point

SC	single carrier
SCR	sustained cell rate
SDMA	spatial division multiple access
SDO	standards developing organization
SDU	service data unit
SFID	service flow identifier
SHA-1	Secure Hash Algorithm 1
SI	slip indicator
S/I	signal to interference ratio
SISO	single-input, single-output
SLA	service-level agreement
SM	spatial multiplexing
SNMP	Simple Network Management Protocol
SNR	signal to noise ratio
SOHO	small office/home office
SPID	subpacket identifier
SR	selective repeat
SS	subscriber station
SSRTG	subscriber station receive-transmit transition gap
SSTG	subscriber station transition gap
SSTTG	subscriber station transmit-receive transition gap
STBC	space-time block coding
STC	space-time coding
STTD	space-time transmit diversity
TCM	trellis-coded modulation
TCP	Transmission Control Protocol
TDD	time division duplexing
TDM	time division multiplexing
TDMA	time division multiple access
TEK	traffic encryption key
TFTP	Trivial File Transfer Protocol
TLV	Type-Length-Value
ToS	type of service

TSS&TP	test suite structure and test purposes
TTG	transmit-receive transition gap
UBR	unspecified bit rate
UCD	uplink channel descriptor
UDP	User Datagram Protocol
UGS	unsolicited grant service
UIUC	uplink interval usage code
UL	uplink
VC	virtual channel
VCI	virtual channel identifier
VLAN	virtual local area network
VoIP	voice over Internet Protocol
VP	virtual path
VPI	virtual path identifier
Wi-Fi®	wireless fidelity
WiMAX	Worldwide Interoperability for Microwave Access
WLAN	wireless local area network
WPAN	wireless personal area network
WRAN	wireless regional area network

Chapter 1 Broadband wireless access (BWA)

Applicable market segments and requirements

The IEEE 802.16 Working Group on Broadband Wireless Access has been developing IEEE Std 802.16™ since 1999. Its original air interface standard was completed in 2001, but it has continued to evolve and will continue to evolve. The basic goal of the standard is to provide specific technologies and protocols to specify the air interface of BWA systems.

The central aim of IEEE 802.16 technology is to support broadband access; that is, to support access (i.e., connection to core networks) at broadband rates (which, according to a popular International Telecommunication Union (ITU) definition, means providing service at a rate of at least 1.544 Mbit/s, although we take the definition somewhat more loosely). Since IEEE 802.16 provides broadband access with a wireless connection, it could perhaps be said to provide "wireless broadband access." In the early days, the IEEE 802.16 Working Group considered this word order, finally settling on "broadband wireless access." It appears that this might emphasize that the work involved a broadband extension of the wireless access concept rather than a wireless implementation of broadband access concepts. In fact, because this standard does both, it exemplifies the ongoing trends toward convergence in telecommunications and data communications. The final choice of word order was rather arbitrary.

As the world's backbone telecommunications networks flourished in the twentieth century, vast data-carrying capacity is now available in fiber-optic lines. Unfortunately, this bandwidth is directly available only to those with direct access to the infrastructure, which is a very small fraction of the potential users. In the developed world, powerful fiber infrastructure may connect large commercial buildings but not the neighboring medium-sized commercial facilities. In many parts of the developing world, connection to the world's fiber infrastructure is still only a plan, or perhaps only a dream. When such a connection does take place, it will initially be at a single port of entry for an entire country. Distribution is critical.

Broadband access is about bridging the gap between the core infrastructure networks and the user's networks. This bridge can itself be built with wires or fibers, but the basic structure of the problem means that this requires a massive proliferation of cabling in order to serve users in a wide variety of places. This is a fundamental barrier to cabling. It is costly to install and maintain. Cabling hung from utility poles requires a series of permissions and rights-of-way. It is costly to deploy and subject to damage caused by everything from weather to traffic accidents. Buried cabling is more reliable (although still subject to cuts), but trenching is expensive and slow, or even prohibitive, due to the disruption to communities and traffic patterns. Doug Lockie, a veteran of the BWA business, has noted many times that "backhoes don't follow Moore's Law." His point is that no amount of exponential improvement in cost and performance of network equipment will enable cost-effective network deployment as long as cable installation is the fundamental limitation.

If broadband access is wireless, many of these construction problems can be bypassed. Compared to wired broadband access, wireless broadband access leans more heavily on technological innovation and less heavily on the physical plant. This is one reason it seems to hold more promise for the future. Of course, the wireless medium has its own limitations and costs; these are associated primarily with spectrum access, restrictions, and limitations. It remains to be seen how the costs and benefits of the wired and wireless cases will balance. It seems clear that cabling will be the leader at the core network side, where data are massively consolidated, and wireless will shine nearer the user, where data are more user-specific. The details of which technologies will be most competitive in various market segments will depend on many details, including the local infrastructure and regulatory environment.

Competition, or course, is driven by demand, and demand cannot always be satisfied by a single technical solution. If all homes and enterprises in an area have direct access to wired broadband, many of the consumers will probably be satisfied. On the other hand, some might still be willing to consider alternative providers based on price, performance, features, or reliability; therefore, alternatives such as wireless broadband access may see an opportunity. But demand in the telecommunications world can also come from another direction: from the top down. Namely, many operators of core networks intend to provide service to retail customers. If these operators lack access networks, they are forced into leasing or partnership arrangements with the access provider. In an area with only one access provider (or only very few), the network operator may be forced into an unfavorable position. In such cases, demand for alternative access may be driven in top-down fashion by the operators.

In many cases, governments are seeking to foster more players and more competition in the broadband network marketplace. The market is sometimes slow to respond to demand, frequently because of natural barriers. For instance, the wired access market is difficult to penetrate, not only because it is slow and costly but, as noted earlier, because the development of local wired access facilities is burdened by regulation, especially because it is disruptive to environments (such as, for example, cities that cannot tolerate streets being trenched, or rural areas in which power lines affect wildlife migration). Barriers to wireless access networks are generally less prohibitive; therefore, wireless networks can be erected relatively inexpensively and also quite quickly. This, of course, depends heavily on spectrum availability and conditions.

One aspect of cabled infrastructure that is hard to avoid is its stationary nature. It is difficult to imagine cable being so densely deployed that it is readily available everywhere in a region the size of a city, or larger. Cable operators pick and choose their locations. Wireless signals, on the other hand, have the ability to cover a metropolitan-sized (or larger) area quite densely so that signals are broadly available, even to moving users. This allows for usage models that simply cannot be accommodated with cabled systems. The great technological and marketplace achievements of the mobile telephone business have helped to demonstrate the appeal of such a system and the flexibility it offers. Meanwhile, the success of portable computers, especially those networked with wireless local area networks (WLANs) (based on IEEE Std 802.11™) have demonstrated the appeal of untethered broadband access. This background points to a future in which broadband access moves toward an increasingly portable, nomadic, and mobile usage model. The possibility of this evolution strengthens the hand of wireless broadband access relative to its wired alternatives.

As already noted, IEEE Std 802.16 is an evolutionary standard. The standard was designed originally for the support of stationary, enterprise-class deployments. However, the long-term goal was always to evolve the standard along with the developing technology to the point at which it could be economically feasible to move deeper into the access network and closer to the user. As of 2003, the standard was first enhanced with technology suitable for residential-class applications. Around the same time, work began to evolve the technology further, toward systems that could support mobile as well as stationary terminals. The IEEE 802.16e amendment, approved in December 2005, brings support for mobile as well as fixed terminals. We do not significantly discuss IEEE Std 802.16e™ in this book. However, in the standard's evolution, it continues to build on the foundations that were previously laid. Therefore, understanding the stationary ("fixed") technology is an essential basis for understanding the fixed/mobile advances.

As IEEE Std 802.16 has grown (see [B13][1] and [B38]), much of the evolution has occurred in the physical layer (PHY), which is the primary arbiter of the physical environment in which the technology can operate. All of this development, however, is based on the essence of the standard: its medium access control layer (MAC) specification. This MAC, which supports all of the standard's PHY options, was originally designed for the first application: carrier-quality, enterprise-based telecommunications services. Because of this history, the IEEE 802.16 standard can support the most demanding service requirements, even as it evolves toward more consumer-friendly applications. This is a key factor in the applicability of the standard. One of the primary MAC-related features of the standard's technology is its support for differentiated quality of service (QoS) among its users. An IEEE 802.16 base station (BS) can simultaneously support a variety of customer service requirements, for example, very demanding services such as real-time video

---

[1] Numbers in brackets refer the reader to additional resources listed in the bibliography of this book.

conferencing along with T1/E1 service, voice over Internet Protocol (VoIP), and best-effort Internet.

Spectrum, as the wireless medium, is a very precious infrastructure commodity. Any standard intended to support wireless communications, in any application, must be critically conservative in its use of this resource. IEEE 802.16 takes pains to conserve spectrum and emphasize spectral efficiency, at the expense of protocol complexity. This approach generally enhances its applicability, although perhaps some applications might be better suited to lower complexity protocols with lower efficiency.

IEEE Std 802.16 is called *the WirelessMAN standard for wireless metropolitan area networks*. "Metropolitan" in this sense indicates not the target geography but instead the target scale. The standard supports networks that are about the size of a city. It is by no means limited to urban applications. Some of the most likely applications are in rural areas in which high-quality broadband access is not readily available. The term *metropolitan area network* (MAN) predates the term *WirelessMAN* by many years. IEEE 802®, under which IEEE 802.16 is chartered, is formally known as the "LAN/MAN Standards Committee." It was called the "LAN Standards Committee" when formed in 1980, but the name grew with the group's portfolio.

IEEE 802.16 is an air interface standard, not a manual for service deployment. It is intended to support transport of any higher layer network requirement and in any application in which someone sees fit to apply it. The goal is flexibility. The developers of the standard may have had some applications in mind, but a good, open standard can be put to a variety of users by people who did not develop it. If a developing area is seeking to install a unified wireless network for voice, data, and video services, thereby obviating the need for a multiplicity of more narrowly focused systems as would be seen in a developed city, then IEEE 802.16 may be a good choice. Likewise, if a city is seeking to enhance the availability of broadband services to its population by constructing a public wireless data network, then IEEE 802.16 may also be an excellent candidate, serving either as a backhaul network supporting the ready deployment of WLAN access points (APs) or as an access network bringing a connection directly to a user. Some operators may seek to provide service to a fixed antenna, either on the outside of a building or indoors. Other operators may seek to communicate directly with portable computers with on-board IEEE 802.16 radios.

In some cases, IEEE Std 802.16 might be a pure consumer system, with no commercial operator. Such use could be, for example, purely inside a home or enterprise, where the standard's QoS support gives it a leg up on alternative technologies. Or perhaps hobbyist or quasi-professional users will set up license-exempt systems for neighborhood communications. The applications are limitless, as they were intended to be.

Given this flexibility, it is difficult to completely address the specific market applications to which IEEE Std 802.16 may be applied. An amalgam of typical deployment scenarios is sketched in Figure 1-1. Many services are supported, consumer as well as commercial, mobile as well as fixed, and backhaul-only as well as directly to the end user. Such a mixed set of services signifies a mature and successful MAN. Perhaps we shall someday see such a network based on IEEE Std 802.16.

Below, we continue this introductory chapter, providing examples of how IEEE Std 802.16 might be applied to some idealized market situations.

COMMERCIAL FIXED BROADBAND WIRELESS: FIBER EXTENSION

In large metropolitan areas throughout much of the developed world, commercial office towers are connected to core networks by high-capacity fiber optic links, with broadband network services provided to the tenants. In the meantime, other smaller businesses, even those located across the street or on the next block, go without proper network facilities for a lack of fiber connectivity. IEEE 802.16 networks can fairly easily extend the reach of the fiber links. Suitable spectrum is available for this purpose in many areas. For example, the United States licensed 1150 MHz of spectrum in the 28–31 GHz range in 1998 for a service called *local multipoint distribution service* (LMDS); since then, additional higher frequency spectrum has become available on a license-exempt basis or with a low-cost license. Supply for spectrum at such frequencies outweighs demand primarily because of the less favorable propagation properties of electromagnetic radiation in this band. In particular, attenuation in air is rather high so that propagation distances are

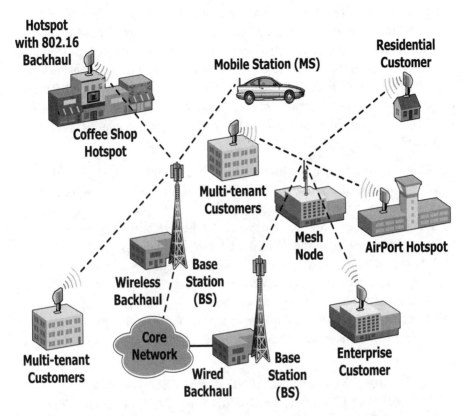

Figure 1–1: Typical IEEE 802.16 deployment scenarios

typically limited to around 3 km to 5 km. Also, attenuation increases dramatically in the presence of rainfall so that power control over a significant range is required. Diffraction is limited so that only line-of-sight (LOS) propagation is practical for communications. Furthermore, electronic components at these frequencies are relatively expensive. On the other hand, spectrum is plentiful enough for massive bandwidth, and short propagation distances and LOS propagation (supported by the compact, high-gain directional antennas available at these short wavelengths) allow for high frequency reuse.

Tall towers, which often have fiber connectivity, are good platforms for LOS BS antennas because the view from the top typically includes hundreds of prospective customers. Commercial buildings allow for rooftop antennas, the cost of which (both the hardware and the professional installation) is spread over a number of commercial customers. In LOS links, multipath propagation is usually minimal; therefore, relatively simple modulation schemes may be effective. The extension of fiber requires the ability to support high data rates (> 100 Mbit/s), with high reliability and service-level guarantees. The ability to transport voice, video, and data traffic with appropriate quality is also an important requirement.

The first version of IEEE Std 802.16, approved in 2001, included a PHY specification, called *single-carrier WirelessMAN* (WirelessMAN-SC), suitable for use in the high-frequency bands associated with fiber extension. At the time, prospects for the deployment of the standard seemed rosy. However, the bursting of the "telecom bubble" early in the twenty-first century put an end to many promising plans. Currently, conditions are becoming more favorable for reconsideration of these applications. If so, then the WirelessMAN-SC air interface will be a good candidate to support such deployments. It supports the requirements, and it allows for stepped investment commensurate with market capacity requirements.

RESIDENTIAL FIXED BROADBAND WIRELESS: DIGITAL SUBSCRIBER LINE (DSL) AND CABLE MODEM ALTERNATIVE

In the developed world, broadband access has become a virtual necessity in the residential market. Homes are frequently the site of full-time or off-duty business, and broadband Internet access is in demand for education, information, voice, video, shopping, and entertainment. Cable modem and DSL networks have become the primary technologies addressing these home markets in many countries. However, such networks do not extend everywhere. Not all areas are served by cable television networks, and not all have been upgraded for cable modem service. While telephone networks are widely deployed in the developed world, not all lines have been upgraded for DSL service, and many homes are too far from the central office for DSL transmission.

Where DSL and/or cable modem service is available, BWA may nevertheless play a role. Customers may be seeking alternatives based on price, performance, and/or reliability. Also, wireless service may have advantages, such as, in some cases, the ability to move a portable terminal throughout a service area. As previously noted, network operators who are dissatisfied with their inability to directly access their retail customers except through a local access provider may also be motivated to deploy a wireless access alternative.

In the developing world, and in many rural areas of the developed world, broadband access is simply unavailable. Wireless access may be particularly attractive in sparsely populated rural areas in which the cost of cabling is prohibitive.

Different design considerations may apply in rural and urban areas. For instance, in rural areas, economic considerations may require that a BS cell size be quite large, up to tens of kilometers. In these cases, LOS propagation with rooftop-mounted directional antennas may be appropriate. On the other hand, urban and suburban applications typically assume non-line-of-sight (NLOS) operation due to the proliferation of obstacles, the high incidence of multipath propagation, and the carrier frequency used (typically under 11 GHz). These networks may use licensed or license-exempt bands. Attractive licensed frequencies are available in the United States in the broadband radio service (BRS) bands [formerly known as multichannel multipoint distribution service (MMDS)] at around 2.5 GHz to 2.69 GHz. Outside the United States, other licensed bands, such as 3.5 GHz, are more common.

License-exempt operation has some additional challenges in MANs. Popular WLANs, which similarly operate in license-exempt 2.4 GHz and 5 GHz bands, are typically deployed in buildings and other areas under the control of a single entity deploying the network. This controlled environment is similar to a licensed deployment, since interference is limited to neighbors normally isolated to some degree by space and walls. However, license-exempt outdoor BWA deployments bring a different set of challenges, as a single entity cannot enforce its rules on other legal transmitters that may interfere. Any BWA solution has to take this into account.

In all of these cases, operators favor solutions that allow self-installation of customer equipment, since professional installation significantly affects the cost. Ideally, the customer equipment is in the form of a portable device that can be placed indoors. Such installation not only reduces the deployment cost but also offers the advantage of portability.

With residential customers becoming more technologically demanding, a competitive residential service offering is becoming suitable to small businesses as well. Because IEEE Std 802.16 is designed to support mixed services, it has no problem supporting such a customer mix. We can easily envision a multitude of applications for such a system. For instance, T1/E1 customers could be provisioned, and commercial WLAN hot spots, which typically require backhaul by 10BaseT (often over DSL), could be made entirely wireless using IEEE 802.16 backhaul. This would allow greatly increased flexibility in the placement of the IEEE 802.11 APs. Backhaul of other devices, such as cellular BSs, is another likely service.

IEEE Std 802.16 added support for a PHY at frequencies below 11 GHz beginning in 2003 with IEEE Std 802.16a™. This work, further refined in IEEE Std 802.16-2004 [B20] (the most recent revision of the standard), includes the specification of three PHY alternatives, known as orthogonal frequency division multiplexing (WirelessMAN-OFDM), orthogonal frequency division multiple access (WirelessMAN-OFMDA), and single carrier below 11 GHz (WirelessMAN-SCa). The first two, in particular, are well suited to NLOS operation in a multipath environment.

While IEEE Std 802.16-2004 specifies a standard for fixed access, the actual applications may allow a significant level of flexibility. For example, the user device may be *nomadic*, meaning that it can move as long as it does not operate while doing so. Such operation, as would be typical of a portable terminal such as a laptop computer equipped with wireless fidelity (Wi-Fi®), is well within the target market of deployments based on IEEE Std 802.16-2004. Even a certain amount of movement during operation is tolerable. Intercell handover, however, is beyond the scope of IEEE Std 802.16-2004. For this, users should turn to the IEEE 802.16e specification that was approved by IEEE as an amendment in December 2005.

At the time of publication, a number of companies have announced products claiming compliance to IEEE Std 802.16-2004.

QUALITY OF SERVICE (QoS)

In order to be competitive in a modern networking environment in which a single BS must support a multiplicity of widely varying transport demands over an inherently fluctuating wireless medium, a BWA network must include rigorous support for differentiated QoS as a fundamental design feature.

This requires support for flexible MAC and PHY framing to optimally use the available airtime adaptively. A centrally controlled MAC is also a necessity for BWA systems because distributed access methods, such as carrier sense multiple access (CSMA), cannot work efficiently in environments where some user devices are unable to hear other ones, due to distance, directional antennas, intervening terrain, etc.

The higher probability of errors in wireless environments introduces another challenge to the wireless QoS problem, especially in NLOS environments. The impact of errors on QoS can be minimized with sophisticated adaptive modulation and coding, error correction, error detection, and retransmission algorithms at both the MAC and PHY.

It is unnecessary, and in fact inappropriate, for a standard to enforce a specific QoS scheduling algorithm. However, a practical standard must define scheduling behaviors that allow the system to enforce uniform QoS. IEEE Std 802.16 does so by use of techniques analogous to those of the asynchronous transfer mode (ATM), where different scheduling behaviors are defined without specifying any particular scheduling algorithm.

The ability to provide QoS in both directions is also a critical requirement for efficiently managing the available spectrum and to support bidirectional applications such as voice and video conferencing.

THROUGHPUT REQUIREMENTS

Apart from basic QoS and reliability requirements, BWA systems must provide support a sufficiently high data transmission rate in order to be commercially successful. For backhaul systems, these throughput requirements are generally higher; depending on the load, systems should be able to carry a significant fraction of a fiber optic network's traffic. Moreover, given the nature of the aggregated traffic over backhaul networks, the throughput requirements may be symmetrical [i.e., identical in both uplink (UP) and downlink (DL) directions], depending on the configuration.

On the other hand, the last-mile access or edge networks have to provide throughput comparable to that of competing technologies, such as DSL or cable. If BWA is used to support business needs, then the networks should be capable of supporting multiple business-class links such as T1, T3, DS3, or OC-3. Most importantly, the ability of the BWA network to scale efficiently with reasonable per-subscriber throughput is a critical requirement.

The increasing popularity of various multimedia, on-demand, and interactive applications is likely to increase the demand for network throughput and scalability in all application scenarios. IEEE Std 802.16 foresees the need for this evolution and is designed to support it.

Chapter 2　IEEE 802.16 standards

The working group and documents

BACKGROUND

IEEE Std 802.16, along with related standards and amendments, is developed and maintained by the IEEE 802.16 Working Group on Broadband Wireless Access. In this chapter, we begin with an overview of the umbrella organizations under which the IEEE 802.16 Working Group performs its activities, and we explain the basic process of developing and maintaining IEEE standards. We follow with a history of the IEEE 802.16 Working Group and review the historical development of its projects.

IEEE STANDARDS ASSOCIATION (IEEE-SA)

The Institute of Electrical and Electronics Engineers (IEEE) <http://ieee.org> is a technical professional society with over 350,000 members worldwide. IEEE has many technical and regional activities, most of which take place in a largely independent fashion.

The development of standards in IEEE is assigned to the IEEE-SA <http://standards.ieee.org>. The business of IEEE-SA is directed by an elected board of governors. The development and maintenance of standards are overseen by the IEEE-SA Standards Board, which mandates the process, approves the initiation of new projects, and approves appropriately balloted drafts as IEEE standards. IEEE-SA operates in accordance with the principles of consensus, due process, and openness defined by the American National Standards Institute (ANSI) and the Code of Good Practice for the Preparation, Adoption and Application of Standards produced by the World Trade Organization (WTO) under its Agreement on Technical Barriers to Trade. IEEE-SA is recognized by important international organizations as an international developer of standards, and IEEE standards are, in many cases, recognized as international standards. One example of particular relevance to

IEEE Std 802.16 is IEEE's international Sector Member status in the ITU's Radio Communication Sector (ITU-R), the same status held by the International Organization for Standardization (ISO).

The IEEE-SA leadership sets policy that directly influences not only the development of IEEE-SA standards but also their use. One critical topic is patents. The IEEE-SA patent policy is similar to that of most of the world's other formal standards developing organizations (SDOs). The key statement of IEEE-SA policy on this issue is "IEEE standards may include the known use of essential patents and patent applications provided the IEEE receives assurance from the patent holder or applicant with respect to patents whose infringement is, or in the case of patent applications, potential future infringement the applicant asserts will be, unavoidable in a compliant implementation of either mandatory or optional portions of the standard." The policy goes on to explain that, if the patent will be enforced, this "assurance" shall be "a statement that a license will be made available without compensation or under reasonable rates, with reasonable terms and conditions that are demonstrably free of any unfair discrimination." The IEEE generally avoids offering interpretations of these somewhat ambiguous statements, and any disagreements need to be settled outside of IEEE processes.

IEEE-SA standards are openly developed with consensus in mind. Participation in their development is entirely voluntary, as is their use. However, history has shown that standards developed in an open forum can produce high-quality, broadly accepted results that can focus companies and forge industries.

IEEE-SA charters over 200 "sponsor" groups to oversee the development of specific standards projects. These are typically organized within one or more of IEEE's technical societies.

While the sponsor groups have significant leeway in how they organize their work and prepare draft standards, the IEEE-SA is particularly assertive in the conduct of the ballot process under which drafts are reviewed as part of their consideration as IEEE standards. Before a prospective standard can be considered for approval by the IEEE-SA Standards Board, it must be balloted in a formal process that is defined by IEEE-SA. This process is, rather confusingly, known as "sponsor ballot" although it would more appropriately be called "IEEE-SA ballot." In any case, balloting is conducted in an open process using a "ballot group" of volunteer individuals. IEEE-SA members are invited to participate in all ballots, regardless of whether they participated in the development of the draft. Balloting is an iterative process in which comments (a polite word for a specific complaint) are solicited and addressed, after which an improved draft is "recirculated" for further comment. As IEEE 802 Executive Committee Member Emeritus Geoff Thompson likes to say, balloting is about "improving" the draft, not about "approving" the draft. More comments lead to a better result. The IEEE 802.16 Working Group seeks comments, and the IEEE 802.16 task groups are experienced at resolving them (up to 500, or 1000, or even 2000 in a weeklong session, if necessary). Provided that a competent, active, and sincere ballot group is engaged in the process, the outcome is a sound and reliable technical document that reflects a broad consensus. While the participants are individual human beings, they often bring with them the technical ideas of their home environments, including national, regional, and corporate viewpoints.

IEEE 802® LAN/MAN Standards Committee (LMSC)

The development of local area network (LAN) and MAN standards with IEEE-SA is assigned to the LMSC, informally known as *IEEE Project 802* or simply *IEEE 802* <http://ieee802.org>. One of the largest, most prolific, and most influential of the IEEE-SA sponsors, IEEE 802 has operated since 1980 under the IEEE Computer Society. It develops and maintains standards addressing the MAC and PHY, each of which fits under a common logical link control (LLC) layer. Taken together, these make up the two lowest layers of the Open System Interconnection (OSI) seven-layer model for data networks (see [B29]).

IEEE 802 oversees a panoply of network standards, using an internal structure based on working groups developing draft standards. IEEE 802's great successes include IEEE 802.3 Ethernet, IEEE 802.11 WLANs, and IEEE 802.15.1 Bluetooth™ personal area networks (PANs). Other significant, although ultimately less successful, projects have included token ring and token bus.

IEEE 802 is a large but tightly managed organization that meets in plenary session each March, July, and November, with recent attendance in the range of 1800 people. The organization is governed by an executive committee composed of the chairs of the active working groups and technical advisory groups (currently 11 individuals) and 7 additional officers.

STANDARDS DEVELOPMENT IN IEEE 802

The IEEE 802 process is designed for quick development of standards with broad consensus. The demand for consensus helps to ensure that standards are technically superior and meet market needs.

The development process in IEEE 802 follows the chronological steps outlined below. The process is overseen by the IEEE 802 Executive Committee and defined by a written set of rules and procedures.

Study group stage

When sufficient interest has been identified in a topic, IEEE 802 may establish a study group to investigate the problem and consider the interest and potential scope of a possible standardization project. Should a study group wish to pursue standardization, it must draft a project authorization request (PAR); this is a form by which all new IEEE-SA projects become authorized. Before the Executive Committee considers approving a PAR for submission to the IEEE-SA Standards Board, it requires a statement addressing IEEE 802's "five criteria for standards development." This statement must demonstrate that the potential standard has broad market potential, compatibility with other IEEE 802 standards, distinct identity within IEEE 802, technical feasibility, and economic feasibility.

Working group development of draft

The Executive Committee assigns each new project to an existing or new working group and charters that group to develop the standard. Technical decisions are made in the working group by vote of at least 75% of its members. Membership in IEEE 802 working groups belongs only to individual people, usually engineering professionals, and is established and maintained by participation in sessions, according to specific rules. Nonmembers participate actively as well, often with significant influence.

The initial draft development method varies among working groups, but the typical process is to delegate a task group to the problem and issue a public call for contributions requesting documentary input. Eventually, the task group develops a first draft, either by adopting a complete contribution or by assembling a collections of inputs. This process is sometimes contentious, as the competing interests of different companies and technology interests are often reflected in meetings. However, it offers a good opportunity for new participants to come to understand the process and become comfortable in discovering commonality among a group with diverse interests. The fact that professional individuals, not companies, are the recognized entities helps to set a tone of collegiality. New participants continuously enter the process. They sometimes appear aggressive at first, only to discover that this approach can be ineffective. The process demands excellent communication and preparation as well as technical skills. Those most effective at furthering their causes are those who clearly state their intent, present well-documented arguments, and look for opportunities to unite with others who have compatible goals. The system is an excellent training ground for bringing out effective communications skills, and many of the most effective participants learn their skills the hard way.

Once a working group has adopted a draft, the process changes subtly but significantly. At this point, if all goes well, the interest of the participants begins to align with the common goals of improving and completing the draft. The typical process is driven by distributing the draft and requesting comments in the form of specific requests to make changes. A strict IEEE 802 policy, atypical in IEEE-SA, is the requirement of a formal ballot process, modeled after the IEEE-SA ballot, before a draft standard may be forwarded to IEEE-SA for sponsor ballot. This dual ballot process is a key factor in the quality control for which IEEE 802 has become known.

In this "working group letter ballot," as in sponsor ballot (both of which are paperless), any vote against the document must be accompanied by specific comments on what changes are required in order to make it acceptable to the voter. This process forces constructive suggestions of change and helps drive the process to quick improvement. Members voting to approve, and nonmembers as well, are also solicited for suggestions. An approval rate of 75% is required for draft acceptance. However, changes made in response to comments, and negative comments that have not been accepted by the editorial team, must be "recirculated" for approval by the voters. This allows for additional reviews and additional improvement. Eventually, however, most ballots reach a terminal period in which a large consensus favors closure. The ballot cannot close until those voting negative have had their say and failed to attract significant support for their argument. The approval margin is typically much higher than 75% at closure, but it need not be.

Following approval in working group letter ballot, drafts are forwarded for IEEE-SA sponsor ballot. This is similar to a rerun of the working group letter ballot except that the ballot group is not restricted to members of the working group. When this ballot is complete, the draft, ballot results, and supporting documentation are forwarded for review by the IEEE-SA Standards Board's Review Committee (RevCom). RevCom's recommendation proceeds to the full board for final action.

Once IEEE standards are approved, they are professionally edited and generally published and offered for sale within about two months, depending on size and complexity and the extent of editing required.

Unique to IEEE 802 within IEEE-SA is the "Get IEEE 802" program <http://standards.ieee.org/getieee802> in which published standards are available for download without charge beginning six months after publication. The cost of this program is subsidized by the individual participants, through the session registration fee, along with a few corporate sponsors.

IEEE 802.16 WORKING GROUP: OVERVIEW

IEEE 802's WirelessMAN work takes place within the IEEE 802.16 Working Group on Broadband Wireless Access <http://WirelessMAN.org>. The working group is a unit of IEEE 802, which serves as sponsor of IEEE 802.16 projects (although, unique to IEEE 802, the IEEE 802.16 projects also have a cosponsor: the IEEE Microwave Theory and Techniques Society).

IEEE 802.16 WORKING GROUP: HISTORY

The activities of the IEEE 802.16 Working Group were initiated by Roger Marks of the (U.S.) National Institute of Standards and Technology (NIST), who organized a meeting on BWA standardization, attended by 45 people, in August 1998 at the IEEE Radio and Wireless Conference in Colorado Springs, Colorado, USA. Following his visit to the IEEE 802 plenary session in July, Marks forwarded an invitation from IEEE 802 Chair Jim Carlo to convene a meeting on this topic at the IEEE 802 plenary session in November. At that session, IEEE 802 approved the formation of the Study Group on Broadband Wireless Access. That study group met twice and drafted a PAR, limited to 10 GHz to 66 GHz, that was endorsed by the IEEE 802 Executive Committee in March 1999. This action (after approval from the IEEE-SA Standards Board) created the IEEE 802.16 Working Group. Following an organization session (Session #0) in May, 106 people became charter members of the working group at its first official session in Montreal, Canada, in July 1999.

The working group has continued to meet at each IEEE 802 plenary session (in March, July, and November) and to hold a working group interim session each January, May, and September. (An additional interim session in August 1999 caused the even/odd numbering of sessions to reverse; since then, even-numbered sessions correspond to IEEE 802 plenaries.) As shown in Figure 2–1, which includes sessions through #39 of September 2005, attendance has grown and waned. Participation interest depends on the current activity. Interest was very high in 2004-2005 as the IEEE 802.16e amendment was being assembled, and attendance peaked at 367 in November 2004.

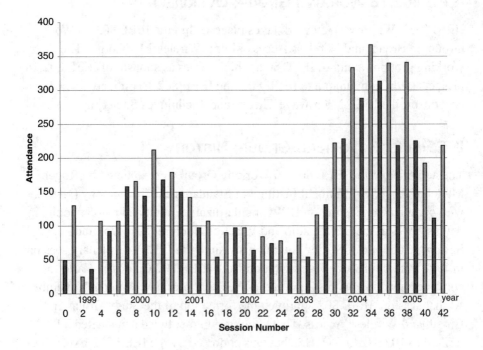

Figure 2–1: IEEE 802.16 Working Group attendance

Beginning in 2004, in recognition of the fact that IEEE 802 plenaries are generally in North America, nearly all sites selected by the working group for its interim sessions have been outside North America, most often in Asia. Like attendance, membership in the working group also fluctuates, lagging behind the attendance figures. As of November 2005, the working group had 310 individual members. According to the addresses they provided, they represent a broad geographical base, as shown in Table 2–1.

Table 2–1: IEEE 802.16 members by geography, as of January 2006

Address	Number of members
Canada	20
China	17
Finland	2
France	3
Germany	7
Ireland	1
Israel	18
Italy	3
Japan	7
Korea	60
Netherlands	4
Romania	1
Singapore	1
Sweden	3
Taiwan	3
UK	11
USA	149

TECHNICAL PROGRESS IN IEEE 802.16 WORKING GROUP

Since 1999, the IEEE 802.16 Working Group has constantly been active in developing standards projects, usually with multiple parallel activities. Although some of its projects have been very large, the group prefers to divide its efforts into specific problems that can be reasonably well-defined and completed within a predictable time. The IEEE process allows for the development of amendments that modify an existing standard. The published amendment, which is designated by a lowercase letter after the primary standard number, is not an independent specification because it includes only the modifications, not the base material from the original standard. Upon approval of the amendment, the applicable standard is no longer the prior version, but the version defined by the application of the amendment. When appropriate, a revision of the standard may be undertaken; in this case, the base standard and its published amendments are editorially merged and reballoted, with the entire document open to comment.

Figure 2–2 shows a timeline of the past and current projects of the IEEE 802.16 Working Group. The designated start date is that of the approval of the project authorization (PAR); in some cases, the chart shows the date of a previous PAR that was later modified before the project was complete. The end date in each case is the actual or anticipated date of approval.

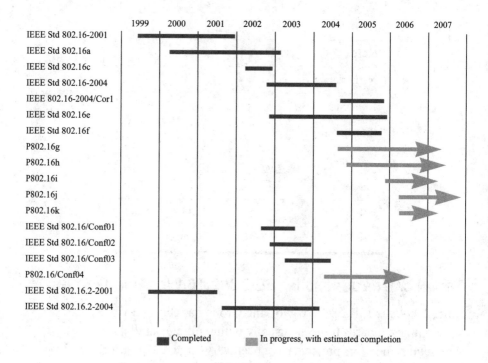

Figure 2–2:   IEEE 802.16 project timeline

Air interface: IEEE Std 802.16

Work under the original IEEE 802.16 PAR, to develop an air interface for 10 GHZ to 66 GHz, began in July 1999. By November 1999, 35 PHY and MAC proposals were considered. By March 2000, two consolidated MAC/PHY proposals were still under consideration. In May 2000, agreement was reached to merge these proposals. A working group letter ballot followed soon thereafter. IEEE Std 802.16-2001, IEEE Standard for Local and Metropolitan Area Networks—Part 16: Air Interface for Fixed Broadband Wireless Access Systems, was approved in December 2001. The MAC protocol is fundamentally based on a time division multiplexing/time division multiple access (TDM/TDMA) protocol supporting time division duplexing (TDD), frequency division duplexing (FDD), and half-duplex frequency division duplexing (H-FDD). The PHY, entitled *WirelessMAN-SC*, is a single-carrier system assuming LOS propagation to fixed antenna terminals.

Because the working group believed that IEEE Std 802.16-2001 allowed too many options for easy interoperability testing, it opened up a follow-up amendment project, IEEE P802.16c, to define a set of profiles that would each define a set of options with sufficient specificity to allow for interoperability. With Ken Stanwood's leadership as task group chair, this project both opened and closed in 2002.

In November 1999, while the 35 MAC and PHY proposals were being discussed, IEEE 802.16 created a study group, under the leadership of Brian Kiernan, to develop a PAR for frequencies below 10 GHz. That PAR was approved in March 2000 and began the working group's activities in NLOS PHY technology.

The working group targeted its higher frequency PHY at carrier frequencies above 10 GHz and its lower frequency work at less than 11 GHz. This overlap simply reflected the fact that some available bands (in particular, 10.5 GHz) seemed suitable for either. In fact, this minor overlap was dwarfed, early on, by more fundamental issues regarding the relationship between these two approaches. Initially, many of those interested in the low-frequency applications favored basing the work on a new MAC, believing that the original MAC was too enterprise-centric and not sufficiently suited to basic residential applications. However, upon further discussion, the working group made the virtually unanimous decision to build all of its PHY specifications upon a single, sophisticated MAC foundation. Another contentious issue was whether to pursue separate PHY projects for licensed and licensed-exempt bands. After starting in this direction, the working group later decided to merge those efforts. As a result of these changing decisions, a number of PARs were revised and renamed in the early years. Eventually, the project became amendment project IEEE P802.16a, and the amendment was approved by IEEE-SA in early 2003.

IEEE Std 802.16a includes three separate PHY specifications:

- WirelessMAN-SCa: single carrier
- WirelessMAN-OFDM: multicarrier with 256 subcarriers
- WirelessMAN-OFDMA: multicarrier with 2048 subcarriers

The debate leading to the decision to include all three modes was long and contentious. The eventual decision was not to everyone's liking, but compromise is a critical element of standardization. In attempting to develop a single standard for worldwide use, it was necessary to recognize a number of different worldwide needs. Since IEEE Std 802.16a was adopted, WirelessMAN-OFDM has become very popular. WirelessMAN-OFDMA, which was more forward-looking at the time of adoption, is increasingly appearing to be the choice of the future, particularly as IEEE Std 802.16 is evolving toward mobile systems. In the meantime, WirelessMAN-SCa has not gained significant industry interest.

Once IEEE Std 802.16a was complete, the working group opened the amendment project IEEE P802.16d. This was intended to parallel the profiles project IEEE Std 802.16c, but oriented toward the lower frequencies. The PAR also allowed for the correction of errors that inevitably plague complex standards such as IEEE Std 802.16. Gordon Antonello agreed to chair the project, which was intended to be short and sweet, like IEEE Std 802.16c™. However, the project quickly grew in complexity, as a number of proposals looked more like enhancements than error corrections. Eventually, the working group decided that it needed to convert the project from an amendment into a revision; this would editorially merge IEEE 802.16-2001, IEEE 802.16a, and IEEE 802.16c, opening the entire result to comments regarding corrections and improvements. The amendment PAR IEEE 802.16d was abandoned in favor of a revision project IEEE 802.16-REVd, and Antonello ended up with a much larger task than he had expected. The work was finally approved in June 2004 and weighed in at nearly 900 pages. Because it was a revision, IEEE Std 802.16-2004 [B20] made IEEE Std 802.16-2001, IEEE Std 802.16a, and IEEE Std 802.16c obsolete. Some people incorrectly refer to this document as the "16d" standard, but letters are used to identify amendments, not revisions.

Once again, bugs and errors turned up. In September 2004, a PAR was approved for a new project to address them. In IEEE parlance, a project allowing corrections but prohibiting new features is called a *corrigendum*. Under the leadership of Jon Labs, the corrigendum IEEE Std 802.16-2004/Cor1 was completed in September 2005 and approved in November 2005. This document put to rest, for the near term, the definition of the IEEE 802.16 air interface for fixed wireless access.

However, a working group in motion seems to remain in motion. As the amendment project IEEE P802.16a was wrapping up in late 2002, the working group opened up a new PAR, IEEE P802.16e, to expand the IEEE 802.16 fixed access system into a combined fixed/mobile system, allowing a single BS to support both fixed and mobile terminals in licensed bands below 6 GHz. The amendment project IEEE P802.16e, chaired by Brian Kiernan, attracted great interest and a great many participants to the working group. The influx of people tended to keep the project unstable; therefore, many decisions were revisited again and again. However, the work did come to a conclusion, and the final draft, at 684 pages, was approved in December 2005. IEEE Std 802.16e amends IEEE Std 802.16. All of the three lower frequency PHY modes are supported, but the WirelessMAN-OFDMA mode is made "scalable" with the addition of new subcarrier counts: 128, 512, and 1024. The details of IEEE Std 802.16e and its content is beyond the scope of this book.

In March 2006, Brian Kiernan was awarded the IEEE-SA Standards Medallion "for steadfast and exemplary leadership of the Task Groups developing the IEEE 802.16a and 802.16e WirelessMAN standards specifying wireless metropolitan area networks for fixed and mobile broadband wireless access systems."

As the working group's attention turned to mobility, it decided that network management would be an increasingly critical issue. The Network Management (NetMan) Task Group, chaired by Phil Barber, was initiated to address such concerns. The NetMan group took on two projects: IEEE P802.16f and IEEE P802.16g. IEEE Std 802.16f™, a management information base (MIB) for fixed systems, was approved in September 2005. IEEE P802.16g is a complex activity on "management plane procedures and services." Approval is not expected until 2007. In the meantime, the working group began planning in late 2005 to follow up IEEE Std 802.16f with a new MIB project for the mobile case. This was launched as IEEE P802.16i in December 2005.

The work to amend IEEE Std 802.16-2004 also continues with IEEE P802.16h, which is attempting to address the long-neglected problem of coexistence in license-exempt bands. The License-Exempt Task Group leading the effort is chaired by Mariana Goldhamer.

In July 2005, following a number of presentations and expressions of interest, the working group created the Mobile Multihop Relay Study Group to investigate the initiation of a new project. The existing standard specifies both the BS and the subscriber station (SS). The study group was chartered to consider the additional specification of a relay station, which would offer a valuable new tool to system operators. The study group, chaired by Mitsuo Nohara, proposed an IEEE P802.16j PAR. Following a tutorial on the topic at the IEEE 802 plenary session in March 2006, the PAR was approved later that month.

Conformance: IEEE 802.16/Conformance0*X*

The IEEE 802.16 Working Group, aware that air interface standards alone cannot specify conformance or interoperability, believes in the importance of conformance test documents. The working group has completed and published three stand-alone conformance test standards, all applicable to WirelessMAN-SC systems and all developed by the Conformance Task Group, chaired by Ken Stanwood. IEEE Std 802.16/Conformance01-2003

[B21], IEEE Std 802.16/Conformance02-2003 [B22], and IEEE Std 802.16/Conformance03-2004 [B23] address the protocol implementation conformance statement (PICS) proforma, test suite structure and test purposes (TSS&TP), and radio conformance tests (RCTs), respectively. Work on a PICS for the PHYs operating below 11 GHz is taking place under the IEEE P802.16/Conformance04 project, chaired by Gordon Antonello.

COEXISTENCE: IEEE STD 802.16.2™

Beginning in 1999, the working group took note of the difficulties that would be faced by operators of systems in licensed bands due to co-channel and adjacent channel interference. It opened a PAR to create a stand-alone *recommended practice*, IEEE parlance for a standard that uses the verb "should" instead of "shall" in its normative statements. IEEE Std 802.16.2-2001, addressing the frequencies important to the WirelessMAN-SC PHY, was approved in 2001. A revision, to include the lower frequencies in IEEE Std 802.16a, was approved in 2004 as IEEE Std 802.16.2-2004 [B26]. Phil Whitehead chaired the Coexistence Task Group for both projects.

Chapter 3 Basic concepts and definitions

Wireless protocol and communication concepts

This chapter introduces some of the basic concepts and definitions related to wireless design, antennas, regulatory issues, MAC, and the PHY. This will form a basis for the more detailed discussions in the rest of the book. Among the countless books on these issues, we refer the reader to a few on the topic of broadband wireless or wireless in general (see [B4], [B7], and [B51]) and on networking (see [B6], [B33], [B42], [B50], and [B54]). These and other references offer a more thorough study of the specific concepts.

In this chapter, we use the term *wireless device* to refer generally to a radio communication terminal, whether a central controller such as a BS, a client SS, or a node in a multihop network. Note that, according to this definition, "wireless" devices may in fact be cabled to a network, such as a backhaul network on the BS side or a user LAN on the SS side, but we call the device "wireless" because it carries out radio communications over a wireless link. The communication between wireless devices is accomplished using protocols and techniques that require an antenna for transmission and reception of radio waves, a PHY for modulation and demodulation of signals, and a MAC for coordinating and controlling access to the medium (the "airwaves"). Other networking, management, and privacy components may be present in a wireless device to support additional services. IEEE Std 802.16 provides support for point-to-multipoint (PMP) systems centered around a coordinating BS and a number of SSs. It also supports, as an option, a multihop mesh mode in which a wireless device receives information and forwards it to another wireless device.

FREQUENCY BANDS

One of the basic parameters that characterizes a wireless system is the frequency band it uses, generally expressed in terms of megahertz (MHz) or gigahertz (GHz).

Channels

A frequency band is often split into multiple frequency *channels* to support independent communication activities. A channel is usually defined with a specific center frequency and occupied channel bandwidth. For example, a 2437 MHz channel with a 22 MHz bandwidth occupies the spectrum from 2426 MHz to 2448 MHz.

Licensed and license-exempt spectrum

Typically, the frequency spectrum in which a specific wireless device operates can be classified into *licensed* or *license-exempt* (or *unlicensed*) spectrum. The allocation and regulation of licensed and license-exempt spectrum are typically controlled by regulatory agencies, such as the Federal Communication Commission (FCC) in the United States.

In licensed bands, individual licensees are typically granted exclusive rights by the regulatory agency to a portion of the spectrum in a specific geographical area, sometimes in exchange for a fee. For example, most mobile telephone operators hold exclusive rights to the spectrum in which they operate; in many cases, the license rights were obtained at auction. Conversely, operation in license-exempt spectrum does not require explicit permission, although generally transmission is permitted only using communication devices that have been certified to meet regulatory specifications. In general, operation in license-exempt situations offers no legal protection against harmful interference.

The distinction between licensed and license-exempt spectrum is, in many cases, overly simplistic. For instance, in some cases, license-exempt devices are permitted to operate in the same bands, and in the same general geographic areas, as licensed devices, provided that the licensed devices are protected from interference in some way. In this case, it is the device operation, not the spectrum, that is properly characterized as licensed or license-exempt.

A licensing model commonly referred to as *light licensing* is coming into use by some regulatory agencies for some frequency bands. In the light licensing model, transmission requires a license, but licenses are nonexclusive. Typically, the license holders are required to register transmitter details with the regulatory agency. The regulator may control subsequent licenses to avoid interference with previous license grants in the same location, or it may leave interference issues to be negotiated by the licensees. With the development of smarter and cognitive wireless devices, many regulators are hoping to move toward less spectrum regulation and let technology address the interference issues.

In most cases, wireless devices, whether in licensed and license-exempt operation, require legal certification by the governing regulatory agency. The certification usually requires that devices meet specific requirements regarding fundamental physical features, such as transmitter power, in-band and out-of-band emissions, and channel selection. In other cases, regulations may specify the type of modulation and detailed protocols, although such an approach is beginning to look anachronistic. Rules may depend on the application; for instance, rules for indoor and outdoor operation may differ.

One relatively new requirement for license-exempt operation in certain cases is *dynamic frequency selection* (DFS). DFS is used by a wireless device to dynamically detect any active transmitters (typically, high-power radar) on the channel and take corrective actions. The actions may include stopping transmission on the occupied channel and switching to another channel. DFS may also be enhanced to support more sophisticated coexistence protocols, including those based on detection of lower powered nonradar devices.

In virtually any regulatory domain, many radio frequencies are unavailable for private terrestrial wireless communications, licensed or not. Many of these bands are allocated to specific public use, such as military or public safety, or to other private services. For a detailed list of allocated bands and specific requirements, please refer to the appropriate regulatory agency in the area of interest.

Spectrum and standardization

While the regulatory agencies do not generally take responsibility for developing standards for wireless communications, wireless standards developers must take regulatory requirements into account and provide solutions that meet those demands, as well as technical and business requirements. In the case of license-exempt operation, protocol designers, including those involved in standardization, must include the support of extra measures for coexistence and interference mitigation. In a license-exempt band, no single entity controls the deployment of devices. Operating devices may well transmit within range of each other. Without defined coexistence mechanisms and interference mitigation techniques, harmful interference may occur. For example, the 2.4 GHz industrial, scientific, and medical (ISM) band is heavily used by IEEE 802.11 WLANs, IEEE 802.15 wireless personal area networks (WPANs, including Bluetooth), cordless telephones, etc. IEEE 802.16 devices are also allowed, provided that they follow the regulatory requirements. The basic rules of operation may provide some protection from interference, and specific protocols can offer additional protection to devices jointly observing those protocols.

Coexistence

In licensed spectrum, the licensee manages and controls the transmitters. This priority control over emitters in the licensed spectrum and geographic area, sometimes combined with coordination with licensees in adjacent geographic areas, adequately ensures coexistence for the emitters of the licensee. Coexistence ensurance hence may take among others the forms of emitter positioning and manual configuration as well as automated configuration through coexistence protocols among emitters. IEEE Std 802.16 addressed the issue of licensed coexistence in IEEE Std 802.16.2-2004 [B26].

On the contrary, license-exempt spectrum usually allows the operation of many incompatible wireless protocols. A network including a set of wireless devices following a specific protocol may successfully share the spectrum. A different protocol might work equally well. However, the two networks, when operating in proximity, may appear to each other as unmanageable and harmful interference. Standards developers with products intended for license-exempt operations are increasingly aware of the need to coexist peacefully with devices based on different protocols. For example, IEEE 802 maintains the IEEE 802.19 Coexistence Technical Advisory Group to "develop and maintain policies defining the responsibilities of 802 standards developers to address issues of coexistence with existing standards and other standards under development." However, practical techniques for coexistence across protocol lines are still in their infancy.

TYPES OF WIRELESS NETWORKS

Fixed and mobile networks

Wireless devices may be classified based on whether the communicating devices are physically fixed at one location or allowed to move during a communication session. For example, the ITU defines fixed service as a "radio communication service between specified fixed points." In *fixed* wireless networks, wireless devices at both ends are stationary during operation. In *mobile* wireless networks, at least one of the devices can move during operation. An example of the mobile wireless network is the cellular telephone network in which the cellular BSs are fixed and the phones or handsets are mobile. Other types of mobile networks offer a range of mobility and support a variety of device velocities.

The regulatory requirements of fixed and mobile wireless networks differ for a variety of reasons. For example, since fixed networks are stationary and not carried by humans during operation, they may be allowed to use higher transmit power in certain regulatory domains. Similarly, mobile devices, especially handsets carried by humans, may have a different set of regulations with respect the transmit power and allowed emission levels. Both fixed and mobile networks exist in both licensed and license-exempt spectrum.

Apart from the regulatory requirements, fixed and mobile devices differ in technical design. Fixed networks may (in some cases) use high-power transmitters and involve high-gain, directional rooftop antennas with a clear LOS. This situation provides for a simple propagation channel. Although some fixed-system designs are based on more challenging links, to account for indoor antennas with minimal installation problems, the channel is relatively static. Mobile wireless networks, on the other hand, may involve a handheld subscriber unit with a small, omnidirectional antenna; therefore, the channel may be quite poor. Furthermore, since the device may be in rapid motion, the system must accommodate a rapidly varying channel. On the PHY level, the system must deal with changes in the radio frequency (RF) signal that are the result of the mobile device's movement, such as large-scale fading and Doppler. These phenomena will be discussed later in this chapter.

Effective mobile communication requires optimization at both MAC and PHY levels. An important additional requirement, with both PHY and MAC ramifications, is fast handover. For example, a mobile device communicating with a fixed BS must establish communication with a second BS, without disconnecting from the first, as it goes out of the range of the first device. Handover usually involves support from components outside the MAC and PHY. However, the MAC must include basic support for handover, working in conjunction with other functions.

Also, fixed devices are often attached to the electrical power grid so that power efficiency is a relatively minor issue. In mobile devices, power use is a foremost factor.

Nomadic and portable networks

Nomadic service refers to a mode of operation in which at least one of the wireless devices in the system is transportable but, during operation, the devices are stationary. For example, most laptops with IEEE 802.11 WLAN connections are operated in nomadic mode. *Portability*, on the other hand, is a physical property of the device itself. Portable devices may be mobile or nomadic from the communication perspective.

WIRELESS NETWORK TOPOLOGIES

The topology of a wireless network depends on how the wireless devices are connected and the role played by each wireless device. When a wireless network consists of only two directly connected devices, the topology is referred to as a *point-to-point* (PtP) network. In a PtP network, the access methods and protocols can be simplified, as there is no need to support multiple access. PtP networks are typically used to connect backhaul networks and are often fixed networks. The wireless devices at both ends of a PtP network usually have similar functionality and capabilities.

On the other hand, a *point-to-multipoint* (PMP) network consists of a single wireless device (a BS) communicating directly with more than one wireless client device (the SSs), thereby representing a star topology. The BS acting as a controller at the center of the network is responsible for performing additional functions to coordinate the actions of other devices. A PMP network requires a multiple access method. The communication from the controller device to other wireless devices is referred to as *downlink* (DL) or *downstream*, and the communication in the reverse direction is referred to as *uplink* (UL) or *upstream*. The DL and UL directions are also referred to as *forward link* and *reverse link*, respectively.

Both PtP and PMP wireless networks are a class of *single-hop* networks, in which the wireless protocol defines the communication across one wireless link. Although the wireless device on both ends may be connected to other wired or wireless networks to support distribution, aggregation, or forwarding services, these networks are considered independent. However, a third type of wireless topology, known as a *mesh*, is in the class of *multihop* networks. A device or node in a mesh network is usually connected to two or more other devices, and the communication needs to be coordinated for more than one hop for a mesh network to operate properly. Apart from basic access and connection management, the mesh networks also need to define additional functions such as routing, fault-tolerance, and interference mitigation. All the devices on a mesh network are expected to have similar capabilities and functions, although some devices may have additional functionality to perform special functions in a mesh network, such as a connection to another network or as an aggregation point.

While these three types of network topology have elements in common, there are some significant differences. A network and protocol designed for a PMP topology may be simplified and optimized to a PtP network. This is often the case in practice, as both networks are single-hop networks. However, the mesh networks require additional functionality that usually requires a different protocol or a set of extensions to support. Since the frame structures of the PtP/PMP and mesh operation can be significantly different, these modes not always interoperable.

The primary benefits of the mesh networks are as follows:

- Extended range with an extra hop
- Increased NLOS coverage when one or more mesh nodes are added to go around obstacles
- Alternate paths in case of failures or performance degradation

Mesh also comes with some disadvantages, including the following:

- Increased delays introduced by multiple hops
- Increased complexity of protocols (i.e., MAC, routing, management, security)
- Increased complexity of planning of initial coverage (network seeding)

If not properly managed, the delays introduced by multihop wireless networks may be unacceptable for certain types of applications.

RF PROPAGATION

LOS and NLOS

Wireless systems are often classified on the basis of whether restricted to LOS operation or are capable of NLOS operation. Strictly speaking, however, the real distinction is not based on visible light, as the propagation behavior is dependent on radio waves of wavelength far greater than that of visible light.

A more accurate description is provided in IEEE Std 802.16.2-2004 [B26], which defines LOS as the condition in which the signal path is more than 60% clear of obstructions within the first Fresnel zone. The first Fresnel zone is defined as the region along the transmission path in which a reflected signal will be less than 90 degrees out of phase. Signals with an absolute phase offset less than 90 degrees add constructively to the received signal, whereas signals with a higher offset actually have a cancelling effect on the total received signal. It is hence desirable to receive the reflected signals of low phase shift.

Figure 3–1 illustrates an obstacle in the transmission path. The first Fresnel zone includes, at each point along the transmission axis [the line connecting transmitter (Tx) and receiver (Rx)], the circular area centered on that axis with radius $R_{F1} = \sqrt{(\lambda \cdot D_T \cdot D_R)/(D_T + D_R)}$, where λ is the wavelength of the RF carrier. The resultant ellipsoid is represented in Figure 3–1 by the dotted ellipse. For a single obstacle, the link is hence said to be LOS only if $R_o > 0.6R_{F1}$, where R_o is the distance of the obstruction from the transmission axis. The dashed line in Figure 3–1 shows an example reflection with exactly 90 degrees of phase offset.

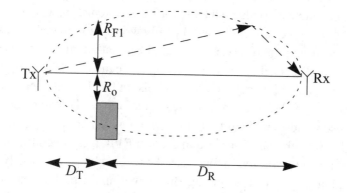

Figure 3–1:   Fresnel zone

The definition of R_{F1} also makes immediately clear that for visible light, where λ is several hundred nanometers, the first Fresnel zone is many times smaller than for the spectrum relevant for IEEE Std 802.16 where λ is on the order of 1 cm to 1 m.

If multiple obstacles are present, the LOS condition is based on the minimum R_o/R_{F1}. Note that the ellipsoid is defined in three dimensions, and obstacles can exist not only to the side of the direct transmission path but also below and, in rare occurrences, above. Specifically, the curvature of the earth (which becomes more important as links get longer) can intrude into the 60% range if the antennas are placed insufficiently high.

Another manner of looking at propagation through reflections is by considering the effect not in the phase domain but in the time domain. This is most easily understood by the realization that reflections have a longer overall transmission path and hence, given the constant propagation speed of the signal, will arrive later and be more attenuated than the direct nonreflected path. This condition is termed *multipath propagation*. As a result, if symbols are transmitted sequentially without interruption, the receiver sees energy from one symbol smeared into the energy for the subsequent symbol.

In general, three methods can be used to mitigate the degrading effect of multipath on IEEE 802.16 networks. The first method is the use of very narrow beam antennas. This is suitable for LOS systems, since by this method the first Fresnel zone reflections will be only weakly stimulated. The second method is the use in the receiver of an equalizer that attempts intersymbol interference (ISI) cancellation by extracting and cancelling the energy of a symbol that has bled into subsequent symbols. In the third method, redundant samples are prepended to the transmitted waveform so that the receiver can discard these samples without loss of information. When sampling is cyclical in nature [as with the fast Fourier transform (FFT)], the prepended samples are termed the *cyclic prefix*. In that case, the receiver need not strictly discard the prefix, but may start its sampling already within the prefix if its duration exceeds the multipath duration. As described in later chapters, the cyclic prefix method is used in the OFDM and OFDMA PHYs of IEEE Std 802.16.

By use of methods like the latter two above, a system may be able to mitigate the effects of multipath propagation and communicate successfully even in NLOS environments. Such a system may be termed an *NLOS system*.

Multipath

When a radio signal is sent from a transmitter to a receiver, depending on the LOS and the reflecting objects, the signal may analytically split into multiple components with different delay, amplitude, and phase. These multiple components of the same signal may have positive or negative impacts on the received signal at the receiver, depending on the relative phases. In general, destructive interference is known as *multipath fading*, also termed *small-scale fading*. This is discussed in the next section.

Apart from the number of potential reflectors, the locations of the transmitter and receiver, and the distance between transmitter and receiver, it is important also to consider the impact of the orientation and movement of the communicating devices on multipath fading. Multipath causes the time offsets at which signals with substantial amplitude are received to lengthen. The energy of a signal transmitted with duration T_b is received with a duration longer than T_b. If the multipath duration is large compared to the symbol duration, the energy of a transmitted symbol will be received during a large multiple of the duration T_b and hence impact many subsequent symbols in a continuous symbol stream. This effect is termed *intersymbol interference* (ISI). A number of methods, know generally as *equalization*, exist to extract the energy of a transmitted symbol out of a received signal, but all of these methods get substantially more complex as the ratio of the multipath duration to the signal duration increases.

A related concept is *delay spread*. The delay spread defines the differences in the arrival time of signals that took different paths from the transmitter to reach the receiver. A good receiver has to be able to tolerate a certain delay spread. The delay spread usually grows with propagation distance (because the radii of the Fresnel zones increase as a function of distance) and, therefore, become significant in the design of outdoor wireless systems, where the distance between transmitter and receiver is typically long and the chances of multipath high as well. As the delay spread approaches the symbol duration, the probability of bit errors increases. The effect of multipath cannot be minimized by simply increasing the transmit power.

The problem of delay spread is minimized in low-frequency (below 11 GHz) IEEE 802.16 systems by either using multicarrier modulation (WirelessMAN-OFDM and WirelessMAN-OFDMA) or using equalizers (WirelessMAN-SCa).

Fading

Fading, which is the reduction of the received signal strength, is generally divided into three categories; large-, medium-, and small-scale. Large-scale fading, which is deterministic, is better known as *path loss* and depends on the separation between the transmitter and receiver. Medium-scale fading is also known as *shadowing*. In general, it is the result of obstructions in the transmission path, such as rain and foliage.

Small-scale fading is in general the result of movement on transmission paths with multipath. Movement causes a shift in the signal's frequency as observed by the receiver, the so-called *Doppler frequency*, and causes rapid variations of the phases of the multipath components, the addition of which causes the rapid fading. The Doppler shift f_d itself can be calculated as $f_d = v \cdot \cos(\theta) \cdot f_c / c$, where v is the mobile's speed, θ the angle of movement relative to the direct line between transmitter and receiver, f_c the carrier frequency of the system, and c the speed of light. For example, for a mobile device traveling at 100 Km/h in a system operating at 2 GHz, the Doppler shift will range from −185 Hz to 185 Hz.

Small-scale fading is often characterized as nonfrequency selective (flat) or frequency selective. A signal is said to be flat faded if the coherence bandwidth of the channel exceeds the signal bandwidth; in other words, the fading of every frequency component is strongly correlated. A more detailed characterization of the above cursory description of fading can be found in [B49].

When a system is installed, its link budget is generally configured according to the large-scale fading with the addition of a margin for the cumulative effect of shadowing and small-scale fading.

ANTENNAS

Antennas are the physical elements responsible for converting the modulated signals into electromagnetic waves at the transmitter and performing the reverse operation at the receiver. The performance of an antenna constructed of usual materials is the same for both transmission and reception due to the fundamental property of reciprocity. A more detailed discussion of antenna technologies can be found in [B4], [B27], and [B47]. In this section, we discuss some of the basic antenna concepts.

Antenna parameters

The characteristics of the antennas used for wireless applications vary based on specific system requirements, such as deployment and coverage. The following is the commonly used set of parameters to characterize antennas:

- *Gain:* The gain of the antenna is the power output in a specific direction compared to that of a hypothetical isotropic antenna. The antenna gain is typically measured in dBi, the logarithmic scale of decibels relative to the isotropic antenna.

- *Radiation pattern:* The radiation pattern of an antenna is the graphical depiction of the antenna gain as a function of the elevation and azimuth angles. The radiation patterns for the horizontal and vertical planes are typically drawn on a polar graph as a function of the azimuth and elevation angles, respectively. A rectangular graph, instead of a polar graph, is sometimes used to illustrate the radiation patterns of narrow beam antennas.

- *Beam width:* The directionality of an antenna is indicated by its beam width. The beam width is measured in degrees between the points on the radiation pattern of the antenna at which the radiated power density is exactly half the maximum.

- *Polarization:* The polarization of an antenna is determined by the orientation of the electrical field component of the electromagnetic waves radiated by the antenna. Antennas usually provide linear, circular, or elliptical polarization. The linear polarization can be further classified into horizontal and vertical polarizations.

- *Bandwidth:* When an antenna is designed to operate in multiple frequency bands, it is important that it maintain uniform impedance across all those frequencies. Antenna impedance matching plays an important role in maximum power transfer between transmitter or receiver circuitry and free space. The band of frequencies in which an antenna is designed to operate is known as the *antenna bandwidth*.

Apart from the system requirements, regulatory rules may also limit the type of antennas that can be used. For example, many regulatory domains have specific limits on antenna gain and/or beam width. Some regulatory domains may limit the effective isotropic radiated power (EIRP), which is a sum of transmit power and antenna gain, minus any loss in the transmit path (all in decibels).

Directional and sectorized antennas

The antennas used in PtP, PMP, and mesh systems usually have different beam widths. Omnidirectional antennas are rarely used in fixed outdoor PtP systems, given the range requirements of the outdoor systems and the need to minimize interference. Directional antennas have important characteristics that are of significance to BWA; i.e., they increase the amount of power transmitted in a specific direction and reject the signals coming from other directions on reception. For PtP systems, highly directional antennas are used, as the goal is to establish communication only between two specific and fixed end points.

PMP systems typically deploy sectorized antennas, with beam width greater than 15 degrees, at the BS to improve frequency reuse in cellular systems. These minimize the interference across cells, as compared to omnidirectional antennas. Sectorized systems use more than one sector to cover all 360 degrees, if necessary. As many as six sectors are used in many practical systems.

Even in the case of PMP, the UL antenna may, like the PtP case, be highly directional and pointed at the BS. The tradeoff for such a high-gain antenna is the need for a LOS and an expensive professional outdoor installation that may call for careful alignment. Consumer-based applications, particularly at the lower frequencies, may call for self-installation and, therefore, require a much different (perhaps omnidirectional) SS antenna.

Diversity

Diversity is a technique used in wireless communication to mitigate the effects of multipath fading and to improve performance by exploiting the fading characteristics of wireless transmissions. For example, the *temporal* and *spectral diversity* schemes make use of the time-selective and frequency-selective fading characteristics of channels, respectively. When neither of these schemes is possible or desired, *spatial diversity* schemes may be used. In spatial diversity, more than one antenna is used in the transmitter and/or the receiver to make use of the spatially independent fading characteristics of the channel. Since spatial diversity requires more than one antenna, it is sometimes referred to as *antenna diversity*. Depending on the use of multiple antennas at the transmitter or receiver, it may be called *transmit antenna* or *receive antenna diversity.*

Multiple diversity schemes may also be combined to provide improved performance. For example, spread spectrum and OFDM schemes essentially make use of spectral diversity of the frequency selective characteristics of the channel. Spatial diversity schemes can also be used in a spread spectrum or OFDM system. The following subsections briefly describe temporal, frequency, space, angle, and polarization diversity.

Temporal diversity

In temporal diversity method, the signal is repeated in time, where the time interval between repetitions is sufficiently large to have different channel characteristics. Therefore, the received signals are expected to have independent fading.

Frequency diversity

As the name implies, frequency diversity involves the transmission of the signals in frequencies spaced sufficiently far apart to cause independent fading. Frequency diversity is typically exploited only in multicarrier systems, since the use of more than one frequency is otherwise not practical. In IEEE 802.16 multicarrier PHYs, frequency diversity is exploited through the use of interleavers and the mapping of subchannels to physical carriers. Frequency diversity may be implemented with one antenna.

Spatial diversity

Spatial diversity schemes can be classified into *diversity combining* and *switched diversity* schemes. With diversity combining, the signals received from more than one antenna are combined to reduce the effect of amplitude fading due to multipath. The receiver typically uses switched (selection) combining, maximum ratio combining (MRC), or equal gain combining (EGC). In switched combining, antennas are placed sufficiently apart that the fading is independent on each receiver antenna. The combining process takes the best signal from the receive antennas. MRC requires the estimation of the channels and weights the signals received at each receive antenna, per the estimation, in order to maximize the signal to noise ratio (SNR). EGC is very similar to MRC except that the weights are independent of the channel estimates. Instead, the signals from each receive antenna are multiplied by the same weight. Since there is no channel estimation in EGC, the SNR gain of EGC is lower than that of MRC.

Diversity combining is sometimes impractical due to its complexity and cost. The simpler switched diversity schemes may be used to switch between one of the available antennas based on known characteristics or some dynamic measurements. The switched diversity schemes can be implemented at either the transmitter or the receiver. The popular IEEE 802.11 WLAN devices often use transmit diversity with two antennas. The receive antennas can also be chosen similarly.

A variation of receiver diversity known as *fast-receive diversity* is sometimes used to adaptively choose the antenna with the best signal for reception. However, fast diversity schemes cannot be supported if the preamble is too short, as there may be insufficient time to detect and measure signals and adaptively switch between antennas. The goal of switched diversity schemes is to mitigate the effects of multipath, not to increase the gain.

All the spatial diversity techniques require at least two antennas. The diversity combining techniques use at least two antennas simultaneously. However, the switched diversity techniques use but one transmit or receive antenna at any given time.

Polarization diversity

Polarization diversity makes use of the orthogonality of polarized transmissions. Since the fadings for horizontally and vertically polarized signals are often uncorrelated due to the phase and amplitude differences, polarization diversity can be implemented with a single antenna that is capable of receiving both vertically and horizontally polarized signals. In practice, polarization diversity is often implemented with two antennas, one horizontally polarized and the other vertically polarized. When space is limited, polarization diversity can be helpful since the antennas can be placed together. It should also be noted that, in a highly reflective environment, the polarization of the transmitted signals may be lost; therefore, polarization diversity may provide only minimal gain.

Angle diversity

With angle diversity, the pointing angles of the antennas are different. This technique makes use of the uncorrelated signals arriving at different angles. Angle diversity requires at least two antennas. As with polarization diversity, angle diversity is useful when the location of the antennas are space constrained.

Multiple antenna systems

Multiples antennas may be used on a single wireless device for a variety of reasons. The most common application of multiple antennas is to support one of the diversity schemes described in the previous section. Adaptive antenna systems (AAS) and multiple-input, multiple-output (MIMO) systems also exploit the properties of multiple antennas, as described below.

Adaptive antenna systems (AAS)

Adaptive antennas can adaptively adjust to meet specific performance goals, such as improved link margin or improved signal to interference ratio (S/I). Although adaptive antennas can be implemented in many different ways, they are classified into one of the following three categories:

- *Switched beam:* This is the simplest form of adaptive antenna design, where the antenna pattern of each antenna is fixed. The best antenna for communicating with a specific remote wireless device is chosen adaptively based on a variety of factors such as past measurements from each antenna. The antenna selection concept of switched beam architecture is very similar to that of switched antenna diversity.

- *Beam steering:* Unlike the fixed beams of switched beam design, the beam steering approach is used to adaptively steer the beams toward the remote wireless device to maximize the gain. The objective is to maximize the gain from and to every wireless device.

- *Optimum combining:* The optimum combining method uses a linear spatial filter to adaptively adjust the antenna, with a periodically adjusted reference signal, through feedback. The linear spatial filter is used to suppress the noise and interference, while separating the desired signal. Optimum combining should not be confused with the aforementioned MRC, as MRC is not spatially discriminatory and is optimal only when the noise is white (i.e., uncorrelated at the different antennas) and there is no interference.

The gain and S/I achieved by switched beam adaptive antennas are limited by the pattern, the angle, the interference, and the LOS conditions of the area covered by the fixed antenna. The complexity of the beam steering and optimum combining approaches also depends on whether the wireless devices are fixed or mobile. For the beam steering approach, maximizing the gain for each wireless device requires that the power be controlled for every beam. While the path loss may not change significantly for a fixed wireless device, a mobile device may require dynamic power adjustments to the beams.

The optimum combining approach depends on a feedback system and requires dynamic update of filter parameters to account for fading. Since the fading rate of mobile devices can be much higher, the complexity of processing the updates in mobile networks is typically higher than that of fixed networks.

Multiple-input, multiple-output (MIMO)

MIMO systems make use of multiple antennas at both ends of the link. In a MIMO system, potentially uncorrelated signals from different antennas can be used to maximize capacity. Given this property, the gain from MIMO in LOS or PtP links will be minimal, as the signals in such environments will be highly correlated. However, MIMO offers many benefits in a typical NLOS environment, where the multipath effect leads to highly uncorrelated signals.

The main difference between adaptive antennas and MIMO is that the adaptive antenna schemes achieve capacity gain by suppressing interference through directional discrimination, allowing simultaneous transmissions to different terminals. MIMO, on the other hand, achieves the same by exploiting uncorrelated signal paths to, in effect, transmit multiple data streams simultaneously to a single terminal.

Impact of antenna technologies on protocol design

In any wireless system, the MAC protocol design must consider the antenna technology. While some MAC protocols may work equally well on a variety of antenna types, some protocols are optimized for or constrained by specific antenna types. For example, if the protocol depends on "listen before talk" schemes such as CSMA, any directionality may work against the efficiency of the protocol because some receivers are inherently unable to detect some transmitters and cannot, therefore, sense ongoing transmissions.

On the other hand, omnidirectional antennas support a broadcast-type network that can be used to minimize the control traffic overhead. The lack of broadcast capability can increase the overhead in multiple antenna systems such as the AAS. Therefore, the MAC protocols designed for multiple antenna systems need to optimize the control traffic overhead without sacrificing other benefits of such a system. There may also be some implementation constraints in multisector systems, where the MAC transmissions and receptions may need to be coordinated across sectors, depending on the frequency plan.

Antenna design for fixed and mobile devices

The choice of antenna technology also varies depending on whether a system or device is fixed or mobile. Since a typical mobile system can include both fixed devices such as BSs and mobile devices such as handsets, the fixed and mobile devices in the same system may use different types of antennas. For example, most mobile devices are too small for directional antennas; even if they were not, orienting such antennas would be problematic. Therefore, most mobile devices use omnidirectional antennas, even if the BSs use directional or sectorized antennas.

The more complex adaptive antennas are also difficult to implement on a mobile handset using current technologies. Similarly, the support for antenna diversity with sufficiently spaced antennas is limited, depending on the frequency of operation, due to the size. Therefore, some of these adaptive antenna and diversity methods are used only in a fixed system or the fixed part of a cellular system. Some types of diversity, such as polarization diversity, are also unsuitable for a mobile device since the polarization is difficult to control.

PHYSICAL LAYER (PHY)

Forward error correction (FEC)

One way to deal with errors in wireless transmissions is to automatically detect and correct symbol or bit errors rather than relying on the detection and retransmission of packet errors [e.g., using checksum or cyclic redundancy check (CRC)]. The most commonly used technique for error correction is called *forward error correction* (FEC), which is capable of detecting and correcting some errors upon reception by adding redundancy to the transmitted signals.

Wireless protocol design must consider the overhead and cost associated with PHY-level error control schemes, such as FEC, and MAC-level error control schemes, such as automatic repeat request (ARQ). For example, while FEC may be able to correct certain types of errors quickly with high probability, the addition of FEC to every transmitted block reduces the efficiency of the channel and could increase the delay of good protocol data units (PDUs) due to longer interleaving. On the other hand, MAC-level ARQ can increase the delays in a channel with high error rate, where the detection, communication, and retransmission can consume a significant portion of the bandwidth. Many systems support hybrid techniques, where a combination of FEC and ARQ parameters may be adjusted, often adaptively, to offer the desired performance.

Single carrier and multicarrier

As described above, the delay spread and fading caused by multipath are challenging problems that any practical outdoor, or outdoor-to-indoor, wireless system must solve. While equalizers can be used to offset the effect of multipath in a single-carrier protocol, the complexity of time domain equalization and associated channel estimation increases with higher data rates and time-varying channels. In most BWA cases, the delay-spread–induced ISI is significant only for high signal rate, typically above 1 Msymbol/s (i.e., symbol durations shorter than 1 μs). If the signalling rate can be reduced by increasing the symbol duration, the effect of ISI on the channel can also be reduced. However, high signal rates are desirable for high data rates. In multicarrier modulation schemes, the available frequency band of a given channel is divided into a large set of (typically, orthogonal) subcarriers, each of which is modulated separately. This leads to a much smaller signal rate per subcarrier compared to single-carrier systems. The data rate achievable over these simultaneously transmitted subcarriers is almost equivalent to the data rate of a high-rate single-carrier transmission, with the difference stemming from the cyclic prefix overhead. Multicarrier modulation schemes, such as OFDM (see [B55]), do come with other drawbacks, including the higher sensitivity to frequency offset (the equivalent of timing sensitivity in other systems) and phase noise. They also require a somewhat larger power amplifier backoff due to higher peak to average power ratio (PAPR).

Instead of time-domain equalization, a single-carrier system may also employ frequency-domain equalization. In this case, the receiver contains an inverse fast Fourier transform (IFFT) and FFT pair, with channel estimation executed between the transforms. The result is some characteristics similar to those of an OFDM transceiver, including similar complexity.

DUPLEXING, MULTIPLEXING, AND MULTIPLE ACCESS

The following subsections provide a brief overview of some of the key duplexing, multiplexing and multiple access concepts. For a detailed discussion on these and related concepts, the reader is referred to [B4] and [B7].

Duplexing

Duplexing defines how bidirectional communication is achieved between two devices or between a BS and a set of client devices in a PMP system. The communicating devices themselves may be capable of half-duplex (transmit or receive but not both simultaneously) or full-duplex (transmit and receive simultaneously) operation. There are two types of duplexing: *time division duplexing* (TDD) and *frequency division duplexing* (FDD). Figure 3–2 illustrates TDD communication in a PMP system, with DL and UL on the same frequency channel. TDD is a half-duplex method by definition; i.e., when one device transmits, the other receives, and vice versa.

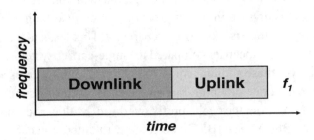

Figure 3–2: TDD illustration

In FDD, the devices use different frequencies to transmit and receive. Therefore, it is possible to support full-duplex communication in an FDD system. Figure 3–3 shows full-duplex operation, where the transmission and reception overlap in time. However, half-duplex frequency division duplexing (H-FDD) operation, as shown in Figure 3–4, is also possible, as is a mix of full-duplex FDD and H-FDD devices in the same system. An H-FDD device is usually cheaper to build as it requires only one transmitter and one receiver. The H-FDD operation is also referred to as *frequency shift division duplexing* (FSDD). Alternate expansions of the abbreviation FSDD include *frequency switched division duplexing* and *frequency simplex division duplexing*.

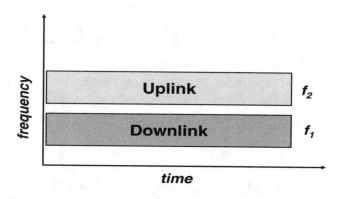

Figure 3–3:   Full-duplex FDD illustration

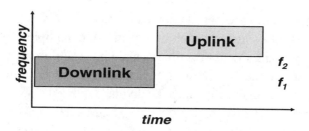

Figure 3–4:   H-FDD illustration

In TDD, the time allocated for each direction (i.e., UL or DL) may be fixed or adaptive. Adaptive time division duplexing (ATDD) is a variation of TDD in which the UL/DL ratio can be dynamically adjusted per frame, based on traffic conditions and other system parameters. ATDD is of great interest to broadband wireless systems, due to its flexibility in supporting asymmetric traffic and efficient use of the spectrum. However, in cellular deployments with multiple ATDD systems, if frequency reuse is not properly employed, ATTD may increase the level of interference. In some practical multisector systems, the ATDD boundaries may need to be coordinated across sectors, even if the sectors operate on different frequency channels due to adjacent channel interference issues.

FDD may require more spectrum than a comparable TDD system, increasing the potential for interference, especially in license-exempt spectrum. The frequency channels used in both directions also need to be sufficiently separated to avoid interference. Moreover, in access networks, spectrum may be wasted if FDD uses symmetric channels in both directions, as the traffic load may be asymmetrical. On the other hand, the QoS scheduling and channel access functions may be less complex in FDD compared to TDD systems.

The duplexing method is independent of the multiple access and multiplexing methods. FDD is commonly used in licensed voice-centric networks such as cellular telephone systems, where most of the traffic has historically been symmetric.

Multiplexing

Multiplexing refers to a mechanism in which a single device transmits to multiple devices on a single channel. In time division multiplexing (TDM), the transmitting device divides the time domain into multiple slots to communicate with multiple devices. Needless to say, TDM requires only one frequency, but the available time must be divided among the potential receivers. In frequency division multiplexing (FDM), the transmitting device uses different frequencies to communicate with multiple devices. Any wireless protocol design has to balance this flexibility against the spectrum usage requirements.

Centralized and distributed multiple access schemes

Multiple access refers to the way that multiple devices access the medium, regardless of whether the communication is many-to-one or many-to-many. Distributed access methods such as CSMA require no centralized coordinator. While this may be simpler to implement, the spectrum efficiency of this method is low; and the potential for collisions affects the overall system throughput and scalability, causing a potentially dramatic decline in efficiency as the system load increases. Collisions also introduce nondeterminism into the system, thereby making QoS guarantees difficult to achieve.

Time division multiple access (TDMA) and frequency division multiple access (FDMA)

In TDMA, multiple devices use their own predetermined time slots to transmit. The number and location of the time slots may be statically determined or dynamically assigned. If dynamically assigned, some form of coordination (centralized or distributed) is necessary. It is possible to provide a deterministic access to the medium using a TDMA scheme. For example, IEEE Std 802.16 relies on the BS to allocate time slots for UL transmissions. The BS also schedules DL transmissions in a similar fashion, using TDMA bursts addressed at specific groups of SSs. Some other PMP systems use continuous TDM in the DL, with the information for each SS multiplexed onto a single stream of data directed to all SSs within the sector. The continuous TDM stream does not require preambles for resynchronization between bursts. The downside, however, is that it does not allow the flexibility of adaptively adjusting the coding and modulation to specific SSs.

FDMA is similar, with multiple devices using different frequencies to access the medium. However, the amount of spectrum needed and the co-channel and adjacent channel interference can decrease the efficiency of the FDMA systems. While traditional TDMA systems used fixed allocations to the existing devices, which may result in wastage of bandwidth if not used, it is possible to allocate bandwidth dynamically in a TDMA system, where the slot allocations are changed and communicated dynamically. Both TDMA and FDMA can be used with FDD or TDD. Hybrid or combination methods are also possible.

Orthogonal frequency division multiple access (OFDMA)

As described earlier in this chapter, OFDM is a multicarrier modulation method, but not a multiple access method. OFDM may be used with many multiple access methods, including CSMA, TDMA, and FDMA. The IEEE 802.11a and IEEE 802.11g amendments use an OFDM PHY with CSMA, and the IEEE 802.16 OFDM PHY option is used with TDMA. However, OFDM's use of multiple subcarriers can be used as the basis of a multiple access method. The extension of OFDM as a multiple access method is called *orthogonal frequency division multiple access* (OFDMA), in which a subset of the mutually orthogonal subcarriers may be allocated to a specific wireless device for access to the channel. The concept is a sophisticated form of FDMA, with the orthogonality of the subcarriers making it easier to separate them without the need for guard bands.

Many variations of OFDMA exist, distinguished primarily on the basis of how the subcarriers are allocated to users. For example, the subcarriers may be allocated in a frequency-hopping pattern. They may also be allocated dynamically by a central controller. The difference between OFDM and OFDMA is illustrated in Figure 3–5. While all the subcarriers will be used by a single user in OFDM, the subcarriers may be shared by a number of users in OFDMA. The number of subcarriers assigned to each user may vary based on system parameters and demand.

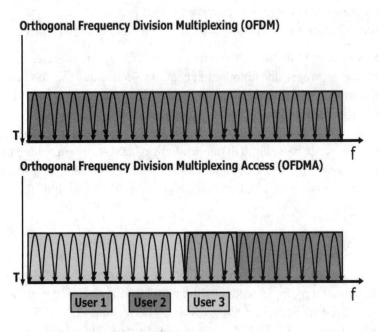

Figure 3–5:   OFDM and OFDMA

Code division multiple access (CDMA)

In CDMA, each user is assigned a unique code used to encode the transmissions on a channel. Instead of transmitting a single data waveform, the transmitter sends the time-domain convolution of the data waveform with the code so that the transmission rate (often referred to as the *chip rate*) is many times higher than the data rate. The high transmit rate, and hence short duration of each chip (code element), results in a spreading of the signal in the frequency domain. This is why these codes are often referred to as *spreading codes*.

At the receiver, the convolution of the signal with the known sequence allows the retrieval of the original data symbol while spreading out any narrowband interference in the channel. CDMA is a popular choice for military systems for the following reasons:

1. The requirement that transmission code be known in order to receive the signal

2. The spreading of narrowband interference

3. The fact that the spreading of the signal in the frequency domain can lower the energy per hertz to well below the thermal noise floor so that its existence is not obvious to third parties

The design of CDMA codes is based on minimizing the autocorrelation of the code and the cross-correlation of the code with all other used codes. The better this design, the easier it is for the receiver to separate the simultaneously transmitted codes, and the more robust the system is to power variations between terminals.

Since multiple codes can be used simultaneously on the same channel, CDMA can be a bandwidth-efficient solution that requires no coordination or synchronization.

In the OFDMA PHY described in Chapter 12, the CDMA concept is also used for contention-based access. In that case, however, the spreading sequence is not transmitted in time, but each chip is modulated on a separate subcarrier before the IFFT.

DATA UNITS

Data units are the basic units exchanged between different layers of the protocol stack. The service data unit (SDU) and the protocol data unit (PDU) are the two fundamental data units used here. The SDU is the data unit exchanged between two adjacent protocol layers of the same device. The service access point (SAP) defines the interface between two adjacent protocol layers, where the services of the lower layers are available to the higher layers. The SDUs are exchanged between two adjacent protocol layers through the SAP, as the lower layers provide the services of accepting the SDU from the higher layer and delivering the SDU to a higher layer. The PDU is the data unit exchanged between the peer entities of the same protocol layer. When a PDU is transferred to a lower layer for transmission, it becomes the SDU of the lower layer.

An SDU received from a higher layer may go through a variety of transformations to become a PDU for transmission to a peer layer. When the MAC or PHY receives an SDU, it may add additional headers and control information to reliably transmit the PDU to the peer. In addition, SDUs may go through fragmentation, in which more than one PDU may be constructed from a single SDU. Similarly, fragments of multiple SDUs or multiple full SDUs may be packed into a single PDU. The only basic requirement is that the inverse transformation must be performed by the same peer layer on the receiving end before delivering the PDU to a higher layer. Figure 3–6 shows the concept of PDUs, SDUs, and SAPs.

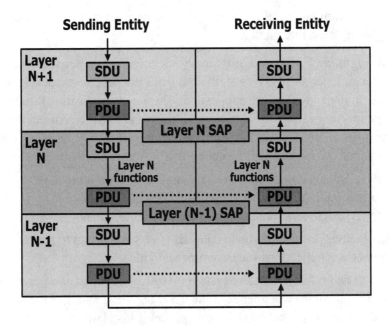

Figure 3–6:   PDUs, SDUs and SAPs

QUALITY OF SERVICE (QoS)

A BWA system supporting multiple services must properly account for the fact that different services require different performance levels in order to maintain the appropriate QoS. A network that supports differentiated QoS can provide services at specified performance levels using a set of qualitative and quantitative measures.

Applications such as voice, video, and data transfer demand various QoS requirements. The QoS requirements of the application traffic can be specified and enforced either as a set of priorities or as a set of parameters. In the prioritized QoS model, the traffic is treated by the network based on relative priority at any instant. However, in a parameterized QoS model, the network attempts to guarantee a set of QoS parameters for the traffic, in conformance with a preagreed or configured service.

The most commonly used QoS metrics in the parameterized QoS model are throughput, delay, jitter, and loss:

- *Throughput:* Throughput is typically specified in bits per second, although it can be also be specified in bytes per second or packets per second, depending on the application. The actual application throughput depends on a variety of factors, including packet size, overhead, and retransmissions. The achievable throughput is bound by the maximum and instantaneous capacity of the network, also specified in bits per second.

- *Delay:* The delay, also known as *latency,* is the time taken for the information to travel from a source to a destination and vice versa. The delay is specified in units of time, such as milliseconds.

- *Jitter:* Jitter is the variation in delay. It is very important to many interactive applications such as voice and video.

- *Loss:* The packet loss in any system is typically defined as a percentage. Data applications can tolerate some delay and jitter, but they cannot readily tolerate packet loss. If a data packet is lost, it must be retransmitted by some layer (e.g., network or application) for the application to work correctly. On the other hand, real-time applications such as voice and video can tolerate a small percentage of loss. Retransmission can be used to compensate for packet loss at the cost of increased latency and reduced available capacity.

The simplest approach to QoS is static provisioning of services and reservation of resources. This is how circuit-switched networks guarantee QoS, where a circuit is established by reservation of specific resources for the duration of a connection or call. However, this is not a practical or efficient solution in packet networks, since not all users are expected to be simultaneously active and not all active users are expected to use a fixed percent of the network resources. The challenge in such networks is to support adaptive QoS techniques that can utilize the available resources more efficiently.

Therefore, in order to fully support QoS at the network layer, the following four functions are required:

- ***Admission control and provisioning:*** Irrespective of how a network solves the QoS problem, networks are constrained by the maximum amount of traffic and services they can support at any particular instant. Therefore, it is important to regulate which users can access the network and how they do so. The control must be dynamic, based on the network load and other parameters. This process is called *admission control* and, without it, networks may become oversubscribed, defeating the purpose of QoS guarantees. *Provisioning* is a related concept. Networks may be statically or dynamically provisioned for specific traffic situations. Static provisioning is often used in conjunction with admission control to over-provision the network and then manage admissions dynamically. While a variety of algorithms can be implemented to perform admission control, the underlying network must provide certain functions to support intelligent admission control functions. A simple admission control algorithm may merely admit or deny new requests based on the number of previously admitted services. However, a more sophisticated algorithm may try to adjust the parameters of admitted services to accommodate the new request. In order to support such admission control functions, a network must provide basic support for modifying admitted service parameters gracefully without affecting the performance or increasing the overhead. Except for such support functions at the network, admission control is primarily an independent function. Therefore, the specification of the admission control algorithm is outside the scope of an air interface standard. Vendors are free to implement specific admission control algorithms to differentiate their products.

- ***Traffic classification:*** In order for the network to provide differential treatment based on QoS requirements, it must be able to classify the incoming packets. The classification process may use some basic header information from a packet to determine how it should be treated. Some QoS-aware networks may also support specific identifying information in each packet to carry the classification or priority information so that explicit traffic classifications at every step (e.g., every hop or every layer of a protocol stack) can be avoided. As we will see in later chapters, the IEEE 802.16 MAC is connection-based and can be used to associate a set of QoS parameters after classification or to map preclassified packets from another protocol to a specific connection.

- ***Shaping and policing:*** The admission control manages the admission of a specific traffic stream or flow into the network. However, the admitted traffic still needs to be regulated dynamically to make sure it conforms to the agreed-upon service specification or profile. Shaping and policing are the two commonly used techniques to monitor and regulate incoming traffic for conformance. *Policing* merely discards or reclassifies the nonconforming packets to a lower priority or class. *Shaping*, on the other hand, may queue the nonconforming packets and delay the release of the packet into the network. The algorithms for shaping and policing are not typically specified in a networking standard. However, a standard may support the necessary signaling and management functions to specify shaping and policing parameters.

- ***Traffic scheduling:*** The traffic scheduler is responsible for determining the transmission order of the packets based on the QoS requirements, when fluctuations in the incoming rate results in queueing of the traffic. Therefore, the scheduler is an important component of any QoS architecture. A network without QoS support treats all types of application traffic alike in a first-in, first-out (FIFO) fashion. A QoS scheduler, on the other hand, has to ensure that the traffic is serviced based on the QoS requirements and not based solely on the order of arrival. However, fairness is still an important criterion for any QoS scheduler, and the higher priority traffic should not completely starve the lower priority traffic. As we will discuss later, wireless QoS scheduling has some important differences compared to wired QoS scheduling. The networking standards may define specific scheduling behaviors and support functions for QoS signaling and enforcement. However, a specific scheduling algorithm is typically not specified by a standard. IEEE Std 802.16 is no exception and leaves the scheduling algorithm as a vendor differentiator.

Per-flow and per-class QoS

QoS can also be enforced at varying levels of granularity. For example, each flow (e.g., every voice conversation) can be classified into a separate stream and handled separately by a network. Alternatively, each class of traffic (e.g., all voice traffic or all video traffic) can be classified into a separate stream.

For a complete discussion on per-class QoS, per-flow QoS, or other ways of differentiating QoS, see [B33] and [B42].

As a general rule, it is practical and sufficient to provide coarse-grain QoS only in core aggregated networks or backbone networks. It would be impractical to maintain state for a large number of flows in a network that carries traffic from thousands of sources. However, fine-grain QoS may be practical and is often necessary at the edge of the network, where the number of unique flows can be smaller and the need for QoS differentiation can be much greater, given the limited availability of resources and other constraints. While some network architectures may support only per-class QoS, others may support both per-class and per-flow QoS.

Wireless QoS

Enforcing QoS can be simply viewed as efficiently reordering transmissions to guarantee throughput, delay, and jitter requirements of all traffic. However, QoS enforcement has fundamentally different challenges in wired and wireless systems. In wireless networks, QoS is not just the reordering of packets. While enforcing QoS in any network in an efficient manner with conflicting requirements can be challenging, most wired networks fall under the class of fixed-capacity networks, where the maximum available capacity does not change over time. Also, the probability of errors in wired networks such as Ethernet or ATM is extremely low. Therefore, a QoS enforcement algorithm in a wired network is primarily responsible for optimal reordering of transmissions. This also simplifies the admission control process, as the instantaneous capacity is fixed and is always known.

On the other hand, wireless networks are varying-capacity networks, where the instantaneous capacity varies based on the conditions of the RF environment. This brings in additional constraints so that scheduling of bytes or packets to enforce QoS may not be sufficient or fair. For example, the airtime required to transmit a byte varies with the modulation and coding scheme. The changing RF conditions often force the modulation and coding rate to be changed dynamically, introducing an additional constraint for the QoS scheduler.

Another important dimension in wireless QoS is the ability of a scheduler to deal with high error rate. For example, if a transmission failed to reach the receiver due to large-scale fading, it may not be advisable to retransmit the data immediately, as transmission is likely to fail again. A high retransmission rate for a specific receiver not only increases the average delay for that receiver but also affects the QoS of others in the system. The QoS scheduler has a challenging task of maintaining fairness in such a system, balancing throughput and the use of over-the-air resources. Dynamic fragmentation is one way to address fairness, fragmenting SDUs dynamically to fairly divide the airtime.

If the wireless system supports ATDD with centralized scheduling (i.e., the BS is responsible for QoS in both directions), the scheduler also has to decide on a frame-by-frame basis how to share the channel between the UL and DL.

MEDIUM ACCESS CONTROL LAYER (MAC)

The MAC function is responsible for controlling access to the medium. MAC protocols may use distributed multiple access techniques, or centrally coordinate the access for all devices, or use a combination of the two, as discussed earlier.

Apart from multiple access, MAC is also responsible for basic functions such as data encapsulation, fragmentation, and adaptive modulation support. A wireless MAC must also define how MAC protocol data unit (MPDU) errors are detected and how, if necessary, the faulty MPDUs are retransmitted. QoS and security are two other functions that must be supported in a wireless MAC, given the limited bandwidth and reduced level of physical security in comparison to wired networks. The MAC, along with the LLC, is Layer 2 of the OSI reference model.

Fragmentation and packing

The MAC is responsible for encapsulating higher layer packets or SDUs in its own PDU format and delivering it to the peer MAC on another device. One simple way to encapsulate a higher layer SDU is to add necessary information and deliver the whole MPDU to the peer MAC. However, there are many reasons for the MAC to fragment a single SDU into multiple fragments before transmission, reassembling those fragments at the other end before delivering the resultant PDU to the higher layers. Fragmentation is necessary to efficiently support higher layer protocols with variable-size SDUs, such as Internet Protocol (IP), if the underlying MAC and PHY are based on a framed architecture (i.e., the frame size and, therefore, the available airtime per frame is fixed for a specific deployment). Some rationale and benefits of fragmentation include the following:

- **Probability of successful delivery**: The probability of successful delivery may depend on the size of the MPDU, especially in random access MAC or lower data rate transmission, where the chances of collision or other failure increases with the size of MPDUs.

- **QoS**: There are many QoS-related reasons for fragmentation. A low-priority, but large, SDU may need to be fragmented so that it does not block a smaller, but higher priority, SDU. The so-called *head-of-line blocking problem* may be alleviated by fragmentation. Another QoS-related reason is to minimize the overall delay in an error-prone channel, reducing the probability of a larger SDU being retransmitted.

- **Spectral efficiency**: Fragmentation in an error-free channel may increase the MAC overhead, thereby adversely affecting spectral efficiency. On the other hand, the cost of retransmitting a large lost MPDU is higher than the cost of retransmitting a smaller fragment of a large SDU. Another important aspect of efficiency is in a framed MAC, where the data have to fit within predefined frame boundaries. In order to fit MPDUs into the available frame time without wasting the airtime, it is important to dynamically fragment SDUs and fit the MPDUs into a frame.

While fragmentation is useful with large SDUs, many communication protocols have smaller SDUs to transport. An example is the widely used networking protocol, Transmission Control Protocol (TCP), in which 40-byte acknowledgments traverse the reverse direction of the data flow and, on average, one acknowledgment is sent for every two TCP packets. Voice application also uses a small payload size to keep the delay at a minimum. Since MAC encapsulation introduces some fixed overhead per packet, it is beneficial for the MAC to allow the packing of multiple higher layer payloads into a single MAC payload, without affecting the delay or fairness of the network.

Automatic repeat request (ARQ)

Any wireless system must define a process for handling MPDU errors. The PHY may support error correction methods to minimize the number of errors. Nonetheless, additional error correction methods are necessary, especially in a wireless MAC, as bit errors are inevitable. The ARQ is a class of retransmission algorithms for supporting reliable delivery in the presence of errors. An ARQ algorithm must define the following:

- Error detection
- Feedback policy
- Retransmission strategy
- Retransmission unit

The detection of MPDU errors is typically handled by adding a checksum or a CRC to every MPDU transmitted. It is important to choose a CRC or checksum that has an extremely low probability of undetected errors and minimal overhead to the protocol. The FEC described in the PHY section earlier in this chapter may also be used as an error detection mechanism, when the FEC detects errors but cannot correct them.

An important part of the retransmission algorithm is determining which MPDUs have not reached the receiver. The design of the ARQ feedback policy needs to consider the overhead associated with the feedback and the potential delay introduced by a specific policy. The ARQ feedback may include positive and/or negative acknowledgments from the receiver, commonly referred to as *ACKs*. Acknowledgments may be sent for every MPDU or for a number of MPDUs in an aggregated form. If the acknowledgment is per-MPDU, then the choice is to send the acknowledgment either immediately following every MPDU or at a later time.

The baseline IEEE 802.11 WLAN MAC used an immediate per-MPDU acknowledgment for all directed MPDUs. While it is possible to make retransmission decisions quickly in this type of acknowledgment strategy, the overhead can be high. A variation of this might be to send only negative acknowledgments instead of positive acknowledgments. If the error rate is low, this might reduce the overhead and keep the delay under control. Of course, the positive and negative acknowledgments can be combined to provide a complete feedback to the transmitter. Another way to minimize acknowledgment overhead is to use a bitmap to indicate positive and negative acknowledgments.

Once an error is detected and the missed MPDU information is communicated to the transmitter, a retransmission policy decides when and how MPDUs are retransmitted. The retransmission policies can be broadly classified into three types: *stop-and-wait*, *go-back-n* (GBN), and *selective repeat* (SR) (see [B6] and [B9]). In stop-and-wait retransmission strategy, the transmitter waits for an explicit acknowledgment before deciding to retransmit the same MPDU or to transmit the next MPDU to the same receiver. While this is a simple algorithm, the strategy can introduce higher delays even if the acknowledgment immediately follows the MPDU.

The GBN algorithm retransmits MPDUs starting from the missing MPDU, including MPDUs that were possibly received correctly. This algorithm is attractive from the perspective of the receiver buffering, as the receiver does not have to buffer out-of-order MPDUs. However, this algorithm makes inefficient use of the wireless medium, and it is possible for an MPDU that was successfully received to get retransmitted multiple times just because an earlier MPDU failed more than once.

The SR algorithm selectively retransmits only the MPDUs confirmed or assumed to be lost. While this algorithm makes efficient use of the medium, the receiver has to buffer out-of-order MPDUs so that they can be delivered to the higher layers in order. SR has been used in many wireless and mobile standards such as European Telecommunications Standards Institute (ETSI) Broadband Radio Access Networks [Technical Committee] (BRAN) HiperLAN/2 (see [B14] and [B36]), GPRS (see [B15] and [B41]), and some IMT-2000 specifications (see [B31]). Since wireless resources are scarce and SR provides higher channel efficiency, one of the many variations of SR has been the preferred algorithm for many wireless systems.

While ARQ can improve the reliability of wireless communication, a certain level of basic reliability is still necessary for a wireless communication to be practical. For example, it is debatable how much ARQ can help time-sensitive traffic such as voice. Depending on the additional delays introduced by a specific ARQ algorithm, ARQ may be unsuitable for voice channels or may be suitable in only limited cases.

Chapter 4 IEEE 802.16 architecture

Overview and key features

In this chapter, we present an overview of the IEEE 802.16 architecture, highlighting the salient features and briefly describing key MAC and PHY components. Detailed descriptions of individual MAC and PHY components of the IEEE 802.16 architecture are given in later chapters. The key aspects of the IEEE 802.16 architecture are summarized as follows:

- *Flexible and extensible with common MAC:* IEEE Std 802.16 defines a common MAC that works with multiple PHY technologies. The MAC is flexible enough to support many existing and future PHY technologies and other MAC extensions, as needed.

- *Modular:* Both the IEEE 802.16 MAC and PHY are very modular, having a set of mandatory and optional features that can be used to realize a variety of fixed and mobile configurations. Many of the optional features are negotiable between an SS and BS.

- *Multiple network topologies:* The support for single-hop and multihop network topologies such as PtP, PMP, and mesh have been built into the IEEE 802.16 MAC and PHY. This enables efficient implementation of any of the supported wireless topologies.

- *Multiple antenna technologies:* The IEEE 802.16 MAC and PHY work with standard antenna technologies such as omnidirectional or directional/sectorized antennas, in addition to multiple antenna technologies such as adaptive antennas and MIMO.

- *MAC CSs:* The IEEE 802.16 MAC is capable of transporting various encapsulated protocol payloads such as ATM, IP, and Ethernet. The core of the IEEE 802.16 MAC is independent of the type of payloads by defining multiple CSs and a standard interface between the core MAC and any CS. This allows the protocol-specific functions such as classification and header compression to be performed outside the core MAC. Additional CSs may be defined in the future without having to change the core MAC.

- *Flexible retransmission policies:* IEEE Std 802.16 provides support for multiple optional retransmission policies, including an SR ARQ and a hybrid ARQ (HARQ).

- *Privacy:* The IEEE 802.16 MAC supports multiple encryption and authentication methods and is capable of supporting additional privacy mechanisms in the future. This is accomplished by a modularized security sublayer that is separated from the core MAC.

- *Subscriber-level adaptive PHY:* The IEEE 802.16 PHY allows a variety of parameters to be changed on a per-connection or per-subscriber basis that can be used to provide varying levels of service by adapting to changing conditions.

- *Integrated QoS:* The integrated QoS support at the IEEE 802.16 MAC defines multiple types of QoS that can be configured and controlled from higher layer. The IEEE 802.16 QoS model defines behaviors for various QoS types that require a QoS scheduler at the MAC. However, the scheduling algorithm is left unspecified to provide full flexibility for the implementor.

- *TDD and FDD support:* The IEEE 802.16 MAC and PHY support both TDD and FDD, with some restrictions as outlined in later sections.

The rest of the book provides an overview of these key features of IEEE Std 802.16 and explains the modules of IEEE Std 802.16 and the flexibility they provide to system designers.

REFERENCE MODEL

The IEEE 802.16 standard describes both the MAC and PHY for fixed and mobile BWA systems. There are two major components of the wireless broadband system: the data/control plane and the management plane. The data plane defines how information is encapsulated or decapsulated in the MAC and modulated or demodulated by the PHY. A set of control functions is needed to support various configuration and coordination functions. While some standards refer to this as "management" (e.g., MAC and PHY management in IEEE Std 802.11), the IEEE 802.16 standard refers to this as the *control plane* to differentiate from the external management system. Every broadband wireless system requires management, including management of the classification, security, QoS, connection setup, and other functions.

Figure 4–1 shows the IEEE 802.16 reference model. The IEEE 802.16 MAC consists of three major components called *sublayers*. The three sublayers are the service-specific convergence sublayer (CS), the MAC common part sublayer (CPS), and the security sublayer.

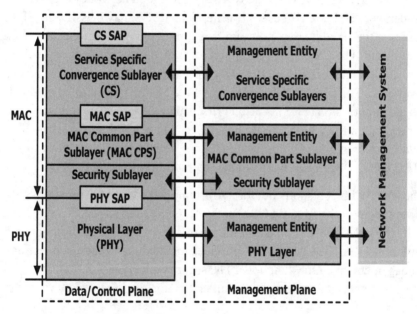

Figure 4–1: IEEE 802.16 reference model

As described in Chapter 3, SDUs need to be classified prior to QoS enforcement. Some higher layer protocols may have their own classification method and deliver preclassified SDUs to the MAC. For example, ATM has its own QoS framework, and QoS identifiers (IDs) are carried in each ATM cell. Similarly, IP also supports QoS protocols such as *differentiated services*. In a differentiated services network, IP packet headers carry differentiated services code points (DSCPs) that are used to identify the QoS class of each IP packet. Such preclassified traffic may be mapped directly into an

IEEE 802.16 MAC connection or may be reclassified based on another set of rules. Other traffic may enter the IEEE 802.16 MAC unclassified, and full classification rules (e.g., based on MAC/IP address or port number) may be needed to classify these packets before applying QoS.

IEEE Std 802.16 currently supports a set of CSs to interface with IP, Ethernet, and ATM protocol layers. In the future, additional CSs may be defined. A service-specific CS performs all functions that are specific to the higher layer protocol it supports, such as classification. The Ethernet CS may classify SDUs based on the MAC address, virtual local area network (VLAN) headers, or IEEE 802.1Q user priorities. The IP CS may classify the SDUs based on the IP addresses, port numbers, protocol types, or other IP-specific parameters. The CS classification process is also responsible for associating a service flow identifier (SFID) and a connection identifier (CID) to the higher layer SDUs, prior to delivering it to the MAC CPS. The optional payload header suppression (PHS) is also performed at the CS, since it is very much dependent on the type of higher layer used. Chapter 5 describes the CS in detail. Following the CS-specific functions, the CS PDU is delivered to the MAC CPS, which is oblivious to the internal format of the CS PDU.

The MAC CPS is responsible for performing the core MAC functions that are independent of the specific CS. The MAC CPS receives MAC service data units (MSDUs) from the CS and transforms them into MAC protocol data units (MPDUs). The MAC CPS provides the medium access, connection management, and QoS functions. The security sublayer is responsible for providing encryption, decryption, authentication, and secure key exchange functions.

The MAC CPS interfaces with the PHYs through the PHY SAP. The MAC CPS may receive MSDUs from multiple MAC CSs. Although the MAC CPS defined in IEEE Std 802.16 can support many PHYs through a PHY-specific PHY SAP, one MAC CPS instance is expected to support only one specific PHY in a specific implementation, whereas it may support multiple MAC CSs in the same implementation. It should be noted that the PHY SAP is not explicitly defined for any of the PHYs in the standard. The QoS scheduling and allocation of bandwidth at the MAC CPS depends on the specific PHY type and parameters, although the MAC itself is PHY independent. The IEEE

802.16 MAC is flexible enough to support additional PHYs that may be defined in the future.

As shown in Figure 4–1, the management plane consists of four management entities corresponding to the CS, CPS, security sublayer, and the PHY. The IEEE 802.16 standard does not define the details of the management plane shown in Figure 4–1 as it is outside the scope of IEEE Std 802.16. However, specific interfaces and messaging may be standardized to support management functions and an IEEE 802.16 amendment. The IEEE 802.16g amendment currently under development is focused on defining management place procedures and services. The approved IEEE 802.16f amendment [B25] defines a MIB for the fixed wireless access systems. Some of the MAC and PHY control messages currently defined in IEEE Std 802.16 may also be used to manage IEEE 802.16 systems through an external management system.

BASE STATION (BS) AND SUBSCRIBER STATION (SS)

The IEEE 802.16 system architecture consists of two logical entities, the BS and SS. Both the BS and the SS have instances of the IEEE 802.16 MAC and PHY, in addition to other support functions. However, specific functions performed by the MAC or PHY differ depending whether it is a BS or an SS, and the IEEE 802.16 standard defines the BS- and SS-specific behavior in detail. In PtP and PMP networks, the BS and SS are in a *master-slave* relationship, where the SS must obey all medium access rules enforced by the BS. The SS in some configurations is referred to as the *customer premises equipment* (CPE) when the SS (or any part of it) is physically located within the customer's premises. Throughout this book, the BS and SS are considered logical entities.

In summary, the BS is responsible for the following functions:

- Enforcing basic MAC and PHY parameters such as frame size, ATDD, and configuration of system parameters
- Performing bandwidth allocation for DL (per connection) and UL traffic (per SS) and performing centralized QoS scheduling, based on the QoS/service parameters configured by the management system and the active bandwidth requests (BW requests) received from the SS

- Communicating the per-frame schedule to all SSs and supporting other data and management broadcast and multicast services
- Transmitting/receiving data and control information to/from one or more SSs within the same frame
- Performing connection admission control and other connection management functions
- Providing other SS support services such as ranging, clock synchronization, power control, and handoff

The SS is responsible for the following functions:

- Identifying the BS, acquiring PHY synchronization, obtaining MAC parameters, and joining the network if necessary
- Establishing basic connectivity, setting up additional data and management connections, and negotiating any optional parameters as needed
- Generating BW requests for connections that require such requests be generated, based on the connection profiles and traffic
- Receiving broadcast/multicast PDUs and unicast PDUs and forwarding them to the appropriate modules
- Making local scheduling decisions based on the current demand and history of BW requests/grants, when a BS allocates bandwidth for the SS
- Transmitting only when instructed by the BS to do so or the SS has some information that qualifies for transmission in one of the slots that may cause "contention" (e.g., ranging and BW requests in contention or broadcast allocations)
- Unless in sleep mode, receiving all schedule and channel information broadcast by the BS and obeying all medium access rules, transmitting data only when the BS allocates slots
- Performing initial ranging, maintenance ranging, power control, and other housekeeping functions

The mobile station (MS) defined in the IEEE 802.16 mobility extension (IEEE Std 802.16e) requires support for additional SS-specific functions such as mobility management, handoff, and power conservation.

One of the basic differences between the BS and SS in a PMP configuration is that the BS, which acts as a centralized controller and a centralized distribution/aggregation point, has to coordinate transmissions to/from multiple SSs, whereas the SS need to deal with only one BS. All traffic originating from an SS, including all SS-to-SS traffic, must go through the BS. Therefore, in a typical IEEE 802.16 system, the BS has to have additional processing and buffering capability (compared to a typical SS) to support a reasonable number of SSs.

The IEEE 802.16 mesh also includes the notion of BS and SS. However, the functionalities of these logical entities in mesh mode are slightly different from those of the BS and SS in PMP. The details of the BS and SS functions in a mesh are described in Chapter 9.

CONVERGENCE SUBLAYER (CS) ARCHITECTURE

The CSs enable the transport of data adhering to some other protocol specification, e.g., ATM, Ethernet, or IP, over an IEEE 802.16 link in a transparent fashion. The CS hides the details of the payload protocol from the IEEE 802.16 MAC, making it "protocol agnostic." Multiple CSs can coexist simultaneously, sharing the same MAC.

The CS for any protocol has to perform the following functions in the transmitter:

- Receive the payload protocol PDU from a higher protocol layer
- Map the payload protocol PDUs to the appropriate MAC service flow
- Optionally compress redundant payload protocol headers
- Deliver the processed packet to the MAC for transmission

In the receiver, the actions taken by the CS are as follows:

- Receive the MSDU
- Restore any compressed payload protocol headers
- Deliver the payload protocol PDU to the higher layer

The payload protocol PDU is mapped to the service flow by a set of configured rules called *classifiers*. The information a classifier considers depends on the protocol being transported. In the case of Ethernet and IP, it is possible to aggregate frames or packets to the same service flow for transport over the air in the CS. The CSs are described in more detail in Chapter 5.

FRAMING AND DUPLEXING

As described in the previous sections, the CS converts upper layer PDUs into MPDUs, and the MAC CPS is responsible for transporting these MPDUs. The data encapsulated in IEEE 802.16 MPDUs are eventually carried in PHY frames. Multiple MPDUs destined for multiple connections and destinations may be carried within a single PHY frame. The duration of the PHY frame is fixed for a specific system configuration and does not change during the operation of the system, as shown in Figure 4–2. The IEEE 802.16 standard supports multiple PHY frame durations for each of the PHY types supported. The amount of higher layer information carried within a single PHY frame of specific duration depends on many factors, including the modulation and coding used, the size of the MPDUs, and other parameters that impact the per-frame overhead. As described in later chapters of this book, some of the per-frame overhead will be fixed for a specific PHY type, while some of the overhead will vary depending on parameters such as the number of SSs and number and types of PDUs carried within a frame.

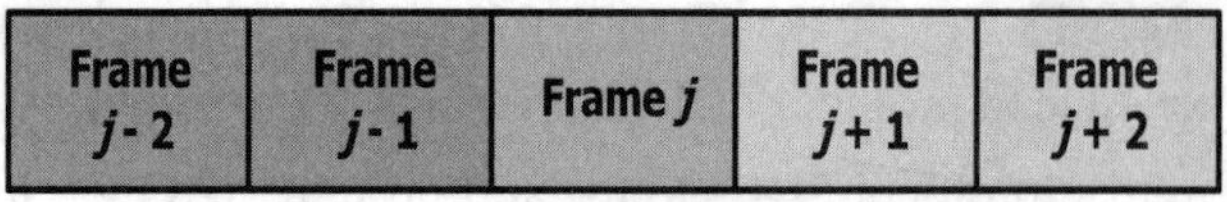

Frame $j-2$	Frame $j-1$	Frame j	Frame $j+1$	Frame $j+2$

Figure 4–2: IEEE 802.16 framing

IEEE Std 802.16 supports both TDD and FDD, with FDD currently allowed only in the licensed bands. By default, TDD systems can be either fixed TDD or ATDD. The standard supports full-duplex FDD and H-FDD operation, and both types of SSs can coexist in the same network. In a mixed network, the BS is responsible for making sure that a half-duplex SS does not receive overlapping allocations for UL and DL.

IEEE Std 802.16 is a framed air interface standard with fixed frames for both TDD and FDD. It is important to understand the following concepts before describing the TDD and FDD operation:

- **Burst profiles:** Each DL or UL burst is characterized by a set of parameters, called a *burst profile*, that includes information such as modulation type, FEC, preamble type, and guard times. The contents of the burst profile is PHY-specific.

- **DL channel descriptor (DCD):** DCD describes the DL PHY characteristics and is broadcast by the BS at periodic intervals. The DCD contains information such as the frame duration code and DL burst profiles. The DL burst profiles define DL interval usage codes (DIUCs) and associated PHY characteristics such as modulation and coding. The contents of the burst profile is PHY-specific. The DIUC is used to identify the burst profile of DL allocations in DL MAPs.

- **UL channel descriptor (UCD):** UCD describes the UL PHY characteristics. The UCD is also broadcast periodically by the BS. The UCD contains UL burst profiles that define UP interval usage codes (UIUCs) and associated PHY characteristics. In addition, the UCD contains backoff parameters to be used for the contention slots in the UL. All the UL and DL transmissions in IEEE Std 802.16 are scheduled; therefore, an SS cannot transmit data at its own discretion unless it is told to do so at a particular time offset. However, there are some cases where an SS may have to communicate some information to the BS without an explicit allocation. This includes a new SS that wants to register and join the network, an active SS that wants to maintain connectivity by requesting a robust DL burst profile, and an active SS that wants to send BW requests asynchronously. UL contention slots are defined to support these types of communication. The SSs are required to follow the specified backoff algorithms along with the parameters advertised by the BS in order to minimize the probability of collision in the contention slots.

- **Downlink MAP (DL MAP):** DL MAP describes the DL allocations. The allocations are specified in PHY-specific DL MAP elements. The DL MAP includes information on DL burst start time and DIUCs to identify the burst profile. While this is the only mechanism used in the WirelessMAN-SC and WirelessMAN-SCa PHYs to describe DL access,

WirelessMAN-OFDM and WirelessMAN-OFDMA PHYs support an additional format called downlink frame prefix (DLFP) to describe part of the DL allocations.

- *Uplink MAP (UL MAP):* UL MAP describes the UL TDMA allocations, which specify the exact time offsets, along with the burst profiles, with which the SSs will transmit data in the UL. The UL contention slots are also defined in the UL MAP. Similar to the DL MAP, the burst profiles are identified by UIUCs broadcast in UCDs.

As we will see in later chapters, there are many PHY-specific compressed or compact DL MAP and UL MAP formats defined in the standard to minimize the MAP overhead.

Physical slots (PS) and mini-slots

PSs and mini-slots are used in IEEE Std 802.16 to express quantities of time slot (bandwidth) allocation. PSs are the basic unit of allocation. The definition of PS depends on the specific PHY alternative. For the single-carrier PHYs, WirelessMAN-SC and WirelessMAN-SCa, PSs are defined to be four quadrature amplitude modulation (QAM) symbols in duration. For the OFDM-based PHYs, WirelessMAN-OFDM and WirelessMAN-OFDMA, PSs are defined to be $4/F_S$, where F_S is the sampling frequency.

Mini-slots are a single-carrier concept, allowing granularity coarser than the PSs in the UL when allocating bandwidth for WirelessMAN-SC and WirelessMAN-SCa. This coarser granularity is used to ensure that certain combinations of bandwidth and frame duration do not cause UL MAP entries to overflow. As such, the definition of a mini-slot varies from one system to another and is broadcast by the BS. A mini-slot is defined as 2^n PSs, where n ranges from 0 to 7. For the ODFM-based PHYs, UL opportunities are allocated in terms of OFDM symbols, and mini-slots are not used.

TDD framing

The structure of the IEEE 802.16 frame is illustrated in Figure 4–3. As shown in the figure, the fixed-length TDD frames are split into DL and UL portions, called DL and UL *subframes*. The TDD frame is repeated at regular intervals, although the duration of the UL and DL subframes (i.e., TDD split) could be adjusted on a per-frame basis based on system parameters and run time parameters such as the amount of existing or expected DL/UL traffic and QoS. The UL and DL subframes carry one or more MPDUs. The DL MAPs and UL MAPs are broadcast at the start of each frame. DCDs and UCDs, if present, are also broadcast at the start of a frame.

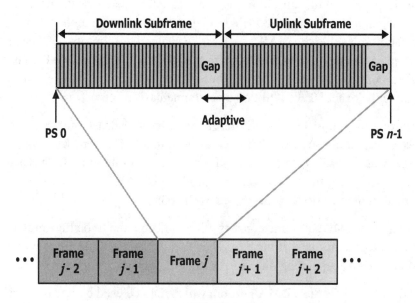

Figure 4–3: PMP TDD frame structure

TDD systems also require a receive-transmit transition gap (RTG) between the UL and subsequent DL subframe (of the next TDD frame). The duration of this gap varies, depending on the PHY type. This gap is used by the BS to transition from receive to transmit mode and by the SS to transition from transmit to receive mode. Similarly, a transmit-receive transition gap (TTG) is required between the DL and UL.

The DL subframe typically starts with information required for PHY synchronization, such as a preamble. The details of the preamble requirements for specific PHY modes are described in the PHY-specific chapters, Chapter 10, Chapter 11, and Chapter 12. The rest of the DL frame consists of multiple DL bursts.

A DL burst may consist of broadcast and/or unicast messages. Since the receiving SSs need to properly decode each burst, the DL MAP is broadcast at the beginning of the DL subframe. The UL MAP is also broadcast at the beginning of the DL subframe. As we will see later, the UL MAPs may refer to the UL portion of the current frame or a future frame such as the next frame, depending on a specific PHY implementation requirements. Other information related to the MAPs, such as the channel descriptors (DCD and UCD), may also be broadcast along with the DL MAPs and UL MAPs. The ATDD is controlled by a BS through these DL MAPs and UL MAPs, which determine the exact DL/UL split on a per-frame basis.

Apart from this frame descriptor information, other broadcast, multicast, or unicast information, such as the higher layer protocol PDUs or MAC management PDUs, is carried in the DL bursts. Each of these PDUs has a MAC header and optional payload that may be protected by a CRC. The construction of the MPDU is described in Chapter 6.

In summary, the DL subframe consists of any information to be transmitted from a BS to one or more of the SSs. The DL subframe is transmitted with TDM or TDMA, depending on the PHY and duplexing requirements.

The UL subframe is the TDMA portion, which may be used by one or more SSs to transmit information to the BS. The UL subframe may consist of contention slots and UL PDUs. The number of contention slots per UL subframe is determined by the BS on a per-frame basis. The UL PHY PDU is similar to the DL PHY PDU, where each UL PDU requires its own PHY synchronization information. However, unlike the DL PHY PDU, each UL PHY PDU can contain only one UL burst. The UL burst is similar to the DL burst, which is used to transport MPDUs and MAC management PDUs with a specific burst profile.

FDD framing

The structure of the FDD frame is shown in Figure 4–4. The structures of the DL and UL subframes are very similar to that of the TDD. FDD also uses fixed frames. The primary difference with TDD is that the DL and UL subframes can overlap in time, as they are transmitted at different frequencies. However, the DL and UL subframes need not necessarily overlap in time. For example, the SS or BS may be incapable of full-duplex operation and, therefore, be unable to transmit and receive simultaneously on separate frequencies. If one of the devices is only half-duplex, the DL and UL subframes are not permitted to overlap in time, as shown in Figure 4–5. The UL MAP information must be provided to the SSs in time for them to initiate UL communication.

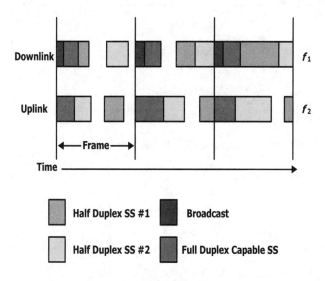

Figure 4–4: PMP FDD frame structure

Full-duplex FDD SSs do not require any gaps between DL and UL transmissions, as they are always listening and can transmit at any time allocated by the BS, as shown in Figure 4–6. However, H-FDD SSs, like TDD SSs, require a subscriber station receive-transmit transition gap (SSRTG) and subscriber station transmit-receive transition gap (SSTTG) between transmission and reception.

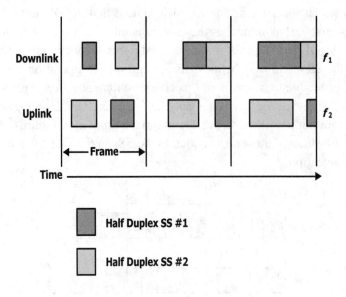

Figure 4–5: H-FDD framing

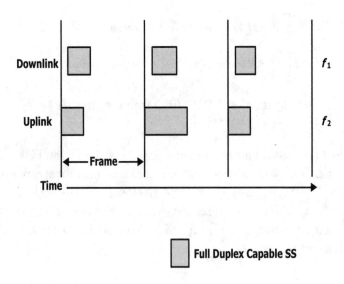

Figure 4–6: Full-duplex FDD framing

Time relevance

The DL MAPs and UL MAPs contain timing information for the SSs on when to receive and transmit. There are some rules on how far ahead in time DL and UL allocations can be broadcast by the BS. Obviously, real-time decisions can be made at the BS and SS if the DL MAPs and UL MAPs always refer to the current frame. However, implementations may not be able to construct and/or decode MAPs in real-time and make transmission and reception decisions based on the current frame's MAP. Therefore, IEEE Std 802.16 defines minimum and maximum time relevance rules that define the minimum and maximum time difference between the MAP broadcast and actual allocations.

The maximum time relevance for the UL MAP and DL MAP for the TDD framing is given in Figure 4–7. In TDD, the DL MAP always refers to the start of the current DL subframe. The UL MAP can be broadcast at most one frame ahead of time. For example, the UL MAP broadcast in Frame n cannot contain allocations for Frame $n + 2$. The minimum time relevance for the TDD framing is given in Figure 4–8. The UL MAP may refer to the start of the current UL subframe, if the SS implementation is capable of decoding the UL MAPs in the current frame and making transmission decisions.

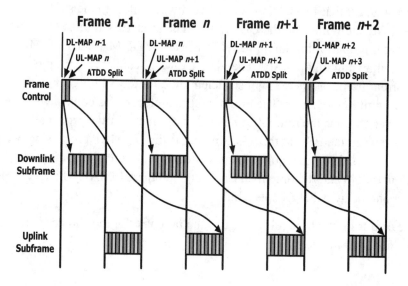

Figure 4–7:   Maximum time relevance of UL and DL MAPs for TDD

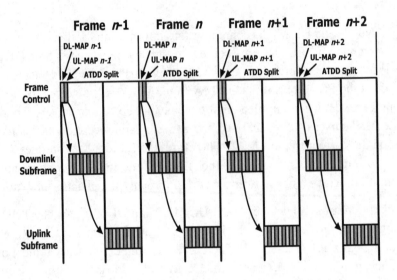

Figure 4–8: Minimum time relevance of UL and DL MAPs for TDD

The maximum time relevance of the FDD UL MAP and DL MAP is shown in Figure 4–9. Similar to TDD, the DL MAP always refers to the start of the current DL subframe, and the UL MAP can be broadcast at most one frame ahead of time. The minimum time relevance of the DL MAP and UL MAP for FDD is shown in Figure 4–10. The FDD UL MAP cannot refer to the start of the current frame, as there is not sufficient time for the SS to receive, decode, and transmit immediately. The minimum time relevance is controlled by two parameters. One is the round-trip delay that accounts for the propagation delay between the BS and the SSs. Another is the UL MAP processing delay, which defines the time required between the arrival of the last bit of the UL MAP in the SS and the time it takes to act upon the UL MAP. The UL MAP processing delay is PHY dependent.

More details on the time relevance and the rationale for the choices are explained later in Chapter 7.

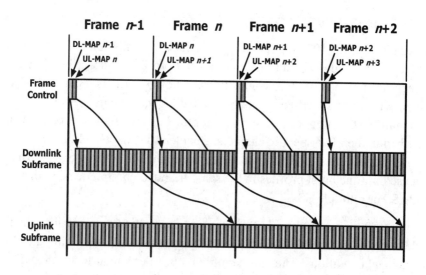

Figure 4–9: Maximum time relevance of DL and UL MAPs for FDD

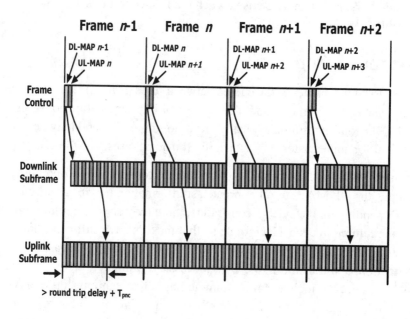

Figure 4–10: Minimum time relevance of DL and UL MAPs for FDD

SUBSCRIBER-LEVEL ADAPTIVE PHY

The adaptive IEEE 802.16 PHY allows for each transmission burst, on each connection, to be described in terms of a burst profile. A subscriber may be characterized by one or more connections. A burst profile is identified by a system-wide code specified by the BS, called the *interval usage code* (IUC). The details of the supported parameters for the burst profiles of each PHY are described in the PHY-specific chapters, Chapter 10, Chapter 11, and Chapter 12. A burst profile is used for both UL and DL transmissions. An BS may use multiple burst profiles in the same frame to communicate with a single SS or multiple SSs.

The selection of a burst profile for a specific connection or subscriber should be determined by the BS based on a variety of constraints, including the QoS and channel conditions. The number of burst profiles supported in the UL or DL depends on the type of PHY. The BS assigns burst profiles based on various factors. For example, a bandwidth-efficient burst profile may be used under favorable link conditions, and a more robust, albeit less spectrally efficient, burst profile may be used when the link conditions are less favorable.

Fixed TDD vs ATDD

As discussed in Chapter 3, the choice between fixed TDD and ATDD depends on the system requirements and other design or implementation constraints. A fixed TDD configuration is typically easier to implement and may be sufficient for some systems with specific traffic patterns.

A minimum DL or UL portion is often required for basic TDD operation. For example, the DL MAPs or other broadcast information may be required in the DL. Depending on the system constraints, there may also be a minimum DL portion required in each TDD frame so that the SSs have sufficient time to switch from receive to transmit mode, after decoding the broadcast information, and transmit in the UL if necessary. Similarly, the UL portion may be required to include a minimum number of contention slots. An ATDD system may make decisions based on these constraints and other real-time constraints to determine the ATDD split. The minimum and maximum values for the ATDD split may also be specified as system parameters that can be controlled from higher layers. ATDD allows a variety of traffic patterns to be supported in the system, including predominantly DL/UL, symmetrical, or bursty traffic.

FRAMED PHY

Early in the development of IEEE Std 802.16, a continuous PHY DL, as found in digital video broadcast (DVB) or data over cable service interface specification (DOCSIS®) systems, was suggested. A continuous PHY has the advantage that interleaving may be applied to increase system gain. In the end, however, the final specification did not include the continuous PHY, for a variety of reasons:

- TDD systems are naturally framed. For FDD and TDD system implementations to be as similar as possible, framed PHYs are more conducive.

- Because of the use of adaptive burst profiles to trade off robustness and efficiency in real time, there was no guarantee a burst would be long enough to take advantage of interleaving. Because of this, the system performance needed to be adequate without interleaving, negating most of the benefit of a continuous PHY.

- Adaptive burst profiles require an easy, natural resynchronization mechanism for SSs that could not stay synchronized during more complex PHY burst profiles. A framed PHY provides this natural resynchronization. A continuous PHY, on the other hand, assumes long-term synchronization and, therefore, complicates synchronization in the short term.

- IEEE Std 802.16 applies true QoS in both the UL and the DL. The framed PHY provides naturally occurring scheduling intervals for some QoS engines to operate at peak performance.

MAC EFFICIENCY

The efficiency of a MAC is determined by a variety of factors. In general, the overhead associated with medium access, retransmissions, and per-MPDU headers are key factors.

The medium access overhead (i.e., the time wasted in trying to access the medium) is almost zero in IEEE Std 802.16, as the centrally scheduled TDM/TDMA MAC can optimally schedule the transmissions without any unused bandwidth. The centrally scheduled MAC differs from distributed medium access methods such as carrier sense multiple access/collision avoidance (CSMA/CA), where nodes make independent transmit decisions, sensing the medium and backing off upon detection of another transmission or collisions. On the other hand, the centrally scheduled MAC introduces additional overhead since BW requests have to be communicated to the central scheduler.

Another source of inefficiency in a MAC is the overhead associated with the retransmission algorithm and associated acknowledgments. If every MPDU is acknowledged immediately, it provides instantaneous feedback and allows automatic retransmission decisions. However, the overhead associated with individual acknowledgments for every MPDU is very high. Moreover, it would not be possible to differentiate between the loss of the MPDU and the loss of an acknowledgment. One of the problems with immediate acknowledgment is that it cannot efficiently deal with PDU loss due to fading that lasts longer than the duration of a PDU. It may be desirable not to retransmit the frame immediately after the detection of the lost MPDU, as the probability of loss is still high for the retransmissions. IEEE Std 802.16 addresses these issues by supporting an SR retransmission algorithm, with efficient acknowledgment mechanisms such as selective and cumulative acknowledgments.

The per-PDU MAC overhead is minimized in IEEE Std 802.16 with the use of small headers. The basic MPDU header (i.e., generic MAC header) is only 6 bytes long and can be used to carry a MAC payload of up to 2041 bytes (when no CRC is included). The generic MAC header supports basic MAC transport functions, and any additional functions such as fragmentation and packing require additional information to be carried in MAC headers. IEEE Std 802.16 defines a set of subheaders, which are extensions to the generic MAC header and may optionally be included in an MPDU, to support such additional functions.

Some real-time applications, such as voice, generate smaller packets. The majority of the data networks today use TCP/IP, a connection-oriented protocol that requires small (typically 40 bytes and 52 bytes with the timestamp option) acknowledgment packets in the opposite direction of the data flow. A TCP acknowledgment is typically sent for every other TCP packet. These small packets are one of the sources of inefficiency in many MAC protocols, if the MAC allows only one SDU per MPDU. Since each MPDU requires a minimum MAC header and PHY header overhead, it is desirable from the efficiency perspective to have the ability to send more than one SDU in a single MPDU. In IEEE Std 802.16, this is achieved by "packing" multiple SDUs into a single MPDU for efficiency.

Another fundamental feature of IEEE Std 802.16 is dynamic fragmentation, where an SDU may be fragmented dynamically based on the real-time conditions, with no requirement for predetermined, fixed-size fragments. Dynamic fragmentation can be used to support efficient QoS as well as to efficiently use the available bandwidth. When the centralized scheduler allocates UL bandwidth for an SS, under no circumstances is the SS allowed to transmit beyond the allocated interval. Dynamic fragmentation allows the scheduler to fit the available frames into these allocated intervals and not leave slots unused.

The concept of ATDD in IEEE Std 802.16 may also be used to efficiently manage the available airtime by reducing wastage in TDD systems. Without ATDD, UL or DL subframes may be underutilized in some frames, as the instantaneous traffic may be asymmetric even in a system with many symmetric flows.

MESH

The IEEE 802.16 protocol includes an option that allows range extension by allowing SS-to-BS communication via one or more intermediate SSs. The IEEE 802.16 specification defines only the MAC-level protocol and assumes that routing is performed on the IP layer above the MAC. Also it is not possible to run the mesh protocol simultaneously with the main PMP protocol.

The IEEE 802.16 mesh protocol offers two possibilities for scheduling transmissions. When centralized scheduling is used, the mesh BS collects data on transmission from the mesh SSs in its domain and computes the schedule for the domain. The schedule is then communicated to the mesh SSs using a special protocol. With distributed scheduling, the mesh SSs compute their own schedule based on information they collect from a two-hop neighborhood.

DIRECTED MESH

Directed mesh is a form of mesh where very directional antennas or beams are used to form a type of repeater network where a node (a combination of some SS and BS functions) acts as a repeater to one or more other nodes in the network. It may physically steer an antenna toward other nodes, or it may use beam forming to create a narrow beam directed toward other nodes. It may do so on the same frequency or on a different frequency. The narrow beams help provide significant frequency reuse and interference avoidance but at the expense of antenna system complexity and network entry complexity. The additional frequency reuse and interference tolerance can in many cases make up for the additional cost caused by this complexity. More details are provided in Chapter 8.

QUALITY OF SERVICE (QoS)

The QoS mechanism supported by IEEE Std 802.16 includes support for DL QoS and UL QoS. The IEEE 802.16 QoS model closely resembles the models defined in DOCSIS (see [B46]), with additional support for wireless-specific issues. As explained in Chapter 3, the QoS scheduling algorithm for wireless QoS has a different set of challenges. Like the DOCSIS and ATM QoS models, the IEEE 802.16 standard defines a QoS behavior and leaves the definitions of the actual QoS scheduler to the implementor.

Any IEEE 802.16 MAC supporting the defined QoS classes should have an integrated QoS scheduler. It is up to the vendor to incorporate a scheduling algorithm that meets the specified QoS requirements. The IEEE 802.16 QoS scheduler is different from a typical packet scheduler that works above the

MAC, as the IEEE 802.16 scheduler resides within the MAC, controlling various MAC parameters and reacting to changing channel conditions.

The IEEE 802.16 QoS support consists of the following:

* Four different QoS classes in the UL and DL
* Support for per-connection QoS and up to 64k connections in each direction
* Generic and PHY-specific mechanisms for UL BW requests and grants
* Flexible QoS models, where most of QoS scheduling can be done centrally at the BS

Since UL bandwidth is always granted per SS, the SS is responsible for making local scheduling decisions and making use of the granted bandwidth fairly and efficiently. In IEEE Std 802.16-2001, per-connection UL bandwidth grants were also optionally allowed. This allowed the BS to schedule UL transmissions on a per-connection basis and support per-connection burst profiles. However, due to the increased complexity at the BS scheduler and lack of flexibility at the BS, this option was later deleted and is no longer specified in IEEE Std 802.16-2004 [B20]. Per-SS grants may have an impact on the system performance when the BS and SS have schedulers with significantly different behaviors.

SECURITY SUBLAYER

The IEEE 802.16 security sublayer consists of two components: an encapsulation protocol for encrypting MAC payloads and a key management protocol called *privacy key management* (PKM). The encapsulation defines the encryption and data authentication method and the rules specifying how to apply these to the MPDU payload. The MAC headers are sent unencrypted. The subheaders, however, are encrypted as they are considered part of the payload. The majority of the MAC management messages are not encrypted, but integrity protection is afforded to management messages to prevent theft of service.

Encapsulation protocols include a data encryption standard (DES) in cipher block chaining (CBC) mode, which only encrypts the data but offers no integrity protection. The CBC initialization vector (CBC-IV) is derived from the frame number, and thus there is no data expansion as a result of applying this protocol. Due to use of the DES algorithm and the way it is used, this encapsulation does not offer a high level of security.

Much better security can be achieved by use of the advanced encryption standard with counter with CBC-MAC (AES-CCM) mode of encapsulating MPDU payloads. This method is essentially the same as the one used in IEEE Std 802.11i™. In addition to strong encryption, it also affords data integrity protection. The drawback is that the size of each MPDU is increased by 12 bytes.

The IEEE 802.16 key management protocol supports secure distribution of keys from BS to SS. The BS may use the key management protocol to enforce conditional access to the network services. The SS uses this PKM protocol to authorize traffic and obtain keys from the BS and performs periodic reauthorization and key refresh. Per-SS unique X.509 digital certificates (see [B19]) and RSA public-key encryption algorithms are used to perform key exchanges between SS and BS. The IEEE 802.16 MAC provides MAC management support for transporting PKM messages between BS and SS. The PKM protocol uses public-key cryptography to establish a shared secret between SS and BS, and the shared secret is used to secure subsequent PKM exchanges.

The IEEE 802.16e amendment introduces support for Extensible Authentication Protocol (EAP) and a new version of the key management protocol (PKM Version 2) that takes into consideration the issues brought into play in a mobile network. Some of the changes include explicit support for mutual and unilateral authentication, where the BS may authenticate the SS, but the SS may not have to authenticate the BS.

AUTOMATIC REPEAT REQUEST (ARQ)

The initial days of the IEEE 802.16 standard development saw support for three classes of ARQ, based on variations of stop-and-wait, GBN, and SR. There was consensus early in the process to not consider a MAC-level stop-and-wait ARQ algorithm. A lengthy debate ensued between the proponents of

GBN and SR variants. The primary argument in favor of GBN was its simplicity, as the SR algorithm would require additional signaling and buffering of out-of-order MPDUs. However, the working group eventually decided to use a SR variant, considering its efficient use of the spectrum. As we will see later, the stop-and-wait algorithm did return, in the form of HARQ.

The group also discussed, at length, the ARQ retransmission unit. Different proposals were floated, one based on the fragment as a retransmission unit and another based on a byte or block. If a fragment is used as a retransmission unit, its size must be decided ahead of time, and once transmitted it cannot be refragmented during retransmissions. On the other hand, a byte- or block-based retransmission would be able to repartition a retransmitted MPDU in any byte or block boundary. While the fragment-based retransmission is less complex, the byte- or block-based approach provided more flexibility to the scheduler in retransmission allocations. Initially, a compromise was reached to use a fragment-based retransmission unit. It was also decided to support transmission of an MSDU in multiple fragments in the first transmission (as a single MPDU), with the additional overhead of ARQ fragmentation headers. This was allowed to support some flexibility in retransmissions, where the original set of fragments could be retransmitted in any order in any number of fragments. However, this decision was also changed in a later revision in favor of only block-based retransmission.

As a result of these machinations, IEEE Std 802.16 currently supports an SR ARQ algorithm that is optionally enabled. The retransmission unit is a *block*, where a block is a contiguous block of at least 1 byte from the SDU, with a maximum size of 2040 bytes per block. The SDU is segmented into a number of blocks, and each transmission or retransmission may consist of any number of blocks. The concept of blocks is used to implement a flexible retransmission algorithm, where the number of blocks in an MPDU may change between retransmissions. The SDUs of an ARQ connection can be thought of as a sequence of blocks, where all blocks, with the possible exception of the last block of an SDU, are of the same size. An MPDU is constructed with an integral number of blocks, which may span multiple SDUs, with appropriate packing subheaders (PSHs).

If the MPDU did not reach the receiver, the retransmission may involve the MPDU as originally constructed, or a new MPDU may be constructed as a subset of the original MPDU (i.e., with fewer blocks). Alternatively, a new MPDU may be constructed by combining the original MPDU (or part of the original MPDU) and other contiguous blocks. The implementation complexity of the block-based retransmission may be reduced by using MPDUs as retransmission units; i.e., keeping the MPDU intact between retransmissions. Other considerations, such as the interaction between encryption and fragmentation, may also force implementations to choose an MPDU-based retransmission.

A compact bitmap-based feedback is used by the ARQ algorithm, which supports multiple types of acknowledgments: cumulative, cumulative with selective, and selective. An implementation may use a mix and match of these acknowledgment types to efficiently use the protocol. It is also important to reduce the overhead associated with ARQ feedback. Therefore, the IEEE 802.16 standard allows the feedback to be sent (in the reverse direction of the information flow) either as stand-alone management messages or as payload piggybacked with other data. The specific algorithm to allocate bandwidth for the ARQ feedback, especially in the UL, is not specified in the standard but left to the developer. For example, the SS may periodically request additional bandwidth for ARQ feedback, or the BS may periodically allocate extra bandwidth for SSs with active ARQ connections. Similarly, bandwidth stealing may be used to send ARQ feedback.

Hybrid automatic repeat request (HARQ)

HARQ is implemented with the cooperation of both MAC and PHY, as opposed to the MAC-level ARQ that is PHY independent. In IEEE Std 802.16, the optional HARQ is supported only in the OFDMA PHY. The HARQ mechanism creates a PHY burst with one or more MPDUs and a per-burst CRC. Four subpackets, each identified by a subpacket identifier (SPID), are generated from the encoded HARQ packet. In the DL, the BS sends one of the subpackets in a burst. Since the subpackets are generated with redundancy, the SS can correctly decode the original encoded packet without having to receive all four subpackets.

As soon as the first subpacket is successfully received, the SS attempts to decode the original encoder packet. The SS sends an acknowledgment to the BS if the decoding of the original encoder packet was successful. If an acknowledgment is received, the BS stops sending additional subpackets of the same encoder packet. If the decoding is unsuccessful, the SS sends a negative acknowledgment. In response, the BS transmits the next subpacket from the four. This process continues until the SS successfully decodes the encoder packet or all attempts fail. Unlike the SR protocol of the MAC-level ARQ, the HARQ is a stop-and-wait protocol with immediate or synchronous acknowledgments. The HARQ mechanism provides a dedicated PHY channel for the SS to transmit positive or negative acknowledgments after a fixed delay. Although the acknowledgments are synchronous, the retransmission can be nondeterministic.

Similar to other stop-and-wait protocols, the IEEE 802.16 HARQ is simple to implement, with no additional buffering at the transmitter or the receiver. The disadvantages of stop-and-wait protocols, such as possible interaction with MAC QoS, also apply to this HARQ protocol.

PHYSICAL LAYER (PHY)

IEEE 802.16 has two different single-carrier PHY specifications:

- WirelessMAN-SC for use from 10 GHz to 66 GHz
- WirelessMAN-SCa for use below 11 GHz

WirelessMAN-SC was the first PHY approved for IEEE Std 802.16. At its high operating frequencies, propagation characteristics dictate that LOS operation is a practical necessity. This severely limits any concept of mobility. As a result, OFDM provides little benefit in this frequency range. A single-carrier PHY provides a simpler and more efficient alternative.

Below 10 GHz, it is a different matter. While the WirelessMAN-SCa PHY does provide a simpler transmit chain and more efficient alternative to WirelessMAN-OFDM and WirelessMAN-OFDMA, it may require a substantially more complicated equalizer in the receiver to combat the additional multipath at these frequencies. This can be either a time-domain equalizer with a substantial number of taps or a frequency-domain equalizer

(FDE). In the latter case, one can abstractly draw a parallel between OFDM and single-carrier, as the Fourier transform in the OFDM transmitter is transferred across the link to the receiver in the form of FDE. At one time during the debate in the working group, a proposal suggested the use of OFDM in the DL and single-carrier in the UL. This would have confined most of the PHY complexity to the BS. Although this combination could have led to attractively simple SSs, the concept was rejected at the time due to the lack of consensus.

Multipath mitigation

In order to mitigate the degrading effect of multipath, the use of an equalizer is recommended with the WirelessMAN-SC and WirelessMAN-SCa PHYs. The duration of the relevant multipath components will typically be multiple symbols long. The cyclic prefix method is used in the WirelessMAN-OFDM and WirelessMAN-OFDMA PHYs. It is not common for the multipath to exceed the duration of the symbols in OFDM and OFDMA PHYs since the symbol duration in these two PHYs is at least hundreds of times longer than those in WirelessMAN-SC and WirelessMAN-SCa PHYs. An equalizer is hence typically not necessary for OFDM and OFDMA PHYs.

MANDATORY AND OPTIONAL COMPONENTS

Before considering the details of the MAC and PHY in later chapters, this chapter concludes with an overview of the major IEEE 802.16 optional and mandatory features. The standard itself is quite complex, with many options. However, specific streamlined system profiles have been defined for vendor interoperability. These are included in Clause 12 of the IEEE 802.16 standard.

Whether a feature is optional, mandatory, or forbidden depends on a variety of factors. Some features, such as support for quadrature phase-shift keying (QPSK) modulation, are mandatory for both BS and SS, irrespective of the PHY type. Many other features depend on the choice of PHY. For instance, MAC-level ARQ is not applicable for the WirelessMAN-SC PHY and is optional for all other cases. Once the PHY is chosen, features may be conditionally mandatory; in other words, implementation of an option makes another feature mandatory. Additionally, adherence to a system profile may require the implementation of certain options, thereby making them mandatory.

Table 4–1 list major features of IEEE Std 802.16, along with the allowed range of values and status as mandatory or optional.

Table 4–1:   Key IEEE 802.16 features and their status

Feature	Parameters and mandatory vs optional status
Channel bandwidths	1.25 MHz to 28 MHz, depending on PHY
FFT size	• 2048 for WirelessMAN-OFDMA (128, 512, 1024 added in IEEE Std 802.16e) • 256 for WirelessMAN-OFDM • N/A for WirelessMAN-SC and WirelessMAN-SCa PHYs
Duplexing	• Mandatory; at least one must be supported • Only one type of duplexing can be enabled at any time; i.e., TDD and FDD are mutually exclusive
Support for H-FDD SS	Mandatory, if FDD is supported
Frame duration	• PHY-dependent options • Must choose a single frame duration
Guard intervals (cyclic prefix)	• WirelessMAN-OFDM and WirelessMAN-OFDMA PHY-dependent options • Must choose a single guard interval
Two-branch transmit space-time coding (STC)	Optional
AAS	Optional
Subchannelization—DL	• Mandatory for WirelessMAN-OFDMA • Optional for WirelessMAN-OFDM • N/A for WirelessMAN-SC and WirelessMAN-SCa

Table 4–1:   Key IEEE 802.16 features and their status (Continued)

Feature	Parameters and mandatory vs optional status
Subchannelization—UL	• Mandatory for WirelessMAN-OFDMA • Optional for WirelessMAN-OFDM • N/A for WirelessMAN-SC and WirelessMAN-SCa
DFS	Mandatory for operation in certain license-exempt bands and regulatory domains
Modulation	• QPSK and 16-QAM mandatory • 64-QAM mandatory in some cases, optional in others • Binary phase shift keying (BPSK) and 256-QAM PHY-dependent
Turbo codes	Optional
CRC	• Optional for WirelessMAN-SC • Mandatory for all other PHYs • CRC may be enabled or disabled on a per-service-flow basis
Data encryption using DES	Mandatory
Data encryption using 128-bit AES-CCM	Optional
BW requests using BW request header	Mandatory
Focused contention BW requests	• N/A for WirelessMAN-SC, WirelessMAN-SCa, and WirelessMAN-OFDMA • Optional for WirelessMAN-OFDM
CDMA BW requests	• N/A for WirelessMAN-SC, WirelessMAN-SCa, and WirelessMAN-OFDM • Mandatory for WirelessMAN-OFDMA
Ability to initiate dynamic service flow creation	• Mandatory for BS • Optional for SS

Table 4–1: Key IEEE 802.16 features and their status (Continued)

Feature	Parameters and mandatory vs optional status
Piggyback BW requests/ Grant management subheader (GMSH)	Optional
Unsolicited grant service (UGS)	Optional, but system must support at least some service type
Real-time polling service (rtPS)	Optional, but system must support at least some service type
Nonreal-time polling service (nrtPS)	Optional, but system must support at least some service type
Best-effort (BE) service	Optional, but system must support at least some service type
Service-specific convergence sublayer (CS)	• Mandatory • Must support at least one variant
Security sublayer	Mandatory
PMP topology	Mandatory
Mesh	• Optional for WirelessMAN-OFDM PHY • N/A for other PHYs
HARQ	Optional
PHS	Optional
Fragmentation	Mandatory
Reassembly	Mandatory
ARQ	• Optional • N/A for WirelessMAN-SC PHY
Classification	Mandatory, within constraints of supported CSs and service types
Unpacking	Mandatory
Packing	Optional

BIT ORDERING

Bit ordering is an often confusing aspect of a protocol specification or implementation. IEEE Std 802.16 follows a simple bit and byte ordering. The fields of MAC messages specified as binary numbers are transmitted as a sequence of bits, starting from the most significant bit (MSB). This rule applies to all fields of the MAC messages and fields of the type-length-values (TLVs), CRC, and header check sequence (HCS). The order of the fields is the same as the order in which they appear in the tables of this book and the standard. The bytes of the SDUs or SDU fragments are transmitted in the same order they were received from the higher layers. Some of the MAC management messages also define strings, which are transmitted in the same order as written. The bits of string symbols and SDU bytes are always transmitted starting with the MSB. The same rule applies to multiple-byte fields, where the MSB is the highest bit. For example, bit 15 and bit 31 are MSB for 16-bit and 32-bit fields, respectively.

Chapter 5 Convergence sublayers (CSs)

Support for multiple protocol transport

The CSs perform the task, at the transmitter side, of transforming a PDU from a higher layer protocol into an IEEE 802.16 MSDU and assigning the MSDU to a particular connection. The CS at the receiver side is responsible for the inverse operation. In the creation of the MSDU, some of the PDU headers may be suppressed. Two CSs are defined in the IEEE 802.16 standard, one for carrying ATM and one for carrying packet-based protocols.

ATM CS

An ATM connection is uniquely identified by its virtual path identifier (VPI) and virtual channel identifier (VCI). A connection may be switched based on virtual path (VP), in which case the cells are forwarded based on the VPI value in the cell header onto the outgoing VP, or virtual channel (VC), in which case cells are mapped to an outgoing connection based on the VPI/VCI pair.

The ATM CS behaves differently depending on whether it is configured into VP-switched or VC-switched modes. When operating in VP-switched mode, the CID of the MAC connection is determined solely based on the VPI field. The VPI field can optionally be suppressed over the air, saving 8 or 12 bits per cell, depending on the flavor of addressing used. The QoS parameters of the MAC connection should be set to match those of the ATM connection at the time the MAC connection is established. The mapping from MAC CID to VPI in the receiver is established at connection setup.

In VC-switched mode, the MAC CID is determined by the VPI/VCI pair, and both of these fields can be suppressed over the air. Again, the QoS and reverse mappings are determined at MAC connection setup time. Since there are fewer than 2^{16} MAC CIDs available, the full range of VPI/VCI values cannot be simultaneously supported. In practice, this limitation should not come into play. Figure 5–1 shows the generic form of an ATM CS PDU. When PHS is not enabled, the ATM CS PDU header is identical to the ATM cell header.

| ATM CS PDU Header | ATM CS PDU Payload (48 bytes) |

Figure 5–1: ATM CS PDU format

When PHS is enabled, the ATM CS PDU header is modified depending on whether the service is VP-switched or VC-switched. In the VP-switched case, the resultant ATM CS PDU header is as detailed in Figure 5–2. In the VC-switched case, the resultant ATM CS PDU header is as detailed in Figure 5–3.

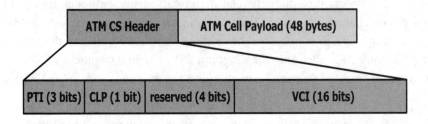

Figure 5–2: CS PDU format for VP-switched ATM connections

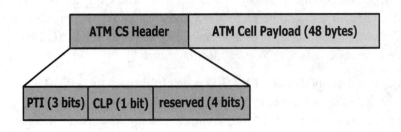

Figure 5–3: CS PDU format for VC-switched ATM connections

While the MAC CPS is allowed to pack ATM CS PDUs into MPDUs, it is not allowed to fragment ATM CS PDUs. This makes packing simpler than in the packet convergence sublayer (PCS) case because the packing process can take advantage of the fact that it has small, fixed-length packets.

However, this is complicated by PHS. The ATM CS can output packets of three different sizes. These correspond to full ATM cells with no PHS, VP-switched ATM cells with PHS, and VC-switched ATM cells with PHS. All ATM CS PDUs from the same service must use the same PHS mechanism. Only ATM CS PDUs from the same IEEE 802.16 MAC connection may be packed into the same MPDU.

PACKET CONVERGENCE SUBLAYER (PCS)

The PCS is designed to cope with any protocol utilizing packets for transporting data. Currently, the service flow signaling used for setting up the PCS supports only Ethernet and IP. There are, however, no fundamental limitations preventing additional support of other protocols such as multiprotocol label switching (MPLS) (see [B43]) either as a part of the standard or as a vendor extension. The PCS PDU is illustrated in Figure 5–4. Note that the PHSI field is always added when carrying Ethernet or IP, regardless of whether PHS is enabled for the connection. If PHS is used, the PHSI field indicates which PHS rule is to be applied by the receiver.

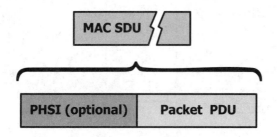

Figure 5–4: MSDU format for PCS

Classification

The PCS contains a classification function that determines on which MAC connection a particular packet shall be carried and which PHS rule applies for that packet. The operation is illustrated in Figure 5–5. The exact rules of classification and the parameters by which the classifier is configured depends on the upper layer protocol and its design. Classifier parameters are configured during dynamic service signaling. Each classifier rule is also associated with a priority. Rules with a higher priority take precedence.

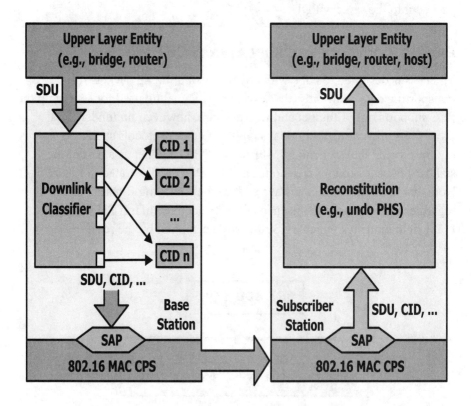

Figure 5–5:   Classification and CID mapping (SS to BS)

Classification of Ethernet packets is based on the following fields in the headers, some of which may be absent:

- Ethernet destination MAC address
- Ethernet source MAC address
- Ethertype/IEEE Std 802.2™ SAP
- IEEE Std 802.1D™ User_Priority
- IEEE Std 802.1Q™ VLAN_ID

The operation performed by the classifier for all of the header fields is an exact match.

Classifying IP packets is more challenging. Often higher layer protocol headers need to be considered in order to enable a meaningful decision. The IP PCS classifiers can be configured to include TCP and User Datagram Protocol (UDP) headers in addition to the IP headers. Both IPv4 and IPv6 can be supported. The IP header fields that can be included in the classification are as follows:

- Source address
- Destination address
- Type of service (ToS)/differentiated services code point (DSCP)
- Protocol

The source address and destination address in an IP packet is matched to a set of prefixes stored in the classifier. Usually a longest prefix match is applied. As no particular classification algorithm is specified, the classifier rule priorities need to be set appropriately to ensure that the classifier behaves as expected. The ToS/DSCP value in the IP header is compared against a range of values after possibly masking away some of the bits. The Protocol field is specified by a set of numerical values checked for an exact match in the IP header.

In case the protocol carried over IP is one that uses "ports" such as TCP, UDP, and Stream Control Transmission Protocol, these can be included in the classification decision. The classifier can be configured to consider nonoverlapping ranges of port values in conjunction with a particular protocol. A rule is not permitted to consider port values alone without considering the protocol.

Classification of IP packets is a relatively complex function. Implementations are not required to support classification on the full set of parameters defined by the IEEE 802.16 standard. When designing the protocol, however, it was agreed that the signaling should not be a limiting factor. Actually, the working group seriously considered the possibility of defining no classifier setup signaling in IEEE Std 802.16, instead relying on Internet Engineering Task Force (IETF) protocols. At the time, however, there was no single de-facto standard protocol for this purpose, and the protocols considered where deemed too "heavy."

Payload header suppression (PHS)

The PHS scheme defined for the PCS is inherited from DOCSIS (see [B46]). The scheme is protocol-independent and relatively simple, trading off some compression performance for ease of implementation. The PHS operation in the sender and receiver is illustrated in Figure 5–6. The BS and the SS negotiate the following PHS parameters constituting a PHS rule:

- *Payload header suppression index (PHSI):* The index to the PHS rule.
- *Payload header suppression field (PHSF):* A string of bytes containing the information allowing the receiver to reconstruct the suppressed information. The sender also keeps this information for purposes of verification (see PHSV below).
- *Payload header suppression mask (PHSM):* A string of bits, each corresponding to a byte in the PHSF. If a bit is set, the corresponding byte is suppressed in the sender, and the receiver will reconstruct the byte based on the content of the PHSF.
- *Payload header suppression size (PHSS):* A parameter indicating the total number of bytes to be processed by PHS. The value is always equal to the length (in bytes) of the PHSF.
- *Payload header suppression valid (PHSV):* A boolean value indicating whether the sender will compare the actual packet header with the version as it will be reconstructed by the receiver. If the two don't match exactly,

the headers of the higher layer packet are not suppressed, and the PHSI field in the PCS PDU is set to zero. Normally, this bit is set. However, some protocols and implementations may be able to cope with the loss of some of the header information, and consequently the verification can be disabled.

The PHS rule to be applied is determined by the sender based on the classification established in the PCS. The receiver, by inspecting the PHSI field, can determine the rule that was used to suppress a particular PCS PDU.

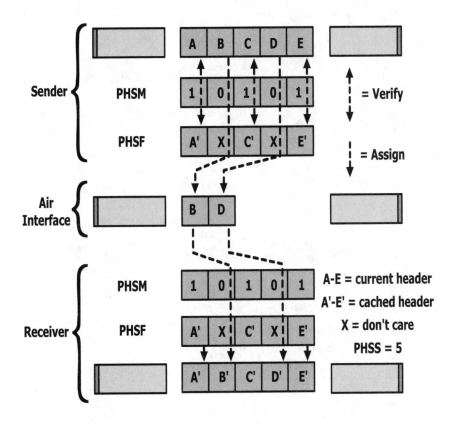

Figure 5–6: PHS operation, with masking

Chapter 6 MAC basics

Concepts, connections, formats, and headers

The IEEE 802.16 MAC specification covers much more than just the protocol for accessing the medium. As with cellular telephony standards, such as those within the Third Generation Partnership Project (3GPP), the IEEE 802.16 MAC specification also contains the associated protocols for radio resource control, radio link control, and security. The IEEE 802.16 MAC does not attempt to separate the control and user planes, but specifies both.

CONNECTIONS AND ADDRESSING

Each IEEE 802.16 MAC instance in an SS has a unique 48-bit MAC address as defined in IEEE Std 802®. An IEEE 802.16 SS may contain more than one instance of the MAC and can, therefore, have multiple 48-bit MAC addresses. One such example is an SS with multiple radios to support multihop or multichannel communication. Throughout this book, the term *MAC* refers to a single instance of the IEEE 802.16 MAC, unless otherwise specified; therefore, we assume there that each SS has a single 48-bit IEEE MAC address. However, this MAC address is used to uniquely identify an SS *only* during initial registration or authentication and as part of some management messages. For example, the initial ranging process, which must be completed before an SS can establish connectivity and gain access to a network, requires exchanging the SS MAC address for identification. On the other hand, it is important to understand that, unlike other wireless protocols such as IEEE 802.11, the 48-bit MAC addresses are *not* carried in every IEEE 802.16 MPDU to identify their source and destination. Instead, the connection-oriented IEEE 802.16 MAC makes primary use of the 16-bit CID to identify all information exchanged between BS and SS, including data, management, and broadcast data. The 16-bit CID is also the basic ID used in bandwidth allocations and bandwidth reservations. This not only reduces the overhead associated with carrying MAC addresses but also provides an easy way to differentiate traffic. All MAC-level QoS functions, such as the classifier and QoS scheduler, use the CID to identify and differentiate traffic in order to maintain the service level and fairness between connections. Despite being connection- oriented, the protocol has been designed primarily to transport connectionless traffic, such as IP.

A maximum of 65 535 connections are supported. Some CIDs, or ranges of CIDs, are reserved for specific purposes. When a new SS wants to join a network, it starts without its own unique CIDs and cannot use any before the initial ranging is completed. A well-known default Initial Ranging CID is provided for this purpose. The details on when and how ranging messages are transmitted by an SS are described in Chapter 7 on MAC operation.

Since the 48-bit MAC address is not primarily used to identify an SS, a Basic CID is assigned to each SS after successful ranging. The Basic CID is used to transport basic MAC control/management messages, to identify an SS, and to communicate some per-SS information that is not specific to a connection. The later sections of this chapter and the next chapter describe various messages and procedures that make use of the Basic CID. Chapter 7 also describes the sequence and the minimum number of CIDs assigned to each SS. Additional CIDs for transporting data and management traffic may be negotiated during SS initialization or added dynamically during operation.

By definition, an IEEE 802.16 MAC connection is unidirectional, although in practice most of the services require a pair of connections, one in each direction. Some exceptions include one-to-many communication, such as multicast and broadcast, which are unidirectional. There are no limits to the number of connections between an SS MAC and BS MAC, subject to the maximum limits specified by the IEEE 802.16 MAC in general and subject to any practical limits that may be imposed by an implementation or to design limits based on scalability and manageability. The composition of the 16-bit CIDs used by the IEEE 802.16 mesh differs from the PMP CID concept as described above. Chapter 9 describes the mesh-specific CIDs and addressing concepts in detail.

The CID space is common between UL and DL. During the early stages of standards development, a separate CID space for UL and DL (i.e., 65 535 connections in each direction) was considered. However, the working group decided against this in order to avoid any potential confusion and complexity. One exception to this rule is management connections, which share the same CID in both directions.

Service flows and service flow identifiers (SFIDs)

Service flows are used by the IEEE 802.16 MAC to efficiently support per-connection services such as QoS. When an SS is added to a system, SFIDs for this SS, along with associated QoS parameters, are provisioned, and each SFID is mapped onto a unique CID. Additional service flows may be defined, or the properties of an existing service flow may be modified depending on the activity and as the needs change. These changes are propagated to the SS or BS using the connection management messages. This model allows static configuration as well as dynamic modification of service flow properties. As we will see, not all CIDs have an associated service flow, as some of the CIDs are used for basic MAC management and control.

All higher layer protocols, connection-oriented or connectionless, are transported over an IEEE 802.16 connection. The actual association of the higher layer SDUs to IEEE 802.16 SFID/CIDs is performed by the CS, as described in Chapter 5. The IEEE 802.16 MAC connections are in no way related to any other notion of connection or lack thereof at the higher layers. It is possible to define a one-to-one mapping between a higher layer connection and an IEEE 802.16 MAC connection. It is also possible to map multiple higher layer connections into a single IEEE 802.16 MAC connection. Similarly, a mix of higher layer connections and connectionless traffic may be mapped onto a single IEEE 802.16 MAC connection. Such mappings are defined by the classifiers used in the IEEE 802.16 MAC CS described in Chapter 5.

CID allocation

The following types of CIDs are defined in IEEE Std 802.16:

- *Initial Ranging CID:* Initial ranging needs to be completed before an SS can establish connectivity to the BS and obtain provisioning parameters, such as other CIDs. Since an SS has no unique CID available before initial ranging, it uses the known Initial Ranging CID for this purpose. The Initial Ranging CID should not be used by an SS if it has already obtained its Basic CID. Since both the BS and SS have to exchange ranging information as part of the initial ranging process, the Initial Ranging CID is reserved in both UL and DL.

- **Basic CID:** Each SS is assigned a Basic CID, during the ranging process, to exchange delay-intolerant and time-critical MAC management messages with the BS. The Basic CID may also be used to identify the SS for the purpose of managing or implementing per-SS functions. The same Basic CID is assigned to both the UL and DL connections used to transport basic MAC information to and from the SS. The use of the Basic CID in specific MAC management messages is described throughout this chapter and the next chapter.

- **Primary Management CID:** Each SS is assigned a Primary Management CID to transport delay-tolerant MAC management messages between SS and BS. The same Primary Management CID is assigned, during the ranging process, to both UL and DL connections for each SS. Appendix A shows the MAC management messages and associated connections used to transport these messages.

- **Secondary Management CID:** The secondary management connection is used to transport higher layer management messages, such as those supporting Simple Network Management Protocol (SNMP), Trivial File Transfer Protocol (TFTP), and Dynamic Host Configuration Protocol (DHCP). The Secondary Management CID is assigned to the SS if, and only if, it identifies itself as a "managed SS" during the registration process following initial ranging. The same Secondary Management CID is assigned in both UL and DL.

- **Transport CID:** Each SS is assigned zero or more transport connections after the completion of ranging and authentication. At least one Transport CID is necessary to transport any user information to or from an SS. The UL and DL Transport CIDs may be assigned independently, since the connections are unidirectional by definition. However, most applications using IEEE 802.16 unicast connections (e.g., TCP-based applications that require TCP acknowledgments in the reverse direction and VoIP that requires the transport of signalling and voice packets in both directions) require bidirectional information exchange. Separate multicast Transport CIDs are not defined. A multicast connection is established by setting up individual connections with the same CID to each SS belonging to the multicast group. An SS may not even be aware that a specific Transport CID is multicast.

- *AAS Initial Ranging CID:* The AAS Initial Ranging CID is used by BSs that support AAS to allocate an initial ranging period for AAS SSs. It is also used by the AAS SSs to initiate the initial ranging process. A separate Initial Ranging CID is required for AAS SSs to prevent the non-AAS SSs from using the slots allocated for AAS SS ranging. Otherwise, the AAS Initial Ranging CID is used in a manner very similar to that for the Initial Ranging CID. Chapter 13 describes the AAS initial ranging process.

- *Multicast Polling CID:* The Multicast Polling CIDs are used to manage the bandwidth (i.e., transmission opportunity) request process. It is important to allocate some UL bandwidth in which the SS can request bandwidth for its connections. This is especially important if the SS has no active UL traffic, which, as we will see later, could allow piggybacked BW requests. It is neither scalable nor efficient for the BS to allocate individual transmit opportunities for all SSs in anticipation of BW requests from specific SSs. To more effectively address this problem, multicast polling groups are defined. An SS may be a member of one or more multicast polling groups. The BS can poll the multicast polling groups by allocating transmission opportunities for the multicast CID so that any SS that is a member of this multicast polling group may use this opportunity to request bandwidth. Since more than one SS may request bandwidth using a single transmit opportunity, collision of BW requests is possible. The multicast Transport CID is used to transmit information in the DL, and Multicast Polling CIDs are used to define UL BW request contention regions. No service flows can be associated with a Multicast Polling CID.

- *Padding CID:* The Padding CID is used to transport padding information by the SS or BS in UL and DL, respectively. Padding information is sent by the SS or BS when it has no other valid information to transmit in an allocated opportunity. The padding byte 0xFF is used in the payload of padding information.

- **Broadcast CID:** The Broadcast CID is used by the BS to broadcast MAC management information to all SSs in the DL. The Broadcast CID is not used in the UL. Apart from the well-known default Broadcast CID, additional Broadcast CIDs may be defined using one of the multicast CIDs to support a variety of additional broadcast functions.

- **Fragmentable Broadcast CID:** The Fragmentable Broadcast CID is used by the BS to transmit fragmented management broadcast information. Management messages sent by a BS in the regular Broadcast CID are not fragmentable. The Fragmentable Broadcast CID can be used by a BS to transmit management information to overcome this restriction. Management broadcast information is typically sent at the most robust modulation and coding so that all SSs can receive it. Therefore, fragmenting some long MAC management messages, in order to maintain fairness and QoS, is advantageous.

Table 6–1 shows the CID allocation in the IEEE 802.16 MAC and the well known CIDs, of which there are five: Initial Ranging CID, AAS Initial Ranging CID (used only by AAS-capable SSs), Padding CID, Broadcast CID, and Fragmentable Broadcast CID. A total of 253 CIDs are allocated to multicast polling groups. The rest of the 65 182 CIDs are reserved for use as a Basic, Primary Management, Secondary Management, and Transport CID, as defined in Table 6–1. The transport connections may include DL multicast connections. The value of m in Table 6–1 depends on the configuration of a specific deployment. Since the Basic and Primary Management CID ranges are defined in terms of m, and there is exactly one Basic and one Primary Management CID per SS, the choice of m is a system parameter that essentially defines the maximum possible number of SSs that can be supported.

Table 6–1: CID allocation and well-known CIDs

CID/Range	Purpose/Type
0x0000	Initial Ranging CID
0x0001 through m	Basic CID
$(m + 1)$ to $2m$	Primary Management CIDs
$(2m + 1)$ through 0xFEFE	Transport and Secondary Management CIDs
0xFEFF	AAS Initial Ranging CID
0xFF00 through 0xFFFC	Multicast Polling CIDs
0xFFFD	Fragmentable Broadcast CID
0xFFFE	Padding CID
0xFFFF	Broadcast CID

MAC HEADERS AND SUBHEADERS

Recall from Chapter 3 that the PDU is the data unit exchanged between the peer entities of the same protocol layer. The IEEE 802.16 MPDUs are the units of data delivered by the MAC to the PHY for transmission and the unit of data delivered by the PHY to the MAC in the receiver. The IEEE 802.16 MAC receives MSDUs (e.g., encapsulated IP packets, ATM cells, Ethernet frames) from the CS above.

There are two classes of IEEE 802.16 MAC information transported between peer entities: stand-alone MAC headers and MPDUs. The 6-byte stand-alone header is the smallest possible information unit that can be transported between two nodes, with the exception of HARQ MAPs, which are sent without the generic MAC headers as we will describe later in this chapter.

Stand-alone MAC headers

Stand-alone MAC headers are used to transport compact control or signalling information between an SS and a BS. None of the stand-alone headers can be used to encapsulate any payload, as they are self-contained and used for a specific purpose. Therefore, it is, technically, a misnomer to call them "headers." All of the currently defined stand-alone MAC headers are 6 bytes in length, as is the generic MAC header used to carry payloads. Future IEEE 802.16 extensions may define additional MAC headers to support other functions.

All currently defined stand-alone MAC headers are relevant only in the UL. Therefore, only the BS is required to process these headers; any SS receiving a stand-alone header should discard it. A header type (HT) bit is used to indicate whether the header is a generic MAC header (HT = 0) or one of the stand-alone headers (HT = 1). Additional fields, such as the Type field, are used to further distinguish the type of generic or stand-alone header being transported or other information that may be present in the payload. In this book, reserved fields are indicated by RSV or Rsv.

BW request header

The 6-byte BW request header (Figure 6–1) carries a full BW request from an SS. Table 6–2 describes the interpretation of the BW request header fields. As shown in the table, there are two types of BW request header: *incremental* and *aggregate*. These are identified by the Type field. The BW request header is protected by a 8-bit HCS.

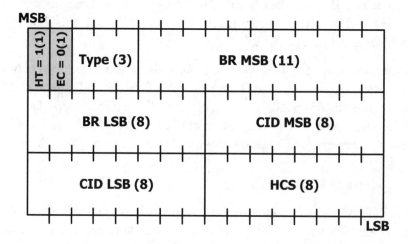

Figure 6–1: BW request header

Table 6–2: BW request header fields

Field	Length (bit)	Description
HT	1	*Header type:* Always set to 1, indicating this is not a generic MAC header.
EC	1	*Encryption control:* Set to 0, indicating no encryption.
Type	3	*Type:* Indicates the type of the BW request. Two types of BW requests are defined, although the Type field is used to demultiplex the different HTs. 000 - Incremental BW request 001 - Aggregate BW request
BR	19	*BW request:* Indicates the amount of UL bandwidth, in bytes, being requested for this CID. This does not include any PHY overhead.
CID	16	*Connection identifier:* The CID for which this BW request is being generated.
HCS	8	*Header check sequence:* The HCS protects the first 5 bytes of the MAC header. It is calculated as the remainder of the division (modulo 2) by the generator polynomial $g(D)=D^8+D^2+D+1$ of the polynomial D^8 multiplied by the first 5 bytes of the MAC header.

MPDU header

Each MPDU begins with a 6-byte generic MAC header, followed by the optional variable-size payload. The payload may consist of MAC subheaders, management messages, special payloads, MSDUs received from a CS, or padding information.

Generic MAC header

Figure 6–2 shows the generic MAC header used to encapsulate all IEEE 802.16 data and management PDUs. Table 6–3 describes the fields of the generic MAC header in detail.

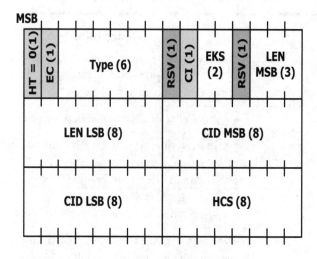

Figure 6–2: Generic MAC header

The type of payload carried within an MPDU is indicated by the CID and the Type field in the generic MAC header. If the CID is a management CID, the payload may consist of one or more MAC management messages. There are some restrictions on fragmentation and packing of some MAC management messages carried on certain management CIDs. If the CID is a Transport CID, the payload may consist of one or more MSDUs or MSDU fragments. The Type field is used to indicate the presence of various MAC subheaders and special payloads.

This generic MAC header may not be very friendly to software implementations, given the rather odd alignment of some fields, but it has all the fields necessary for basic MAC communication. Optional subheaders may be used to support additional MAC functions. Like the BW request header, the generic MAC header is also protected by an 8-bit HCS. If subheaders are included, they are considered part of the payload and not protected by this HCS. The payload may be protected by the optional CRC.

Table 6–3: Generic MAC header fields

Field	Length (bit)	Description
HT	1	*Header type:* Indicates if the header is a generic MAC header or a stand-alone header. Always set to 0 for the generic MAC header.
EC	1	*Encryption control:* Indicates whether the payload is encrypted. 0 = Not encrypted 1 = Encrypted
Type	6	*Payload type:* Indicates the type of subheaders or the type of special payloads present in the payload, if any. See Table 6–4 for the type encodings.
CI	1	*CRC indicator:* Indicates whether a CRC-32 has been included after the encryption, if applicable. 0 = No CRC included 1 = CRC included
EKS	2	*Encryption key sequence:* Indicates the index of the traffic encryption key (TEK) and initialization vector (IV) used to encrypt the payload. These bits are ignored if EC = 0.
LEN	11	*Length:* Indicates the length of the payload in bytes, including the MAC header and CRC if present.
CID	16	*Connection identifier:* Indicates the CID of the connection to which the payload belongs. See the description on packing for more details on packing payloads from multiple connections in the same MPDU.
HCS	8	*Header check sequence:* Same as described in Table 6–2.

Table 6–4 shows the interpretation of the 6-bit Generic MAC HT field. Each bit of the Type field corresponds to an optional subheader or a special payload type. If the subheader or the special payload is present in this MPDU, the corresponding bit position in the Type field is set to one; otherwise, it is set to zero. The description and usage of the subheaders and payload types are given in later sections.

Table 6–4:   Generic HT encodings

Bit	Corresponding subheader or payload
0	*DOWNLINK:* Fast-feedback allocation subheader (FFSH). *UPLINK:* Grant management subheader (GMSH).
1	Packing subheader (PSH).
2	Fragmentation subheader (FSH).
3	Extended version of fragmentation and/or PSH, used when ARQ is used. The presence of the fragmentation or PSH is indicated separately. This bit is set to 0 if ARQ is not used.
4	ARQ feedback payload.
5	Mesh subheader (MSH).

MAC header demultiplexing

Table 6–5 shows how various MAC headers are demultiplexed by a BS based on the HT, EC, and Type header fields. The HT field is used to distinguish a generic MAC header from a stand-alone header. Some of the reserved Type field values and the EC field may be used in the future to define additional stand-alone MAC headers. Since 0xFF is the stuff byte used to fill allocated but unused slots, the MAC header formats are defined so that no MAC header starts with 0xF.

Table 6–5:   MAC header demultiplexing

HT	EC	Type	Header
0	▮		Generic MAC (see Table 6–4 for type encodings)
1	0	000	Incremental BW request
		001	Aggregate BW request
		010	Reserved
		011	Reserved
		100	Reserved
		101	Reserved
		110	Reserved
	1	▮	Reserved

MAC subheaders

The MAC subheaders are used to extend the basic functionality supported by the generic MAC header. All subheaders are defined per PDU, except for the PSH, which is defined per SDU. As we will see in later subsections, the PSH is used to pack more than one SDU or SDU fragment into a single MPDU. Since the PSH serves as a superset of the FSH, the PSH and FSH are mutually exclusive. The FSH can be used only when the MPDU consists of only one SDU or SDU fragment.

As shown in Table 6–4, the Type field of the generic MAC header indicates the type(s) of subheaders present in the MPDU. Since the Type field indicates only the presence or absence of subheaders, an ordering of the placement of subheaders after the generic MAC header is enforced in order to properly decode the subheaders whenever multiple subheaders are present in a single MPDU.

The following subsections describe in detail the subheader formats and the MAC operations supported by these subheaders.

Fragmentation subheader (FSH)

FSHs are used to transport encapsulated MSDU or SDU fragments in a connection. Therefore, they carry sufficient information for the receiver to properly identify and reassemble fragments. Once a connection is provisioned with fragmentation enabled, all MPDUs transported on that connection should have a FSH, irrespective of whether a specific MPDU carries a fragmented or full SDU.

Three different types of FSHs are defined in IEEE Std 802.16, as illustrated in Figure 6–3 and Table 6–6.

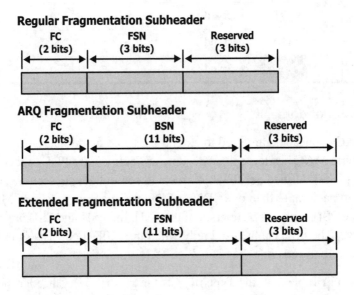

Figure 6–3:   FSHs

Table 6–6:   FSH fields

Field	Size (bit)	Subheader type	Comments
FC	2	Present in all FSHs	Indicates the type of fragment: 00 = No fragmentation 01 = Last fragment 10 = First fragment 11 = Continuing or middle fragment
FSN	3	Regular	Sequence number of this fragment. Increments by one, modulo 8, for every SDU or SDU fragment.
	11	Extended	Increments by one, modulo 2048, for every SDU or SDU fragment.
BSN	11	ARQ	Sequence number of the first block in this SDU or SDU fragment.

All FSHs include the 2-bit FC (fragment control) field that indicates the fragmentation state of the current fragment. The regular FSH is 1 byte in length, with a 3-bit fragment sequence number (FSN). The extended FSH is similar to the regular FSH, except that it is 2 bytes long and supports an 11-bit FSN space. The 2-byte ARQ FSH, on the other hand, uses a 11-bit block sequence number (BSN) instead of the FSN. The concept of blocks and the BSN are described elsewhere in this book.

Packing subheader (PSH)

Packing allows efficient transmission of multiple MSDUs or MSDU fragments in a single MPDU. As we will see later, the ARQ mechanism requires the SDUs or SDU fragments be transmitted out of sequence when retransmitting blocks lost during previous transmissions. The packing mechanism is designed to support such transmission of noncontiguous ARQ blocks or fragments. Three PSHs, similar to the FSHs, are defined, as shown in Figure 6–4.

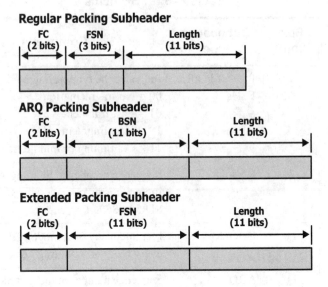

Figure 6–4: PSHs

The three PSHs parallel the corresponding FSHs, with an 11-bit Length field in place of the three reserved bits. This was the motivation for defining an extended FSH, instead of using the three reserved bits of the regular FSH to extend the FSN.

The Length field indicates the length of the unit of data being packed. This includes the length, in bytes, of the SDU/SDU fragment plus the PSH itself. The definitions of the rest of the PSH fields are the same as the corresponding fields in Table 6–6.

Grant management subheader (GMSH)

The 2-byte GMSH is used by the SS to request additional bandwidth and transmit other signalling related to bandwidth allocation. This BW request is essentially piggybacked onto an ongoing connection, in contrast to the stand-alone BW request using a BW request header. The IEEE 802.16 QoS model defines a set of rules for the use of the BW request header and GMSH for specific types of services. Figure 6–5 shows the two GMSHs.

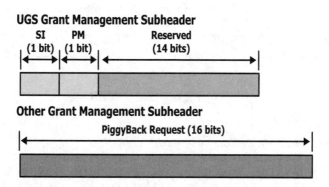

Figure 6–5: GMSHs

In the QoS section of Chapter 7, we describe the different types of per-connection QoS supported by IEEE 802.16. One such QoS type is called *unsolicited grant service* (UGS), in which a connection may be set up with a periodic grant. The UGS GMSH is used by such connections to indicate the status of the current allocation. The slip indicator (SI) bit indicates that the UGS grants are not keeping up with the backlog of the connection queue. The poll-me (PM) bit is used to request a bandwidth poll. The GMSH is always piggybacked on an MPDU of the connection for which additional bandwidth (or signaling) is being requested.

All non-UGS connections use the simple 16-bit GMSH that indicates the number of bytes requested, not including the PHY overhead. Recall that the stand-alone BW request header can be used for aggregate or incremental BW requests. The piggybacked requests are always incremental. Since the size of the GMSH is 2 bytes, it can be used to request up to 65 535 bytes of incremental bandwidth. The stand-alone MAC header allows incremental or aggregate BW requests of up to 524 287 bytes.

Mesh subheader (MSH)

The 2-byte MSH (Figure 6–6), used only by a mesh network, carries the mesh Node ID discussed in Chapter 9.

Transmit Node ID (16 bits)

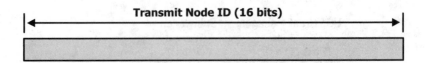

Figure 6–6:   MSH

Fast-feedback allocation subheader (FFSH)

The FFSH (Figure 6–7) is used by the BS to request feedback from an AAS SS. The support for AAS functionality is PHY-specific and described in Chapter 13. The allocation offset is defined in units of slots. This is the offset from the beginning of the fast-feedback UL bandwidth allocation of the slot in which the SS must send a fast-feedback message. The allocation applies to the UL subframe of exactly two frames after the frame including the FFSH. The Feedback Type field identifies the type of the feedback, as shown in Table 6–7. Chapter 13 describes these feedback types in detail.

Allocation Offset
(6 bits)

Feedback Type
(2 bits)

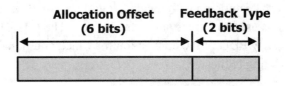

Figure 6–7:   FFSH

Table 6–7:   Fast-feedback allocation feedback type encodings

Feedback type	Description
00	Fast DL measurement
01	Fast MIMO feedback, antenna #0
10	Fast MIMO feedback, antenna #1
11	MIMO mode and permutation mode feedback

ARQ feedback

ARQ feedback refers to the ARQ control information, such as acknowledgments from a receiver to the transmitter. Since this control information flows in the direction opposite to that of the actual ARQ data flow, efficient mechanisms are required to transport the feedback as a piggybacked payload or as a separate management message. In order to send a separate ARQ feedback management message, especially in the UL, the BS has to make specific allocations for the feedback, or the SS has to steal bandwidth from another allocation. An alternate approach is to piggyback on another connection, which may effectively steal some of its bandwidth.

IEEE Std 802.16 allows such piggybacking of ARQ feedback on any MPDU, subject to system policies. This is similar to the way in which subheaders are transported with MPDUs. However, ARQ feedback is a special payload, not a MAC subheader. The format of the ARQ special payload and rules for piggybacking are described in later sections.

DATA AND MANAGEMENT PDU CONSTRUCTION

Simple MPDU

A simple MPDU consists of the generic MAC header and optionally the payload and CRC. Although CRC is optional, it is a good practice to include it in real-world implementations, unless perhaps in the case of an LOS system with a very low bit error rate (BER). The IEEE 802.16 standard mandates the implementation of CRC if the PHY type is WirelessMAN-SCa, WirelessMAN-OFDM, or WirelessMAN-OFDMA. Figure 6–8 shows the simple MPDU structure.

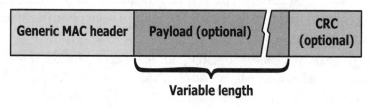

Variable length

Figure 6–8:   MPDU structure

The MPDU format is flexible enough to carry both small and large payloads with minimal overhead. With the appropriate choice of packing and fragmentation and an intelligent scheduler, the overall overhead can be low. When subheaders and special payloads are present, all per-PDU subheaders are placed just after the generic MAC header. The optional subheaders, if present, are considered part of the payload and hence protected by the same CRC that protects the rest of the payload. The format of the MPDU, with subheaders and ARQ special payloads, is shown in Figure 6–9. As shown in the figure, all per-PDU subheaders precede per-SDU subheaders.

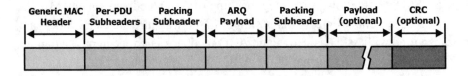

Figure 6–9:   MPDU with subheaders

Subheader ordering

As discussed in earlier sections, the generic MAC header indicates the presence or absence of subheaders. However, in order to properly decode MPDUs and subheaders, an ordering of the subheaders and special payloads is defined. The following rules define the use of subheaders in UL and DL:

- ***PSH and FSH:*** The PSH and FSH are mutually exclusive. The FSH is used only to transport a single fragment. The PSH contains the necessary information to support both packing and fragmentation. Therefore, the PSH is used to transport more than one full SDU and/or fragment in the same payload.

- ***ARQ feedback payload:*** The ARQ feedback special payload, if present, is always inserted as the first payload (i.e., after the subheaders, if any) when packing is enabled. This is illustrated in Figure 6–9. Since ARQ feedback is a payload, a PSH must precede it, if packing is required.

- **MSH:** The MSH, if present, precedes all other subheaders.

- **FFSH:** The FFSH will always be the last per-PDU subheader. Therefore, only the PSH can follow this subheader.

- **GMSH and FSH:** The GMSH precedes the FSH.

Based on these rules, the ordering of the subheaders and the special payload is defined as shown in Table 6–8.

Table 6–8: Subheader and special payload ordering

Order	Subheader
1	MSH (mesh subheader)
2	GMSH (grant management subheader)
3	FSH (fragmentation subheader)
4	FFSH (fast-feedback allocation subheader)
5	PSH (packing subheader)
6	ARQ special payload

Figure 6–10 illustrates the ordering of the subheaders in UL and DL. Note that some subheaders are relevant in only one of these.

ARQ blocks

It is important to explain the concept of ARQ blocks before we discuss the fragmentation and packing in detail. The ARQ block is the basic unit of transmission or retransmission for an ARQ-enabled connection. An MSDU is logically partitioned into blocks of a fixed length, negotiated during connection setup. When the length of the SDU is not an integer multiple of the negotiated block size, the final block of the SDU is formed using the remaining SDU bytes after the final full block has been determined.

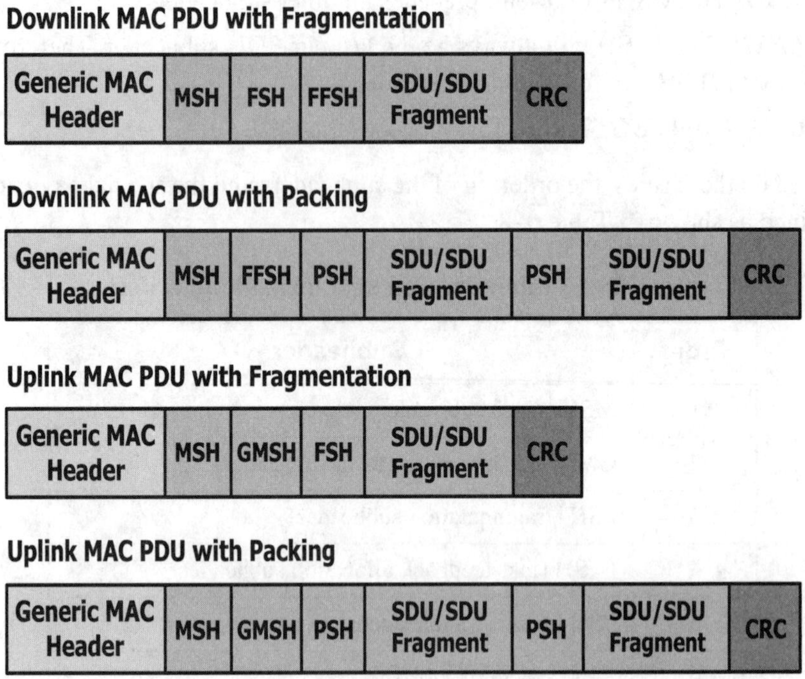

Downlink MAC PDU with Fragmentation

Generic MAC Header	MSH	FSH	FFSH	SDU/SDU Fragment	CRC

Downlink MAC PDU with Packing

Generic MAC Header	MSH	FFSH	PSH	SDU/SDU Fragment	PSH	SDU/SDU Fragment	CRC

Uplink MAC PDU with Fragmentation

Generic MAC Header	MSH	GMSH	FSH	SDU/SDU Fragment	CRC

Uplink MAC PDU with Packing

Generic MAC Header	MSH	GMSH	PSH	SDU/SDU Fragment	PSH	SDU/SDU Fragment	CRC

Figure 6–10: Ordering of MAC subheaders

The size of the ARQ block can be of any value from 1 to 2040 bytes. The appropriate choice of the ARQ block size will depend on the ToS carried over the connection, the underlying PHY properties, and the expected feedback delay of the ARQ protocol, among other factors.

Each ARQ block of an SDU is assigned a sequence number called the *block sequence number* (BSN). Fragmentation is allowed to occur only on ARQ block boundaries. Therefore, an ARQ fragment is formed with an integral number of contiguous ARQ blocks. The BSN value carried in the FSH is the BSN for the first ARQ block appearing in the segment.

Figure 6–11 illustrates the concept of ARQ blocks with two consecutive SDUs of the same connection. SDU #1 consists of seven blocks starting from BSN 5. The seventh block, with BSN 11, is of a size smaller than the negotiated ARQ block size. Similarly, SDU #2 consists of five ARQ blocks, from BSN 12 through BSN 16, where the last ARQ block is smaller than the rest. Both the SDUs are divided into two fragments each: <Frag /SDU#1> (3 blocks, BSN = 5), <Frag 1/SDU#1> (4 blocks, BSN = 8), <Frag 0/SDU#2> (3 blocks, BSN = 12) and <Frag 1/SDU#2> (2 blocks, BSN = 15).

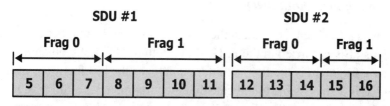

Two consecutive SDUs presented to MAC for the same connection

Figure 6–11: ARQ blocks

Fragmentation

The fragmentation function is negotiated on a per-connection basis during connection setup. There are two choices of FSH for non-ARQ connections. The 11-bit FSN is used by all management connections that support fragmentation. The support for the 3-bit FSN is optional in IEEE Std 802.16. If a specific connection has fragmentation enabled, all SDU fragments on that connection are preceded by a FSH. The FSN is incremented by 1 (modulo 8 or modulo 2048) for each SDU fragment and unfragmented SDU.

The fragment size may be chosen dynamically, based on the real-time conditions and other system requirements. This is often called *dynamic fragmentation*. If the regular FSH is used, up to eight fragments are supported per SDU to avoid ambiguity. However, with the extended FSH or when ARQ is enabled, an SDU may be fragmented into any number of fragments to support QoS and other system requirements. There is no restriction of the size of the fragment.

As described in the previous subsection on ARQ blocks, fragmentation works differently for an ARQ connection. For both ARQ and non-ARQ connections, the FSH is of the same size, but the sequence number fields are interpreted differently. On connections with no ARQ, fragmentation can be done on any byte boundary, and the FSN counter is incremented for every fragment. For ARQ-enabled connections, fragmentation occurs on ARQ block boundaries, and the BSN counter is incremented for every ARQ block.

Fragmentation without ARQ

As soon as a fragment is created, the next valid sequence number is assigned to it. This FSN is carried in the FSH, along with the FC field indicating whether a fragment is the first, an intermediate, or the last fragment from an MSDU. It is also possible to indicate that the "fragment" is a full MSDU (i.e., FC = 00), mostly for convenience of implementation. Although there is no requirement to include a FSH in an MPDU containing a full MSDU, the FSN counter must be incremented even if the FSH is omitted. Figure 6–12 shows the structure of an MPDU with fragmentation.

Generic MAC Header	Other Subheaders	Fragmentation Subheader	Payload (One SDU or fragment of an SDU)	CRC-32

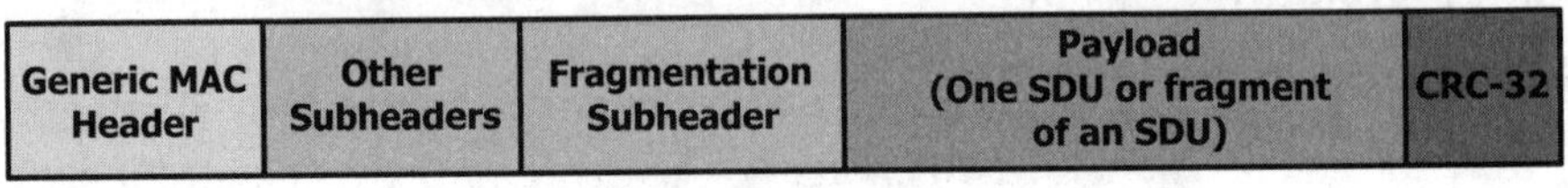

Figure 6–12:   MPDU with fragmentation

The reassembly process for non-ARQ connections in the receiver is fairly straightforward, since fragments are always transmitted in order and can be transmitted only once. If the receiver discovers that a fragment has been lost, it discards any fragments waiting to be reassembled and all fragments until it encounters (a) a fragment indicating it is the first or (b) an MPDU containing a full MSDU. Therefore, the receiver has to buffer at most one full MPDU per connection.

Fragmentation with ARQ

Each MPDU sent on an ARQ-enabled connection must include either the ARQ FSH or the ARQ PSH, even if fragmentation is not enabled, as the BSN must be carried to the receiver. If only fragmentation is used, the MPDU structure is similar to the one shown in Figure 6–12. The only difference is the type of FSH used for ARQ connections.

Packing

There are three types of packing supported in IEEE 802.16: packing of fixed-length SDUs, packing of variable-size SDUs, and packing of ARQ connections.

Packing of fixed-length SDUs

Fixed-size SDUs, such as ATM cells, can be packed and transported on a connection without using PSHs. During connection setup, the SDU parameters can be negotiated for fixed-size packing with no fragmentation. This allows the fixed-size SDUs to be packed efficiently without the PSH overhead. The number of SDUs per MPDU can still be varied dynamically based on QoS and other requirements. The Length field in the generic MAC header implicitly defines the number of packed SDUs. Figure 6–13 shows an MPDU packed with fixed-size SDUs.

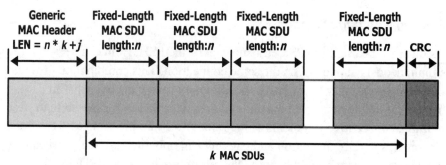

Note: j is the length of the MAC Header plus the CRC, in bytes.

Figure 6–13: Packing of fixed-length SDUs

Packing of variable-length SDUs

Variable-length SDUs can be packed using one of the PSHs. Although all PSHs have all the necessary fields to simultaneously support packing and fragmentation, packing can be used without fragmentation. The regular PSH is sufficient to support packing without fragmentation. A packed MPDU, with each SDU preceded by a PSH, is shown in Figure 6–14.

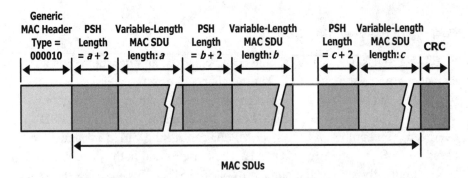

Figure 6–14:   Packing of variable-length SDUs

Packing with fragmentation

As discussed earlier, the PSH is a superset of the FSH. Both can be simultaneously supported on the same connection. This combination provides the utmost flexibility to the MAC to efficiently utilize the over-the-air bandwidth and to maximize overall system performance. It is possible to pack multiple consecutive fragments of the same SDU into the same MPDU. However, such fragmentation and packing can only increase the overhead unnecessarily, with no real benefit. Therefore, typically an MPDU with packing and fragmentation will have at most two fragments of two different SDUs and zero or more unfragmented SDUs.

Figure 6–15 shows an example of a packed MPDU with two fragments and some unfragmented SDUs. The first packed payload is the last fragment of an SDU and the last payload is the first fragment of another SDU. Another example of a packed MPDU with one fragment and some unfragmented SDUs is shown in Figure 6–16.

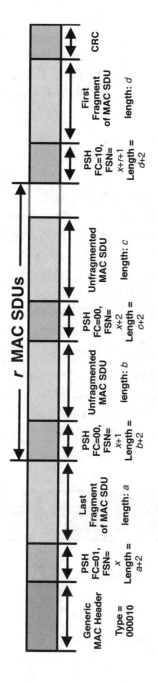

Figure 6–15: **Packed MPDU with two fragments**

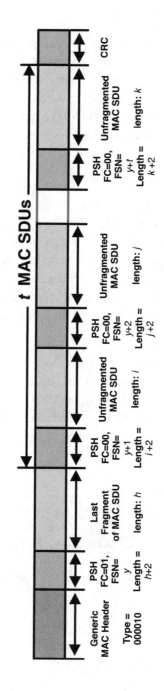

Figure 6–16: **Packed MPDU with one fragment**

Packing and ARQ

An ARQ connection must use either a FSH or a PSH, as the ARQ algorithm requires the BSN to keep track of ARQ blocks. If ARQ is used with no fragmentation or packing, the ARQ FSH is used to carry the BSN. The same subheader can be used to add fragmentation function to an ARQ connection without packing. If the ARQ connection requires packing (with or without fragmentation), the ARQ PSH must be used.

Packing of ARQ payload

The ARQ feedback payload consists of one or more ARQ Feedback information elements (IEs), the format and contents of which will be covered elsewhere in this book. When the ARQ feedback is piggybacked onto another MPDU with packing enabled, it shall always be the first packed payload. The FC and FSN fields are irrelevant to the ARQ feedback and are set to zero. Since the length of the ARQ feedback payload is variable, the Length field in the PSH is used to correctly decode the ARQ payload from the packed MPDU.

Although only the Length field of the PSH is of significance for packing ARQ feedback payload, all PSHs used within an MPDU should be of the same type (i.e., regular, extended, or ARQ), as negotiated during the connection setup. The ARQ feedback payload can be piggybacked on a connection that may or may not have ARQ enabled. Figure 6–17 shows an example MPDU with an ARQ feedback payload.

Figure 6–17: ARQ feedback and packing

Concatenation

Further efficiency can be achieved by reducing the per-PDU PHY overhead through concatenation. This is achieved by transmitting multiple MPDUs in a single PHY burst. Although concatenation is not a MAC feature, it is covered in this section because it is related to other efficiency functions, such as packing and fragmentation. Concatenation is allowed in both the UL and DL.

For some of the IEEE 802.16 PHY specifications, an intermediate sublayer, called the *transmission CS*, is specified to efficiently support concatenation. This sublayer inserts a pointer byte at the beginning of a payload to help the receiver separate the MPDUs in a burst. This can be used to correctly receive some of the concatenated MPDUs in a PHY burst even if part of the PHY burst was received incorrectly (e.g., if a part of the burst was corrupted). The transmission CS is mandatory for the WirelessMAN-SC PHY and optional for the WirelessMAN-OFDM PHY.

Figure 6–18 shows the concept of concatenation for an UL burst transmission. As shown in the figure, any type of MPDU or message may be concatenated into a single PHY burst. The figure shows two UL bursts, one that concatenates a BW request and a user data MPDU and another that concatenates a MAC management message and user data MPDUs.

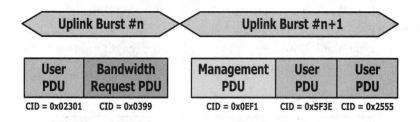

Figure 6–18:   Concatenation of MPDUs

MPDU encryption and CRC

The details of the MPDU encryption are covered in Chapter 8. A summary of MPDU encryption and CRC rules are listed here:

- CRC is optional in the IEEE 802.16 MAC, although it is a good practice to include it.
- CRC is mandatory for ARQ connections, as it is used to detect errors in MPDU transmissions.
- If CRC is included, it covers both the MAC header and payload. The generic MAC header is also protected by HCS.
- If encryption is used, the generic MAC header is not encrypted.
- If both CRC and encryption are enabled, CRC is calculated after encryption.

MAC management

The format of the MAC management messages is shown in Figure 6–19. The MAC management messages are carried as (a part of) MPDU payload. All MAC management messages begin with an 8-bit Type field, followed by the message payload. An example management message payload definition is shown in Table 6–9.

The first part of the payload follows no particular set of encoding rules. In the example, it a string of 3 bytes, the two first of which are to be interpreted as the transaction ID and the third one as the confirmation code. The remainder of the message is defined as *TLV-encoded information*, which follows an encoding that closely resembles ASN.1 Basic Encoding Rules. For each parameter encoded as a type-length-value (TLV), the first byte identifies the parameter type, the following byte/bytes[2] gives the length in bytes of the value field, and the actual parameter value follows.

| Management Message Type (8 bits) | Management Message Payload (Variable) |

Figure 6–19:   Format of the MAC management message

---

[2]   If the length of the Value field is less than 128 bytes, a single byte suffices. For longer parameters, more than 1 byte is used to express the length.

Table 6–9:   DSA-RSP message format

Syntax	Size	Notes
DSA-RSP_Message_Format() {		
Management Message Type = 12	8 bits	
Transaction ID	16 bits	
Confirmation Code	8 bits	
TLV Encoded Information	*variable*	TLV-specific
}		

ARQ

The following subsections describe the basic IEEE 802.16 ARQ concepts of block-based retransmissions and bitmap-based ARQ feedback. The actual SR ARQ algorithm, with its operational parameters and state machines, are described in Chapter 7.

ARQ block-based retransmissions

The concept of ARQ blocks and block-based retransmissions is illustrated in Figure 6–20. The figure shows the transmission, loss, and subsequent retransmission of two consecutive SDUs, shown in Figure 6–11, on an ARQ-enabled connection with packing. The number of blocks per SDU is based on the prenegotiated block size. During initial transmission of the SDUs, both are fragmented into two fragments each. The number of fragments per SDU and hence the number of blocks per fragment are dynamically decided based on a variety of factors. For example, the scheduler may limit the fragment size or airtime allocated to a specific connection. Even if the allocated airtime is fixed, the adaptive modulation may change the number of blocks that can be transmitted within the allocated time.

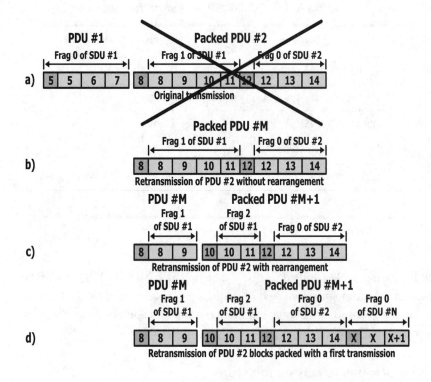

Figure 6–20:   ARQ blocks and retransmissions

Figure 6–20 a) shows the construction and transmission of two MPDUs, where PDU #1 consists of fragment #0 of SDU #1, and PDU #2 consists of fragment #1 of SDU #1 and fragment #0 of SDU #2. The transmission of fragment #2 of SDU #2 is not shown in the figure. Assume it is determined (by mechanisms discussed in the next subsection) that PDU #2 did not reach the receiver. The block-based approach offers many possibilities for retransmission of the lost PDU #2. We assumed that, by the time the loss of PDU #2 is detected, fragment #1 of SDU #2 (consisting of blocks 15 and 16) has already been transmitted successfully and that SDU #*N* of the same connection, where *N* is greater than 2, is ready for first transmission.

Out of the many possibilities for the retransmission of PDU #2, three examples are shown in Figure 6–20. The first example, in Figure 6–20 b), shows the retransmission of PDU#2 with no change or rearrangement as a new PDU #*M* (*M* > 2). The second example, in Figure 6–20 c), shows the retransmission with a rearrangement where the original PDU #2 is split into two MPDUs, *M* and *M*+1. The third example, in Figure 6–20 d), is similar to the second one, except the PDU #*M*+1 consists of the blocks to be retransmitted and the first or original transmission of fragment #0 (with BSN *X*) of SDU #*N*. As shown in Figure 6–20 d), an MPDU on an ARQ connection may contain blocks that are transmitted for the first time as well as those being retransmitted.

It is up to the transmitter's policy to decide whether a set of blocks once transmitted as a single fragment should be retransmitted as a single fragment. Similarly, it is up to the transmitter to decide whether a set of fragments once transmitted in a single MPDU should be retransmitted as a single MPDU. The advantage of keeping the number of blocks per fragment and fragments per MPDU the same between retransmissions is simplicity. On the other hand, the flexibility of refragmentation and rearrangement can significantly improve efficiency and QoS in certain traffic conditions. If fragmentation is not enabled for this connection, the ARQ fragmentation or PSH would still be used with the same block size, as if the fragmentation or packing were enabled. In this case, each "fragment" would contain all the blocks of a single SDU.

ARQ Feedback information element (IE)

Apart from the flexible retransmission, IEEE Std 802.16 defines compact ARQ feedback that is used by the receiver to provide feedback on successfully received and missing blocks. The ARQ Feedback IE contains an ACK Type field and an optional variable-size bitmap indicate positive and/or negative acknowledgments. Four types of acknowledgments are defined to support flexible feedback. Figure 6–21 shows the format of the ARQ Feedback IE.

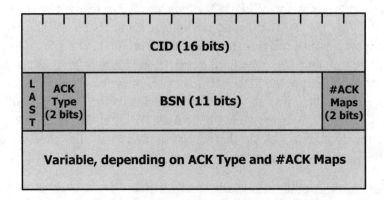

Figure 6–21: ARQ Feedback IE

The fields of the ARQ Feedback IE are described below:

- **CID:** The CID of the connection for which this feedback is being sent.
- **Last:** The Last bit is used to pack multiple ARQ Feedback IEs to create an ARQ feedback payload. The Last bit is set to zero for all ARQ Feedback IEs, except the last IE in the payload.
- **ACK Type:** This field indicates the type of acknowledgment being transported in this feedback IE. Table 6–10 shows the acknowledgment types and their descriptions.
- **BSN:** The BSN is interpreted differently based on the acknowledgment type as described in Table 6–10.
- **# ACK MAPs:** The value of this field plus one represents the total number of 16-bit acknowledgment bitmaps included in this feedback IE. This field is not valid for acknowledgment type 0b01 and, therefore, is set to zero.
- **ACK MAPs:** The number of acknowledgment MAPs is variable, and the acknowledgment type 0b01 does not have an associated acknowledgment MAP. The acknowledgment MAPs are treated as a bit stream, where the MSB is the MSB of the first acknowledgment MAP and the least significant bit (LSB) is the LSB of the last acknowledgment MAP.

Table 6–10:   Acknowledgment types

Value	Type	Description
00	Selective	The BSN carried within this IE corresponds to the MSB of the first 16-bit acknowledgment bitmap. Each subsequent bit corresponds to the BSN corresponding to the previous bit plus one. No assumptions should be made about the status of any of the BSNs prior to the first BSN or the last BSN represented by the acknowledgment MAPs. If the bit is set, it indicates that the receiver has successfully received the corresponding ARQ block.
01	Cumulative	The ARQ block indicated by the BSN and all ARQ blocks prior to this within the transmission window have been successfully received. No ARQ MAP is included with this acknowledgment type.
10	Cumulative with selective	This is the combination of the previous two acknowledgment types (0b00 and 0b01). Combines the functionality of types 0x0 and 0x1. The first bit of the first acknowledgment MAP, which corresponds to the included BSN, is always set to 1, indicating cumulative acknowledgment for the BSN.
11	Cumulative with block sequence	This acknowledgment type combines the functionality of the cumulative acknowledgment type 0b10 with the ability to acknowledge a sequence of ARQ blocks in a compact format. If a set of ARQ blocks with consecutive BSN values all have the same reception status (i.e., either positive or negative), a single bit indicating the status, together with a sequence length, is used to indicate the status of the whole sequence of BSNs.

Since all the bits in a selective acknowledgment MAP must refer to an ARQ block that was successfully received or an ARQ block that was presumed lost or received incorrectly, it is important to choose the right BSN for the Selective ACK Feedback IE. The receiver must assign a BSN so that all bits in a MAP either indicate a positive acknowledgment (ACK) or a negative acknowledgment (NACK). One way to achieve this is to choose a lower BSN so that all bits are valid. For example, if only 12 ARQ status bits are available (either ACK or NACK) starting from BSN N, the receiver may send a selective acknowledgment with a single 16-bit MAP and a BSN of $N - 4$, where all the first 4 bits will be set to one.

The format of the block sequence acknowledgment MAP is shown in Figure 6–22. There are two block sequence formats. The first (and most significant) bit of the block sequence acknowledgment MAP indicates the format. The first format includes a 2-bit sequence acknowledgment MAP, where the bits are interpreted similarly to the selective acknowledgment type, except that each bit corresponds to a BSN sequence instead of a single BSN. Since this is also a cumulative acknowledgment, the BSN carried within this ACK Feedback IE indicates that the ARQ block with this BSN and all ARQ blocks with lesser BSN values within the transmission window have been successfully received. However, the first bit of the sequence acknowledgment MAP corresponds to the first BSN of a sequence that is being selectively acknowledged, which is the BSN carried in the feedback IE plus one. Therefore, the first bit may be a negative acknowledgment, unlike the acknowledgment type 0b10, where the first bit must always be set to one, as it corresponds to the cumulative acknowledgment BSN. The length of the each BSN sequence is indicated in the two 6-bit sequence length fields that follow. The first format can be used to indicate the selective acknowledgment status of up to 126 ARQ blocks.

The second format shown in Figure 6–22 is very similar, except that it has a 3-bit sequence acknowledgment MAP, where the length of each sequence is indicated by the 4-bit sequence length fields. Therefore, a total of up to 45 ARQ blocks can be selectively acknowledged using this 16-bit compact format.

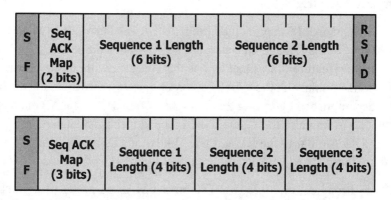

Figure 6–22:   Block sequence acknowledgment MAP formats

ARQ feedback payload

The ARQ feedback information can be either sent as a stand-alone MAC management message on the appropriate basic management connection or piggybacked as a special payload on an existing connection. ARQ feedback sent in a basic management connection cannot be fragmented. When transported in a management connection, more than one ARQ Feedback IE may be sent, and the Last bit is used to properly decode the contents.

HYBRID AUTOMATIC REPEAT REQUEST (HARQ)

The HARQ is an optional part of the MAC. HARQ is supported only for the OFDMA PHY. HARQ, along with associated parameters, is negotiated during initialization of the SS. HARQ defines four basic concepts to efficiently support retransmissions:

- Compact MAP IEs that define access information for both UL and DL
- HARQ packet construction
- Reduced connection identifiers (RCIDs)
- HARQ UL/DL acknowledgments

Compact MAP IE

The Compact MAP IE may include both DL/UL IEs that define the access information for the DL and UL burst of HARQ–enabled SS. One distinguishing feature of this message is that it is sent without the generic MAC header, unlike all other MAPs, in order to save a few bytes of overhead. This is accomplished by setting the first 3 bits of the HARQ MAP message to a value that is not valid for a generic MAC header encapsulated management or data PDU. Similar to the DL and UL MAPs for TDM and TDMA operation, these HARQ MAPs are broadcast by the BS.

A BS may broadcast multiple HARQ compact MAP messages using multiple burst profiles (i.e., modulation and coding rate). The HARQ compact MAP always follows the DCD and UCD messages, if present. Otherwise, HARQ compact MAPs are sent after other DL/UL MAP messages. When multiple HARQ compact MAP bursts are present, they are ordered based on the modulation and coding used, starting from the lowest rate to the highest. A HARQ MAP Pointer IE is used by the BS to indicate to the SS the number of slots allocated to the OFDMA burst containing HARQ MAP messages.

HARQ Control IE

HARQ Control IEs that are part of the Compact MAP IEs carry the necessary encoding/decoding information of the HARQ–enabled DL/UL bursts. This information is used to properly decode the HARQ subpackets. The information contained in the HARQ Control IE includes the following:

- A 1-bit HARQ sequence number, toggled between 1 and 0
- A 2-bit SPID that identifies the subpacket as described in the following subsection
- A 4-bit HARQ channel ID that identifies the HARQ channel

Since HARQ is a stop-and-wait ARQ, a single bit sequence number is sufficient for proper operation. However, more than one HARQ channels may be defined per terminal, where each connection may have an encoded packet pending. The HARQ Control IE may also be used by the BS to request a retransmission of a subpacket.

Construction of HARQ packets

The construction of an HARQ packet involves the concatenation of one or more MPDUs, as shown in Figure 6–23. A CRC is added to this concatenated packet to form an HARQ PHY burst.

After the construction, an HARQ packet is encoded according to the PHY specification, and four subpackets are generated from the encoded result. The subpackets are identified by an SPID. The encoding and generation of four subpackets introduce redundancy among the subpackets. Therefore, the subpackets can be transmitted one at a time, in sequence, and the receiver can correctly decode the original encoder packet without having to receive all four subpackets.

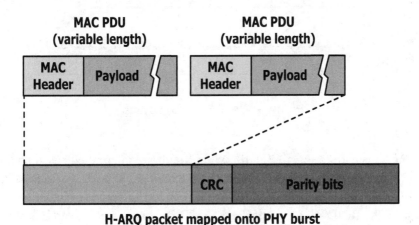

H-ARQ packet mapped onto PHY burst

Figure 6–23:   Construction of HARQ packet

Reduced connection identifier (RCID)

Figure 6–24 shows the format of a RCID. An OFDMA BS may use RCID instead of a Basic CID or multicast CID to reduce the size of MAP messages, including HARQ MAP messages. There are three types of RCIDs defined in IEEE Std 802.16: RCID-11, RCID-7, and RCID-3. The type of RCID used in a specific system is determined by BS considering the range of Basic CIDs of the SSs connected with the BS. The RCID type to be used is specified in the

RCID_Type field of the Format Configuration IE that is used in the OFDMA PHY to select format configuration.

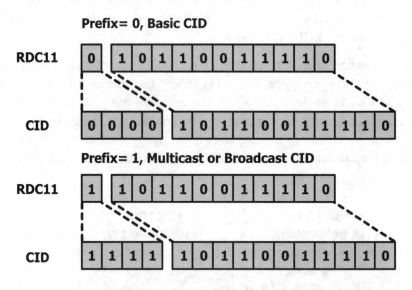

Figure 6–24:   RCID decoding

The RCID is carried in a RCID_IE, whose fields are interpreted based on the RCID_Type field (from the Format Configuration IE) and the Prefix field in the RCID_IE itself, if applicable. When the RCID_Type field value is nonzero, the RCID_IE is composed of 1 bit of prefix and n-bits of LSB of the SS CID. The prefix is set to one for the Broadcast CID or Multicast Polling CID and set to zero for the Basic CID. If the RCID_Type field value is zero, then the RCID_IE consists of a regular 16-bit CID. Therefore, an RCID_IE can be either 16 bits, 12 bits (RCID-11), 8 bits (RCID-7), or just 4 bits (RCID-3). Table 6–11 shows the interpretation of the RCID_IE fields, where the [R]CID length and RCID_IE length are specified in bits.

Table 6–11: RCID field interpretation

RCID_Type	Prefix	[R]CID length (bit)	RCID_IE length (bit)	Comments
0	NA	16	16	This is the same as normal CID
NA	1	11	12	11 LSB of Multicast, AAS, or Broadcast CIDs
1	0	11	12	11 LSB of Basic CID
2	0	7	8	7 LSB of Basic CID
3	0	3	4	3 LSB of Basic CID

The RCID cannot be used in place of the Transport CID, Primary Management CID, or Secondary Management CID.[3] For the purpose of per-SS HARQ, an SS is always identified by its Basic CID (or a corresponding RCID for the Basic CID).

HARQ acknowledgments

The HARQ acknowledgments are handled differently in the UL and DL. The BS uses a Compact MAP IE to send HARQ acknowledgment bitmap information for the HARQ UL bursts from the SS. The BS specifies a "frame offset" that indicates the offset between the UL HARQ burst and the corresponding HARQ bitmap. This frame offset f is communicated in the HARQ_ACK_Delay_for_UL_Burst field of the DCD messages. The SS uses the frame offset to correlate acknowledgment bitmaps to the HARQ bursts. For an HARQ UL burst in the ith frame corresponding to the jth HARQ burst in the UL MAP, the BS uses the jth bit (LSB) of the HARQ bitmap sent in the $(i + f)^{th}$ frame to indicate the status. The DL Compact MAP IE (type = 5) that

---

[3] For other uses of RCID in OFDMA, an exception to this rule is made for the Multicast Polling RCID in the DL. If a DL CID with prefix 1 and RCID 11 is in the range of the Multicast Polling CID (0xFF00–0xFFFD), then the DL CID is interpreted as a DL Transport CID by subtracting 0xFF (0xFE01–0xFEFE).

carries the HARQ bitmap consists of a Length field that indicates the length of the bitmap and the actual bitmap.

The DL HARQ bursts are acknowledged in the HARQ region allocated by the BS in a UL Compact MAP IE. The OFDMA subchannels in the HARQ region are divided into two half-subchannels, where the $(2n)^{th}$ half-subchannel is the first half-subchannel and the $(2n +1)^{th}$ half-subchannel is the second half-subchannel of the nth subchannel. An HARQ DL burst at the ith frame is acknowledged by an SS through the half-subchannel in the HARQ region at the $(i + f)^{th}$ frame, where the frame offset f is defined by the HARQ_ACK_Delay_for_DL_Burst field in the UCD message. The order of DL HARQ burst defined in the HARQ MAP determines the half-subchannel offset in the HARQ region. For an HARQ–enabled burst at the ith frame that is the nth HARQ–enabled burst in the HARQ MAP, the SS transmits the HARQ acknowledgment at the nth half-subchannel in the HARQ region that is allocated by the BS at the $(i + f)^{th}$ frame.

Chapter 7 MAC operation

Radio control, QoS, and ARQ

NETWORK ENTRY AND INITIALIZATION

Once the SS has powered up, it begins the network entry and initialization process. After completing the steps of the process, the SS has all the addresses and parameters it needs to communicate with the rest of the network.

Scanning and synchronization to the DL

Before the SS can do anything else, it has to first find a BS transmitting a signal that it is capable of decoding and understanding. To accomplish this, the SS consults the frequency lists with which it shipped, chooses one of the channels mentioned in the list, and starts to search for the periodically occurring frame preamble. If the SS finds the preamble, it tries to determine the DL transmission parameters by looking at the frame control header (FCH) and the DCD message, which is transmitted periodically but not in every frame. Also, the SS learns the specified UL parameters from the UCD messages. Afterwards, it has sufficient knowledge to transmit on the UL, and initial ranging may begin.

Initial ranging

Initial ranging accomplishes the following:

- The timing advance of SS transmissions is adjusted to make the SS appear collocated with the BS.
- The transmission power of the SS is adjusted for optimal reception at the BS.
- The SS is allocated its Basic and Primary Management CIDs.
- In some systems, the frequency of SS UL transmissions is adjusted.

Initial ranging occurs in a part of the frame that is set aside for this purpose and partitioned into initial ranging slots. The slots need to be large enough to accommodate the ranging transmission and twice the propagation delay to the edge of the cell, since the distance between SS and BS is not known at this time. The exact details of the initial ranging procedure depend on the underlying PHY, but in short the process goes as follows: First, the BS signals the available initial ranging slots in the UL MAP. After determining the location in time of the initial ranging slots, the SS randomly selects a slot and sends an RNG-REQ (ranging request) message. If this is received correctly, the BS replies with a message specifying the power, timing, and possibly frequency adjustments the SS needs to make prior to its next transmission. This message can also instruct the SS to repeat the initial ranging process. However, the BS would normally want to instead fine-tune the SS transmission parameters using one of the maintenance ranging procedures, as they are less resource-consuming than initial ranging.

Now, there is always the possibility that two terminals may send their RNG-REQs in the same slot, in which case the BS may be unable to decode either transmission. Also, it might be that the SS transmission fails because its power is too low (noting that the SS will initially use low power so it does not disrupt transmission in neighboring cells). In any case, if the SS does not hear from the BS within a certain interval, it will increase the power by a notch and try again after waiting for a randomly chosen number of initial ranging slots. In addition, for some PHYs, if the BS can detect energy but fails to decode the initial ranging message, it can signal for the SS to retransmit and make adjustments, identifying the SS in the RNG-RSP (ranging response) message by the transmit slot.

SS basic capability negotiation

Once the SS timing and transmit power have been adjusted, the SS needs to tell the BS which optional functionality it supports. Also, the BS needs to tell the SS which of those options the SS is allowed to use. The protocol by which the support and use of options are agreed upon is called *SS basic capability negotiation*. As this name suggests, this basic negotiation covers not all options, but only those having to do with fundamental MAC and PHY features, such as maximum transmit power, modulation schemes, FEC codes, and support of MAC bandwidth allocation schemes. The idea with exchanging these parameters at an early stage is to make the rest of the network entry process more resource-efficient.

The protocol itself is a straightforward request-grant exchange, as shown in Figure 7–1. The timer T18 has a default value of 50 ms and should be significantly less than the registration timeout (i.e., the timeout at the BS for the maximum time between a RNG-RSP to an SS upon successful ranging and the receipt of an SBC-REQ (SS basic capability request) message from the same SS). In the request message (SBC-REQ), the SS will list all the "basic capabilities" it supports. The BS inspects the list of capabilities and, in the SBC-RSP (SS basic capability response) message, informs the SS of those it does not support or does not want to use.

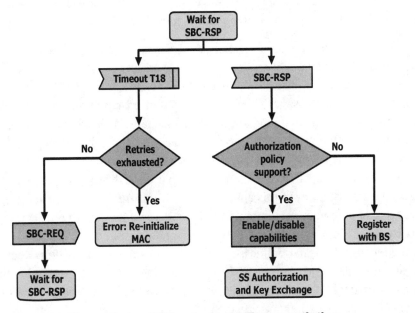

Figure 7–1:   SS basic capability negotiation

Note that some of the most fundamental parameters of operation are not subject to negotiation with this protocol. For example, with different PHY system profiles, establishing communication would be challenging or impossible.

Authorization, security association (SA) establishment, and key exchange

At this stage of the network entry process, the SS should be capable of communicating efficiently with the BS. The BS, however, does not yet know with certainty the identity of the SS with which it is dealing. The SS has up to this stage provided its IEEE MAC address in the RNG-REQ, but has offered no credentials verifying its identity. The authorization phase establishes the identity of the SS, the authorization key (AK), and the list of SAs that the SS is authorized to use. To achieve this, well-known public key cryptographic methods are used. For a treatise on these, consult [B45].

The authorization protocol starts with the SS sending the authentication information privacy key management (PKM) message. All PKM messages are encapsulated in either the PKM-REQ (SS to BS) or the PKM-RSP (BS to SS) management message. The authentication information message is strictly informational, but is included to provide a mechanism by which a BS can learn the X.509 certificate of the SS manufacturer in a standard way. However, the IEEE 802.16 standard says nothing about how the BS can authenticate the certificate. It is envisaged that the industry will set up the needed infrastructure for this authentication if this method is to be used in practice. Meanwhile, the assumption is that the manufacturer certificate ends up, by some secure means, in the BS. Immediately after sending the authentication information message, the SS sends the Authorization Request message encapsulated in a PKM-REQ management message containing the X.509 certificate of the SS together with a list of the SS security capabilities. The BS proceeds to authenticate the certificate by checking the signature of the manufacturer. Assuming the certificate is in order, the BS generates the AK and encrypts it using the public key of the SS contained in the X.509 certificate. Additionally, the BS determines which SAs the SS is entitled to use. The AK, the key lifetime, and the list of SAs along with their parameters are sent by the BS to SS in the Authorization Reply message. Also, the BS has the option to send any PKM protocol parameters that are different from the default values given in the IEEE 802.16 standard.

After receiving the Authorization Reply message and decrypting the AK, the SS becomes authorized to operate normally in the network. However, it still lacks the capability to transmit anything besides MAC management messages. Before becoming fully operational, the SS still needs to (a) acquire a key for each SA given by the BS using the protocol given in the SA parameter perform registration, (b) possibly acquire an IP address, and (c) establish the MAC connections. Each SA's keys, called *traffic encryption keys* (TEKs), are established by the SS sending a digitally signed Key Request PKM message and the BS returning the encrypted key in a Key Reply message, also signed with a key derived from the AK. The algorithm used in the encryption of the TEK is triple data encryption standard (3DES), AES, or RSA, depending on algorithm support of the SS and the parameters of the SA.

Registration

During the registration phase, the SS and BS negotiate additional operational parameters of the MAC. Also, the SS informs the BS whether it will be part of the managed network. If the SS is to be managed, the bidirectional secondary management connection is established between the SS and BS. The registration message exchange also offers the possibility to negotiate the version of IP and the QoS parameters for the secondary management connection. The registration process is shown in Figure 7–2. The SS may wait for up to 3 s (timer T6 in Figure 7–2) for the registration response before resending the request or reinitializing the MAC.

Establish IP connectivity

A managed SS also needs to acquire an IP address (using DHCP), pull down a configuration file (using TFTP), and establish the time of day (using the Internet Time Protocol). Afterwards, the SS sends a message to report that the tasks were successfully completed.

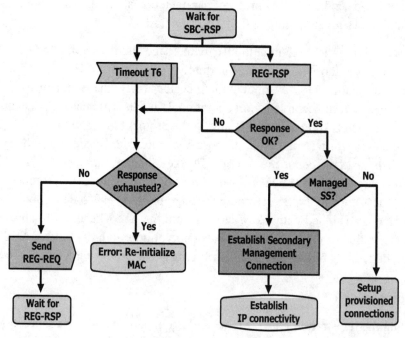

Figure 7–2:   SS registration

Dynamic service establishment

The final step before actual data communication can begin is dynamic service establishment. In the case of a managed SS entering the network, the reception of the TFTP-CPLT ([configuration file] TFTP complete) message triggers the BS to start connection setup. When dealing with unmanaged SSs, the successful completion of registration serves as the trigger. The service flows for a given SS are assumed to have been preprovisioned by the BS. To activate them or take them to the admitted state, the BS initiates a dynamic service addition (DSA) exchange, as illustrated in Figure 7–3. After at least one service flow has been activated, the SS is capable of sending and receiving user data. The DSA signaling can also support setting up the classifiers and PHS on the CS.

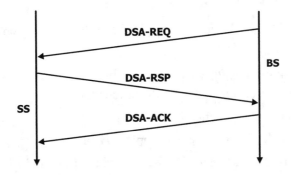

Figure 7–3: DSA message flow (BS-initiated)

PHY MAINTENANCE

Using IEEE Std 802.16, the PHY parameters used to communicate between the BS and an SS may change rapidly. These changes may be caused by myriad events that may differ depending on the operational frequency and choice of PHY specification. For instance, at 28 GHz the WirelessMAN-SC PHY is susceptible to rain fade, while at 2.5 GHz the WirelessMAN-OFDM PHY is susceptible to changes in multipath. How often PHY quality needs to be assessed and how often PHY parameters may need to change often depend on the environment of the system. For instance, at 28 GHz, the rain region may dictate the maximum rain fade, while at 2.5 GHz, population densities and urban or rural physical settings may impact multipath change. Many PHY adjustments are possible. For instance, power control may be applied, or the FEC or modulation scheme may change. Adjustments may occur in real time. Some of these changes, most notably FEC and modulation, impact the availability of bandwidth in real time. This aspect of IEEE Std 802.16, coupled with a need for true QoS, has blurred the lines between the classical partitioning of the MAC and the radio link control (RLC) layer. In the IEEE 802.16 standard, little has been done to prevent this blurring. In fact, the standard does not specifically discuss an RLC layer, but merely handles it as an extension of the MAC. Some messages, such as MAPs, span both layers by simultaneously allocating bandwidth and modifying PHY parameters.

MAPs and channel descriptors

In IEEE Std 802.16, all four currently defined PHYs share some common attributes:

- They are all framed PHYs, providing deterministic, periodic opportunities for resynchronization and centralized scheduling.
- They can be used in either TDD or FDD mode.
- In FDD mode, SSs may be half-duplex (H-FDD).
- They all use MAPs to describe the bandwidth usage and allocations in both the DL and the UL.
- They all use channel descriptors to associate sets of PHY parameters to DIUCs and UIUCs allowing a shorthand to make the MAPs more efficient.

In both the TDD and FDD cases, the PHY frame duration is selected by the BS from one of the allowed frame durations. Typical durations, depending on the PHY, range from 0.5 ms to 20 ms, with 1 ms being common for WirelessMAN-SC systems and 5 ms being typical in the case of WirelessMAN-OFDM. The duration is expressed to the SSs or discovered by them in a PHY-dependent way.

The PHY frame is further divided into a DL subframe and an UL subframe. In the TDD case, the frame starts with the DL subframe and ends with the UL subframe. Between the two is a time gap (TTG) allowing devices the time to shift from transmit to receive, or vice versa. Similarly, another time gap (RTG) falls between the end of the UL subframe and the start of the next DL subframe. When scheduling the end of the DL subframe and the beginning of the UL subframe, the BS must take care to respect the half-duplex nature of TDD devices. Two additional time durations besides the TTG and RTG are of importance, as well, as described in the following two paragraphs.

First, the BS must be aware of round-trip delay because SSs advance their transmissions so that they appear to be collocated with the BS from a timing point of view. The BS has two choices, depending on the complexity of the scheduling within the BS. Either it may use the concept of a maximum round-trip delay based on the planned cell size, or it may track the round-trip delay of each SS individually and use this when scheduling. The maximum round-trip delay is a useful simplification in fixed systems, but it can cause problems if nomadic or mobile stations are allowed. It may also complicate adding an additional subscriber just beyond the current cell boundary. The use of a tracked round-trip delay is more robust to changing conditions, but it can add complexity to the BS scheduler.

The second important time duration is T_{proc}. This is the amount of time allowed for an SS to process the UL MAP after receipt. In systems that allow the UL MAP to apply to the current frame (WirelessMAN-SC and WirelessMAN-OFDM, but not WirelessMAN-SCa or WirelessMAN-OFDMA), the BS further needs to ensure that the SS has had sufficient time to receive and process the UL MAP before it is required to transmit in the UL. More on the time relevance of MAPs will be given in a few paragraphs. An example of a TDD PHY frame for a TDM system, taken from the WirelessMAN-SCa PHY, is shown in Figure 7–4.

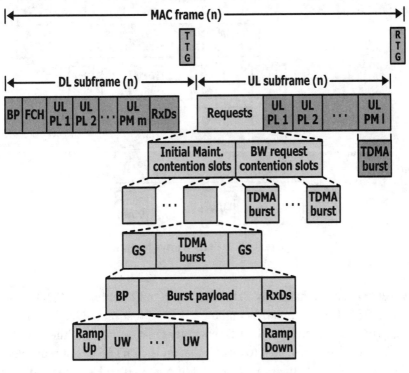

Figure 7–4:   TDD frame format

In the FDD case, the PHY frame can be viewed as being the same as the TDD case, except that the DL subframe is stretched to end later and the UL subframe is stretched to start earlier. Therefore, the two effectively overlap. The structure of each subframe stays basically the same as the TDD case. In FDD, the standard allows two types of SSs: full-duplex and half-duplex. From the BS point of view, full-duplex SSs are relatively easy to schedule since there is no need to worry about receive-transmit exclusions. An example of a FDD PHY frame for a TDM system, again taken from the WirelessMAN-SCa frame, is shown in Figure 7–5.

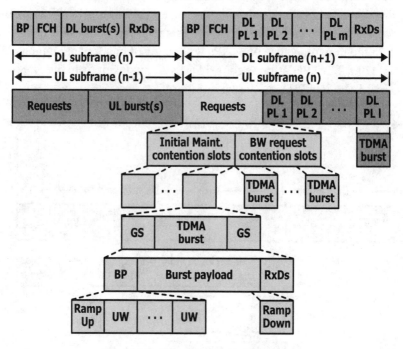

Figure 7–5: FDD frame format

However, we expect that many FDD SSs will be half-duplex (H-FDD) to save cost. The BS scheduler must, as in the TDD case, respect the half-duplex nature of the SSs. Therefore, the BS scheduler must be aware of the round-trip delay, turn around times, etc., to ensure that an H-FDD SS is not requested to transmit and receive at the same time. Additionally, if an H-FDD terminal is scheduled to receive later in the frame than it transmits, it may lose synchronization during transmission and, therefore, may require a method to resynchronize with the DL prior to reception. Due to certain benefits of the OFDM symbol structure, this is usually of concern only for WirelessMAN-SC and WirelessMAN-SCa. For these two single-carrier PHYs, resynchronization points in the form of additional preambles provide the midframe resynchronization. Figure 7–6 shows the modified DL frame structure, taken from WirelessMAN-SC, for FDD with a preponderance of H-FDD SSs.

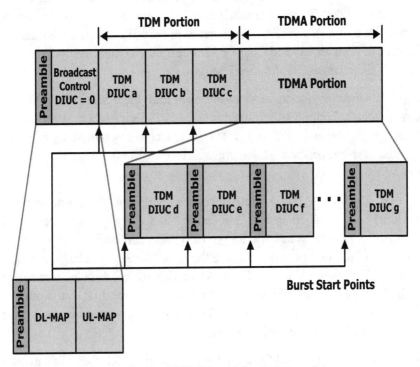

Figure 7–6: FDD DL subframe structure

For both the DL and UL subframes, the BS transmits MAPs that define the allocations of bandwidth in the subframe. The MAPs can take a variety of forms depending on the PHY and the options enabled. In the simplest form, there is a single DL MAP and a single UL MAP for a PHY frame. Certain

PHYs may add additional concepts. For instance, private MAPs are used for AAS zones, and a number of other PHY- and option-specific MAP variants are specified. Some of these MAP variants are caused by PHY-related issues concerning whether an SS can hear broadcast MAPs. In all cases, the basic concept is that the DL and UL subframes are described by MAPs.

At first glance, this appears to be merely a scheduling exercise determining the timing of the transmissions. But the content of the MAPs reaches beyond simply scheduling the time (and subchannels of OFDM-based PHYs) on which an SS sends or receives. The MAPs contain content impacting the PHY. Exactly what PHY information may be contained in the MAPs depends on the PHY and is described in more detail in the PHY-specific chapters, Chapter 10, Chapter 11, and Chapter 12; but it may contain fast power control, FEC changes, modulation changes, etc.

In the UL, since most transmissions are from individual SS's, the MAP elements state which SS is to transmit. In every PHY, the SS is also instructed which modulation and FEC to use. The SS reacts to any parameter changes, using the new information for the time (and subcarriers) covered by the MAP entry.

In the DL, the DL MAPs for some PHYs may address MAP elements to individual SSs, in which case the allocation can act as a quick change of DL burst profile. In these cases, the SS typically has negotiated with the BS a most robust DL burst profile to ensure the signal quality it needs for reliable operation. For some PHYs, the DL MAP entries are not directed to a specific terminal. In these cases, the SSs have a negotiated current DL burst profile and listen to all DL portions they can, including those at that negotiated burst profile and those that are more robust.

DL MAPs are always valid for the DL subframe in which they are carried. This can require some very fast processing, and PHY-specific features may be specified to allow the processing to occur fast enough. For instance, the WirelessMAN-OFDM PHY uses a DLFP in each frame, followed in only some frames by a DL MAP message. For the WirelessMAN-SC PHY, the DL MAP message is always transmitted at the beginning of the DL subframe. The start of the WirelessMAN-SC DL subframe is structured so that at least two noninterleaved FEC blocks of a well-known modulation and FEC will be present. The first DL burst profile transition that is expressed in the DL MAP is always included in the first of these FEC blocks, allowing the SS the duration of the second FEC block for decoding, processing, and taking action.

The timing of the UL MAP is more complicated and is related to the burst profile, the duplexing scheme, and possibly the network setup. At the simple level, the UL MAP describes the UL allocations for a period starting at a point called the *allocation start time*. This is relative to the start of the DL subframe in which the UL MAP was received.

For WirelessMAN-SC and WirelessMAN-OFDM systems, the limits on the allocation start time are slightly different, depending on whether the system is TDD or FDD, but they basically have the same result: The UL MAP can describe the allocations in the UL subframe of either the current or the next PHY frame. For TDD systems, in other words, the MAP allocates the use of the time just after the TDD split of either the current or next frame, as shown in Figure 7–7.

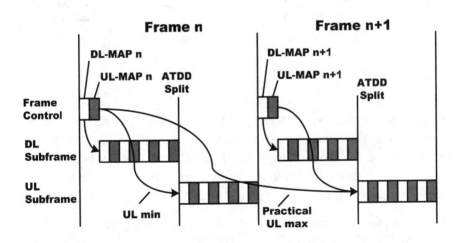

Figure 7–7: Practical TDD UL minimum-maximum relevance

For FDD systems using one of these two PHYs, the situation is a bit more complicated. The minimum delay is as small as the round-trip delay plus

T_{proc}. The latest allocation start time is the beginning of the next frame. Using this latest possible start time, the UL subframe period is aligned with the DL subframe following the one carrying the relevant UL MAP. With an earlier start time, the UL subframe is offset in time from the DL subframe. This is shown in Figure 7–8.

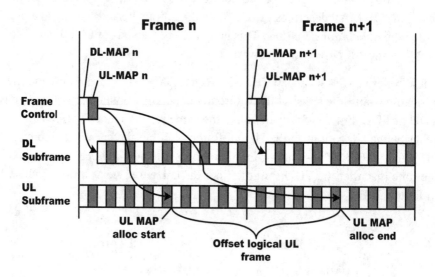

Figure 7–8: FDD logical offset

For WirelessMAN-SCa and WirelessMAN-OFDMA, the concept is similar, except the UL MAP messages are relevant for the next frame in TDD systems and the next one or two frames later in the FDD case. It is also possible that the UL MAPs sent in two different frames overlap. In such a case, the most recently received UL MAP takes precedence.

The reasons for the differences in UL MAP relevance between the PHYs are historical. The WirelessMAN-SC and WirelessMAN-OFDM systems can offer a shorter turnaround time on protocols requiring responses and can offer a shorter request/grant cycle. On the other hand, the BS scheduler and the implementation of UL MAP processing at the SS must be correspondingly more complex in able to respond with due speed.

Because there are so many options for how to configure the PHY in a given implementation, a shorthand was developed to ensure that the length of the UL and DL MAPs did not get out of hand from repeating the PHY parameters. As mentioned above, the MAPs contain codes that indicate a set of PHY parameters in the compact form of DIUCs or UIUCs. Periodically, the BS broadcasts DCD messages and UCD messages. These messages associate the chosen set of PHY parameters with a DIUC or UIUC, respectively. Then, these DIUCs and UIUCs can be used in MAP messages as if the MAP message contained the entire set of PHY parameters. While the sets of PHY parameters for a particular network are generally not expected to change, a protocol is defined for distributing new sets of parameters and coordinating the change from the old set to the new set.

Periodic ranging

In order to ensure effective UL communication, ongoing maintenance is required regarding the parameters adjusted during initial ranging, such as the transmit time advance, the power level, and the burst profile. This process is called *periodic ranging* and is used to compensate for clock drift, changing environmental conditions, etc. For most IEEE 802.16 PHYs, this involves the BS periodically granting UL bandwidth to the SS. If the BS senses that the SSs PHY parameters need to be changed, it can generate an unsolicited RNG-RSP message that can direct certain changes. Timing advance and power adjustments are fairly straightforward. The actual process for changing the burst profile is a bit more complex and is described in the next section. The process generally proceeds as follows:

- The BS must determine how often it needs to hear an UL transmission from each SS in order to maintain the UL within limits that allow continued communication. This maximum time between UL transmissions can be based on a number of factors, such as frequency band of operation; the likelihood of change in environmental conditions (e.g., rain region for WirelessMAN-SC); whether the setting is urban, suburban, or rural; etc.
- Any UL transmission from the SS should suffice. Therefore, as long as the SS has UL activity for any reason (e.g., user data, other management messages), there is no need for the BS to grant bandwidth specifically for

periodic ranging. The BS merely takes measurements from the scheduled UL transmissions, resetting its periodic ranging timer for the SS with each UL transmission. If, however, the SS has no other need for a UL transmission opportunity within the time necessary for UL PHY maintenance, the BS grants it UL bandwidth anyway. Since this bandwidth is not in response to a direct request to transmit user data or management messages in the UL, it is termed an *unsolicited grant*. The process of using normally granted UL bandwidth and unsolicited grants to manage the timing of unsolicited grants is shown in Figure 7–9.

- If, upon analysis of a UL transmission, the BS determines that the SS's UL transmission parameters need to be adjusted, it sends an RNG-RSP message to the SS instructing it to adjust its parameters. If this RNG-RSP indicates "continue," the SS responds with a RNG-REQ to keep the iterative PHY maintenance process in this quick response state until the BS issues a RNG-RSP message indicating "success."

- If the SS has not been granted UL bandwidth in a sufficient amount of time or if the BS sends it a RNG-RSP message indicating "abort," the SS must fall back to periodic ranging to avoid detrimental impact on the UL transmission of other SSs until it gets its UL transmission parameters back with a safe window.

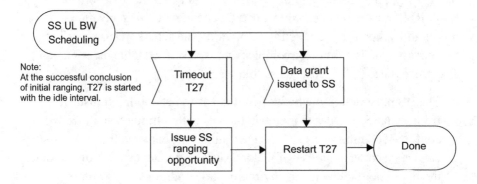

Figure 7–9:   Periodic ranging opportunity allocation at BS

The BS side of the UL periodic flow is shown in Figure 7–10. The SS side is shown in Figure 7–11.

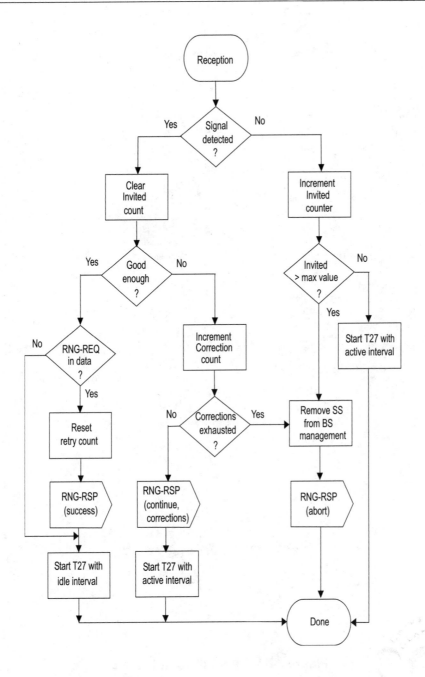

Figure 7–10: Periodic ranging receiver processing at BS

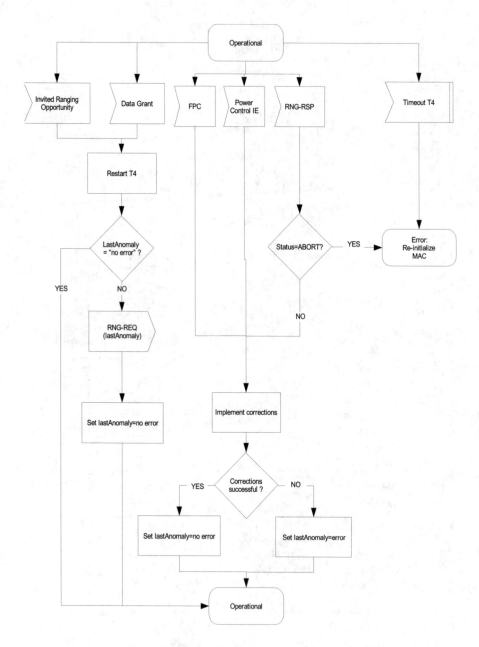

Figure 7–11:   Periodic ranging at SS

For the WirelessMAN-OFDMA PHY, while the BS can initiate ranging as described, the SS also has the option of initiating CDMA-based ranging. This allows a distribution of the timer function to the SSs, easing BS requirements while taking some control away from the BS in the management of the system. In CDMA periodic ranging, the SS sends a randomly chosen ranging code in a randomly chosen contention-based ranging slot. Since the BS does not know which SS sent the code, it broadcasts a RNG-RSP advertising the code rather than directing the RNG-RSP to the specific SS. This method can be used to adjust the SS's power and timing advance. However, since the BS does not know the identity of the SS, it must still use measurements on the UL allocations to determine whether a burst profile change is necessary. The UL CDMA periodic ranging process is shown in Figure 7–12.

Burst profile changes

When the BS is communicating with an SS, it may use a UL burst profile differing from that of the DL. The burst profile used for communication with a particular SS is selected by the BS based on a number of considerations, such as the carrier to interference-plus-noise ratio (CINR), or the bit error rate (BER), or even the packet error rate (PER). The burst profile is always selected from the set described by the most recent DCD and UCD messages.

The initial burst profile is normally negotiated during the initial ranging process. The BS and SS determine the initial quality of the link, with the SS requesting a DL burst profile using the RNG-REQ message and the BS commanding a UL burst profile using the UL MAP. As time goes by, the burst profile used for transmissions to or from a particular SS may need to change. This may be due to changing environmental conditions such as rain or changes in multipath. The change may be either for the better—the conditions have improved to where a more efficient burst profile is practical—or for the worse—the conditions have degraded to the point where a more robust burst profile is necessary.

How these changes are implemented is different for the UL and the DL. In the UL, the BS monitors the UL quality from each of the registered SSs. The BS must give each SS a transmit opportunity at least often enough to adjust the parameters fast enough to keep up with environmental changes. For instance,

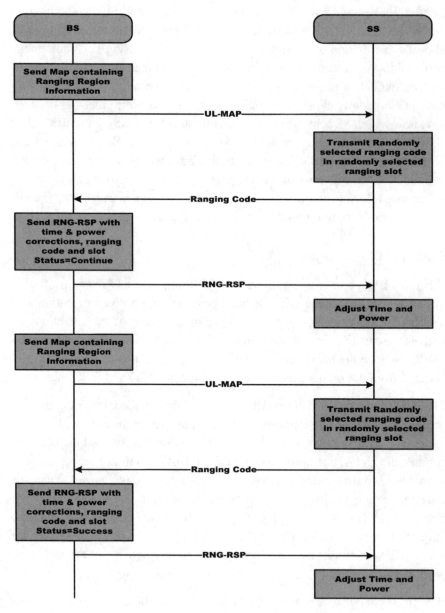

Figure 7–12: CDMA periodic ranging

for WirelessMAN-SC, where rain fade is the biggest physical degradation culprit, the BS generally needs to receive something from an SS every 80 or so milliseconds; otherwise, it may drop back to the most robust burst profile to ensure it can be heard. This monitoring constraint is a small burden, but the actual change is easy. Since each allocation in a UL MAP specifies the SS and the UIUC, the BS commands a change by simply allocating bandwidth for the SS using a new UIUC, from the set previously defined in the UCD message, in the UL MAP.

The DL can be more complicated. In the DL, while the SS has the advantage (except in certain AAS cases) of always hearing the broadcast portion of the DL, it constantly updates its measure of the DL quality from its point of view. The problem is informing the BS of a need for a change. Typically, the SS will request a change via either a downlink burst profile change (DBPC) message, if it has bandwidth allocated on the UL, or using an RNG-REQ message in a ranging interval, if it does not have its own UL allocation. For some situations with some PHYs, the BS may also periodically request PHY metrics from the SS to fine-tune such issues as which subchannels to use. For MAP efficiency purposes, the DL MAP, unlike the UL MAP, for most IEEE 802.16 PHYs does not direct a DL interval to an individual SS. The SS merely decodes all data that are as robust or more robust than the last negotiated DL burst profile. This is an important point. It allows the BS to send data to the SS using a more robust burst profile before first communicating its intent to switch, as shown in Figure 7–13. It also allows the BS to tell the SS to start listening to a less robust burst profile and then immediately begin sending data with that profile, as shown in Figure 7–14. This avoids complicated synchronization of the messages and the burst profile changes. In some cases, with some PHYs, the DL bursts are identified in the DL MAP as being directed to a specific SS. In these cases, the process works much like the UL process.

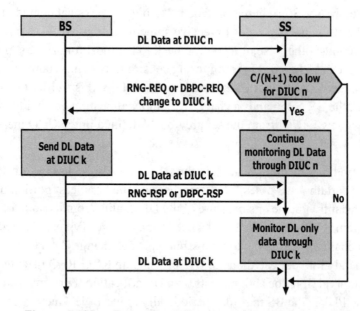

Figure 7–13: Transition to more robust burst profile

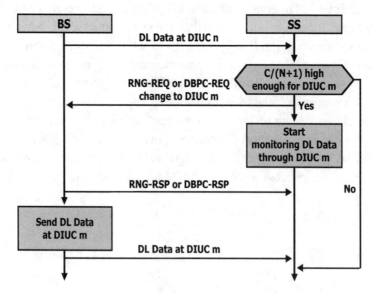

Figure 7–14: Transition to less robust burst profile

QoS AND SERVICE FLOWS

Dynamic service establishment and deletion

Besides the BS-initiated DSA described in the network entry and initialization section at the beginning of this chapter, other mechanisms are defined for SS-initiated DSA, as illustrated in Figure 7–15.

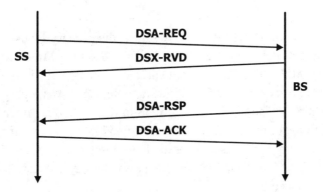

Figure 7–15:   DSA message flow—SS initiated

The standard does not go into details on what actually triggers the DSA; triggering is just assumed to happen, stimulated by the upper layers when needed. For example, a nomadic device may be powered on in a network, similar to Wi-Fi networks that might be found in an airport or other public place today. The SS, being unknown to the network, asks for service. Authentication includes validating the user's credit card. Looking forward to IEEE Std 802.16e and mobility, another example would be a mobile VoIP call initiated at the MS.

One noteworthy detail in the procedure is the DSX-RVD (dynamic service addition, change, or deletion received) message that only acknowledges the reception of the DSA-REQ (dynamic service addition request) message by the BS. This allows the BS to take its time determining whether to admit the service flow without triggering unnecessary retransmissions of the DSA-REQ.

There may also be a need to change the parameters and state of the existing service flow in response to external circumstances. This can be done with the dynamic service change (DSC) procedure that is almost identical to the DSA.

Service flows are deleted using the DSD-REQ (dynamic service deletion request) message and the DSD-RSP (dynamic service deletion response) message. Deletion can be initiated by either BS or SS. Only one service flow can be deleted per message exchange.

QoS model

When IEEE Std 802.16 was in its early stages, the primary focus was from companies targeting last-mile extension of fiber and the PMP backhaul of cellular sites. At that time, IP core networks were considered interesting, but ATM core networks were the norm. Furthermore, ATM was highly advanced in its QoS support. The result was that QoS in IEEE Std 802.16 was modeled after ATM. Some of the concepts were modified based on input from proprietary system proposals and based on proposals for "wireless DOCSIS." As it turned out, ATM, DOCSIS, and the proprietary systems are compatible in their QoS concepts.

While extension of ATM networks was an early market driver, IP was considered to be the future, and the developers of IEEE Std 802.16 did not want to tie the standard too closely to ATM. DOCSIS, as it turned out, was forward-looking in its application of QoS and traffic parameters and was compatible with both ATM and IP services. Because of this, the QoS description and naming conventions from DOCSIS were chosen as the basis for describing QoS in IEEE Std 802.16. Other parts of the DOCSIS model were adopted as well, including the three-state model in which a service can be provisioned, admitted, or active. The state impacts the actual reservation state of the resources necessary for the service.

In order for QoS and traffic parameters to be fully applied to services, IEEE Std 802.16 MAPs services, including connectionless services, to an IEEE 802.16 MAC connection. This is done in basically the same manner as in ATM and in the UL of DOCSIS. Similar concepts are employed in Ethernet virtual circuits in metro Ethernet. Typically, a service is bidirectional, resulting in the mapping to separate UL and DL IEEE 802.16 MAC connections.

The benefit of connection-oriented mapping of services is that it provides a "handle" with which the MAC, in particular the scheduler and QoS engines, can associate a packet with the full realm of QoS and traffic parameters, instead of just simple individual packet parameters such as a priority. Since no packets are expected until a service is active and no resources are reserved until a service is at least admitted, services in the provisioned state are not mapped to IEEE 802.16 connections. The services are mapped only when they become admitted or active.

QoS and traffic parameters

A number of QoS and traffic parameters, which the IEEE 802.16 standard calls *service flow encodings*, are available to help in the implementation of QoS within an IEEE 802.16 system.

SFID and CID

The SFID and CID are important concepts at the foundation of IEEE 802.16 QoS. The SFID identifies a service, which in turn identifies the right of an IEEE 802.16 SS to certain system resources, and also defines which of a user's packets will be mapped to the corresponding IEEE 802.16 MAC connection. The CID is then used internally in the IEEE 802.16 MAC for determining which SS receives the data on the DL, which service it maps to on the UL, and, most importantly, what QoS and traffic parameters are associated with the traffic. The mapping of potentially connectionless traffic to MAC connections is called *classification*. It typically involves some amount of packet inspection and can work much like a combination of a router and a bridge. Classification of IP, VLAN, Ethernet, and ATM to IEEE 802.16 MAC CIDs is currently defined in the standard.

Service class name

The service class name is a parameter that implies a predefined set of other QoS parameters. There is not much demand to define such common service types yet. Once specific methods for handling certain services is agreed upon (if ever) by industry, service classes will most likely be used as an efficient shorthand for sets of parameters.

QoS parameter set type

The QoS parameter set type indicates whether the current set of parameters is the provisioned, admitted, or active set, as shown in Figure 7–16 and Figure 7–17.

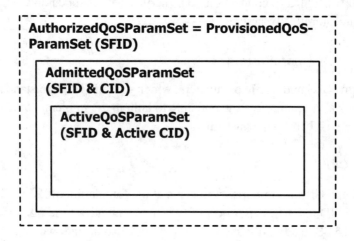

Figure 7–16:   Provisioned authorization model

The differences between parameter sets are as follows:

1. The provisioned set is one that is allowed; i.e., when a service-level agreement (SLA) was set up, this is what the users paid for. But the provisioned set is neither reserved nor allocated. Depending on the implementation, the provisioned set may be ignored by the BS's call admission control (CAC) function. For instance, in a mobile system, the users have paid for a service and have a right to certain resources. However, if their phones are turned off, they have no resources allocated, and the BSs ignore them from a CAC point of view because they may never turn on or roam into their field of view.

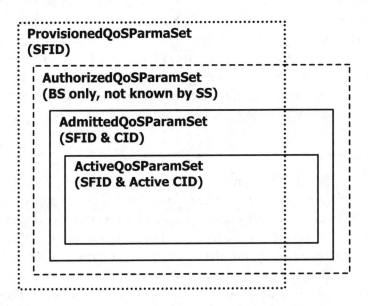

Figure 7–17:   Dynamic authorization model

2. The admitted set is one that is allocated but not being used. IEEE Std 802.16 is a bandwidth-on-demand system in which users gets their guaranteed bandwidth only if they ask for it. Allocated but unused bandwidth will cause CAC to keep the resources for this service in reserve so they are available, but will allow other services that are peaking above their guaranteed bandwidth to use the resources on a BE basis. This could be useful in a handover situation, with the necessary resources "admitted" on the new BS prior to the handover.

3. The active set is one that is currently reserved and being used by a service. For bursty services, there is little to distinguish it from the admitted set, since they both consume resources from a CAC point of view, but bandwidth is not used until requested. Technically, the service can't ask for the bandwidth until it transitions to the active state; therefore, the admitted set would rarely be used for bursty services.

These parameters have some interesting potential above and beyond simple resource reservation. For instance, SIP-based VoIP could be modeled as a low-bandwidth (but guaranteed), bursty, nonreal-time polling service (nrtPS) for the SIP signaling, coupled with a fixed-bandwidth UGS [more likely extended real-time polling service (rtPS) in IEEE Std 802.16e] that toggles between provisioned and active, where the active bandwidth may be modified from the provisioned due to choice of codec.

Traffic priority

Traffic priority is a value from 0 to 7 that describes the priority of traffic. Priority alone, in a situation of frequent congestion, does not provide QoS but rather causes starvation. Therefore, this parameter should be used sparingly. It should not override the minimum reserved traffic rate of a connection. It can be useful for determining how to distribute excess bandwidth among connections that peak above their minimum reserved traffic rate. Since the parameter is on a service-by-service basis, rather than a packet-by-packet basis, it cannot be used to differentiate between two packets from the same service for the purpose of discard on congestion.

Maximum sustained traffic rate

The maximum sustained traffic rate is the peak data transmission rate allowed for a service. In the absence of bandwidth demands from other services, a service would expect to be able to transmit at this rate continuously. Since the time base over which this is statistically guaranteed is not specified, it is left to vendor design to determine how smoothly this is guaranteed. It is expected that client devices will police traffic at the ingress and not request more bandwidth than the maximum sustained traffic rate, again statistically determined over some period. BSs can generally assume that the traffic has been policed before it is received across the backhaul from the core network. If the bandwidth demand is greater than the link can bear, services should first receive their minimum reserved traffic rate. Excess bandwidth above the sum of the minimum reserved traffic rate should be distributed proportional to the difference between the minimum reserved traffic rate and the maximum sustained traffic rate. Traffic between the minimum reserved traffic rate and the maximum sustained traffic rate is not guaranteed to be transmitted and may be discarded. Intelligent discard algorithms that understand the relative importance of individual packets of a service can enhance the QoS perception over general discard methods like weighted random early detection.

Maximum traffic burst

The maximum traffic burst is the largest burst expected at the ingress port for the service. Its main purpose is to ensure adequate buffering in case of temporary congestion or delays between receiving user traffic on the ingress and transmission over the air.

Minimum reserved traffic rate

The minimum reserved traffic rate is statistically guaranteed to a service over time. QoS algorithms should be designed to give services their minimum reserved traffic rate before any services receive bandwidth in excess of their minimum reserved traffic rate, even if they are a higher class of service (CoS) and the traffic is within the bounds of their maximum sustained traffic rate. This parameter is used to prevent starvation of data from connections that are less real-time than others but have guaranteed rates as part of their SLA. Some systems may chose to oversubscribe the minimum reserved traffic rates when performing CAC, assuming that not all services will request bandwidth simultaneously. Typical QoS algorithms use either a credit or a debit scheme with some finite duration memory in determining the bandwidth available to a service under this parameter. For instance, a service that is idle for hours typically cannot expect to wake up and instantly transmit data equivalent to the minimum reserved traffic rate accumulated during those hours.

Minimum tolerable traffic rate

The minimum tolerable traffic rate is related to the Minimum reserved traffic rate. For certain services, it may be convenient to discuss the service paid for—the minimum reserved traffic rate—and the worst service the user will tolerate—the minimum tolerable traffic rate. While CAC should normally use the minimum reserved traffic rate, there may be circumstances in which resource availability changes after a call has been admitted. For instance, the signal quality may degrade due to some interference or, in a mobile system, due to the client moving to an area of degraded coverage (e.g., farther away or indoors). Alternatively, a mobile call accepted in one cell may move into the neighboring cell that may already be at capacity with respect to minimum reserved traffic rates. The minimum tolerable traffic rate allows calls to be retained at a degraded level even when the call would not normally be accepted by the CAC algorithm.

Vendor-specific QoS parameters

The vendor-specific QoS parameters provide a QoS backdoor allowing two devices to share proprietary information with the expectation that they can do a better job distributing bandwidth without using general algorithms based on generic QoS and traffic parameters. Care must be taken, however, to maintain the ability to interoperate with devices from other manufacturers that do not use the same parameters. In general, an unknown device is handled by dropping back to exclusive use of the parameters defined in the IEEE 802.16 standard.

Service flow scheduling type

The service flow scheduling type defines the CoS.

Converted to ATM terminology, UGS is equivalent to constant bit rate (CBR) and usually has a fixed packet size and periodicity. The BS allocates bandwidth at the desired rate automatically, without any BW request from the client device. Keep in mind that the BS may "clump" the allocation into a portion of the PHY frame (unavoidable for TDD links), increasing the jitter and possibly causing the need for jitter buffering.

Real-time polling service (rtPS), nonreal-time polling service (nrtPS), and best-effort (BE) service are all variable-rate packet-based service classes. The main reason for differentiation is that there are different rules governing how a client device is allowed to request bandwidth for the different CoS. Also, when resources are tight, rtPSs may receive their minimum reserved traffic rate in preference to nrtPSs, which in turn may have priority over BE services.

CoS may impact the allocation of bandwidth above the minimum rate (generally policed to the maximum sustained traffic rate).

Generally, a UGS will have minimum reserved traffic rate set equal to maximum sustained traffic rate. A BE service will have its minimum reserved traffic rate set to zero, but have a nonzero maximum sustained traffic rate. rtPSs and nrtPSs will generally have nonzero minimum reserved traffic rates that are less than, but not equal to, their maximum sustained traffic rates.

Request/transmission policy

The request/transmission policy parameter is slightly misplaced in being considered part of the service flow encodings. While these parameters can be set on a per-service basis, most are really device capabilities and would normally be set to the same values for all services of the same CoS to or from an individual client device. They include the use of broadcast BW request opportunities, piggybacked requests, fragmentation, PHS, packing, and MAC CRC.

Tolerated jitter

The tolerated jitter parameter indicates the jitter that is allowed for a given service. It is quite often assumed that the receiving application has jitter buffering or other means to mitigate the problem. It is not clear from the IEEE 802.16 standard whether the tolerated jitter is meant to be total jitter from all sources or that due to the IEEE 802.16 link of the data's path. It could be used as the basis of jitter buffering but, if other points on the path do jitter buffering as well, unnecessary delay can result. This is a worrisome parameter because certification bodies may impose tests that would require jitter buffers where they may be of little value in the end-to-end system.

Maximum latency

Maximum latency is the maximum delay between receipt of a packet from the network and forwarding it on the air interface. A system is generally designed to minimize latency anyway. This parameter can, however, be used in times of congestion to give packets of similar CoS/QoS relative priority with respect to each other. Since the other QoS and traffic parameters have more influence on which packets are transmitted and when, it can also be used to age out old packets.

Fixed-length vs variable-length SDU indicator

The fixed-length vs variable-length SDU indicator specifies whether the service carries fixed-length SDUs (like ATM cells) or variable-length SDUs (like IP packets). This parameter is usually meaningful only for QoS in the case of UGS connections, where the periodic grants should be a multiple of the natural fixed packet length of the CBR service, such as an ATM T1. It can, however, be useful in applying packing and fragmentation more intelligently for variable-rate services that use a fixed packet size.

SDU size

If the fixed-length vs variable-length SDU indicator parameter is set to fixed, the SDU size must be used to indicate the size, allowing scheduling algorithms to take advantage of the known fixed size in whatever way is appropriate.

One interesting omission in IEEE Std 802.16 is the equivalent of the cell loss ratio (CLR) of ATM systems. This omission probably stems from the fact that this parameter is used not so much in the scheduling aspects of QoS, but in the link quality aspects. Typical systems would calculate and report a BER or PER that could be used to determine the need to transition to a more robust PHY mode if and when the loss rate is too great for the ToS. Unfortunately, without a parameter, this decision is left to the vendor or to vendor-specific parameters.

A common fallacy is that IEEE Std 802.16 describes only UL QoS. This is an unfortunate artifact of the naming conventions for services and the QoS model description coming from DOCSIS. DOCSIS assumed that since the head end (the cable modem equivalent of a BS) was in control, the cable modem didn't need to know how QoS was handled on the DL. This works in general but, for some issues like jitter, it's best if the receiving end knows what's going on as well. Unlike in DOCSIS, service setup in IEEE Std 802.16

associates QoS and traffic parameters with the DL as well. Therefore, even though IEEE Std 802.16 uses service names like "rtPS," the QoS concepts apply equally to the DL and the UL.

Interactions between QoS, CAC, and adaptive PHY

Determining available bandwidth

BWA systems are normally intended to provide very reliable service, as an extension of fiber optics or some other high-reliability network. When BWA systems are installed, frequency planning is normally performed to minimize interference while maximizing frequency reuse and, therefore, available bandwidth. Cell sizes are chosen based on some availability goal such as 99.99% or 99.999% (often referred to a "four 9's" or "five 9's").

Furthermore, in systems with a subscriber-level adaptive PHY, each SS can be assigned a "planned" PHY mode. The SS is expected to be able to operate at or above the target BER of the system using the planned PHY mode, or a more efficient one, a fraction of time equal to or exceeding the desired availability. For example, if the system availability goal is 99.99% and an SS's planned PHY mode is 16-QAM, then the SS is expected to operate at 16-QAM or 64-QAM the entire year except for 52 min and 33.6 s. The choice of planned PHY mode for a given availability is generally a function of the distance from an SS to the BS and the LOS conditions. Additionally, certain SSs, such as those at the edges of sectors, may be capped so that they cannot use certain more efficient PHY modes due to interference issues.

As a system operates, the physical environment may also change. Most of the time, most SSs will be able to operate in both the UL and DL in not just their planned PHY mode but also in a more efficient one. For LOS systems in the 10-66 GHz frequency range, rain fade is the main variable in link quality. During rain, an SS may not be able to operate above its planned PHY mode. In fact, during heavy rains, an SS may need to drop to a PHY mode more robust than its planned PHY mode to maintain sufficient link quality to communicate with the BS. For lower frequency systems, multipath variation, due for instance to moving objects such as vehicles, is the main culprit.

The CAC algorithm for modern, third-generation BWA systems must deal with the issues created by these dynamic environments. Many fixed BWA services are established as permanent connections and remain active (in operation) for a long time—months and sometimes years. Rain fades, on the other hand, tend to be relatively short in duration. Therefore, the BWA CAC need not consider short-term changes in system capacity. Those are best handled by the queuing, scheduling, and discard algorithm. CAC must, however, consider the planned PHY modes of the SSs, as these determine the amount of bandwidth that can be guaranteed while still meeting the system's availability goals. The BWA CAC algorithm must address three issues:

- What aspects of a service's QoS and traffic parameters are mapped to the bandwidth availability determined by planned PHY modes to guarantee contractual and QoS requirements?

- What should be done with the additional bandwidth available when a terminal can operate at a more efficient PHY mode, which may be most of the time?

- What should be done when a terminal is unable to operate at its planned PHY mode and must use a less efficient one?

Bandwidth on demand setting the basis for CAC

The solutions to the first two of the problems above are based on the concept of bandwidth on demand. While some BWA systems will be expected to carry legacy T1/E1 TDM voice traffic, demand for variable-rate, bursty services such as Internet access is growing. Even for voice, there is a growing interest in using compressed voice, which results in variable bit rate rather than CBR used by traditional TDM connections. These variable-bit-rate services allow more users and more voice calls to be multiplexed into the same bandwidth by taking advantage of not only compression but also statistical multiplexing.

While third-generation BWA systems can interface to a plethora of user services, such as Ethernet, IP, TDM voice, and ATM, it is ATM that currently has the most mature concept of QoS and the richest QoS and traffic parameter terminology. Therefore, while the concepts apply equally well to IP-based systems, they will be explained using ATM terminology and mapped to the more generic WirelessMAN terminology.

The ATM service architecture defines six CoSs, as shown in Table 7–1:

- **CBR:** Constant-bit-rate services require a static amount of bandwidth that is available for the lifetime of the connection and impose stringent control on delay and jitter. CBR services are intended to emulate legacy TDM services, such as voice and video, and do not generally support statistical multiplexing.

- **rtVBR:** Real-time variable-bit-rate services are bursty, but like CBR also require tight control of delay and jitter. rtVBR services are a good choice for VoIP or streaming video. Because of the bursty nature of rtVBR services, statistical multiplexing may be supported.

- **nrtVBR, GFR, UBR, and ABR:** Nonreal-time variable bit rate, guaranteed frame rate, unspecified bit rate, and available bit rate are categorized as nonreal-time services. These services may be very bursty and have a large difference between average and peak usage rates. GFR, nrtVBR, and ABR offer a minimum guaranteed rate while UBR is simply a BE service.

Table 7–1: Mapping of ATM CoS concepts to WirelessMAN

ATM CoS	WirelessMAN CoS
CBR	UGS (unsolicited grant service)
rtVBR	rtPS (real-time polling service)
nrtVBR, GFR, ABR	nrtPS (nonreal-time polling service)
UBR	BE (best-effort service)

These service categories relate traffic descriptions and QoS parameters to network resources. Queuing, scheduling, and CAC are structured differently for each service category. The purpose of CAC is to ensure that the system has the necessary resources (e.g., bandwidth and buffers) to support a new connection without impacting any preexisting connections. CAC uses the following QoS parameters and traffic descriptors to calculate resource availability.

- ***QoS parameters***
  - ***End-to-End Transit Delay:*** This parameter is the amount of delay the service can tolerate. Information delayed longer than the specified delay is considered to be less useful to the application and thus may be discarded by the network. In WirelessMAN, this is the maximum latency parameter.
  - ***Peak-to-peak delay variation (or jitter):*** This parameter is used to size buffers at various multiplexing points within the system. It also impacts the overall delay experienced by the application. In WirelessMAN, this is the tolerated jitter parameter.
  - ***CLR:*** The cell loss ratio parameter is used in sizing the buffers allocated to the connection and in the level of statistical multiplexing gain that can be supported by the BWA system.

    It should be noted that each BWA system has inherent end-to-end transit delay and peak-to-peak delay variation characteristics that are directly related to system design choices.

- ***Traffic description parameters***
  - ***PCR:*** Peak cell rate (or peak bit rate for non-ATM systems) is the maximum data rate allowed for the service. In WirelessMAN, this is the maximum sustained traffic rate parameter.
  - ***SCR***: Sustained cell rate is the average data rate expected to be available to the service. The minimum cell rate of the GFR service would map to this as well. In WirelessMAN, this is the minimum reserved traffic rate parameter.
  - ***MBS:*** Maximum burst size is used in buffer allocation and statistical multiplexing calculation. In WirelessMAN, this is the maximum traffic burst parameter.

The focus of this section is the impact of adaptive PHYs on CAC. Therefore, we will ignore for clarity and brevity the impact of CLR and MBS, since they affect mostly buffer allocation calculations. Likewise, as long as a system is not oversubscribed, the delay requirements of a service are impacted much more by the bandwidth allocation scheme and the implementation of fairness algorithms.

So what does it mean for a system with a subscriber-level adaptive PHY to not be oversubscribed? The answer to this question lies in the definition of what is guaranteed to a service. For a CAC algorithm to work, certain rules should be obeyed. In particular, policing should be performed at the ingress point to the network. Services should not, in general, be allowed to transmit more data than their maximum sustained traffic rate. Granted, this must be done in a constructive way, since the service should not be penalized for the fact that the physical line rate of its input port is greater than its maximum sustained traffic rate. For BE services, it is not uncommon for the maximum sustained traffic rate to equal the line rate of the port.

SLAs typically stipulate the grade of service the end customer is getting from the service provider. The SLA's attribute that is of interest to CAC is the minimum reserved traffic rate. The customer is typically guaranteed that this rate will be available. This guarantee is, however, a statistical guarantee, meaning that, over time, the customer will get the minimum reserved traffic rate, but that there may be short periods in which that rate is not available. In general, the following holds:

- For UGSs, minimum reserved traffic rate = maximum sustained traffic rate.
- For rtPSs, minimum reserved traffic rate < maximum sustained traffic rate.
- For nrtPSs, minimum reserved traffic rate << maximum sustained traffic rate.
- For BE services, minimum reserved traffic rate = 0.

Bandwidth demands in excess of the minimum reserved traffic rate can be viewed as not guaranteed and are candidates for statistical multiplexing of services. For instance, as any cable modem user knows, BE customers can be continually added until they complain.

CAC can determine whether the minimum reserved traffic rate of a service can be guaranteed, by comparing that rate against the bandwidth available in the airlink after subtracting the minimum reserved traffic rates of the other services already admitted. This is easier said than done. First, there is the issue of jitter. If a service is at all bursty, even if it uses exactly its minimum reserved traffic rate, it will need less at one instance and more at another.

Therefore, some additional bandwidth may be necessary to compensate for jitter. Alternatively, an operator's policy may be to oversubscribe somewhat the minimum reserved traffic rate. Therefore, CAC should base a service's guaranteed bandwidth on a function of minimum reserved traffic rate and jitter. The actual function used can be tailored to take into account oversubscription policy and may also account for the responsiveness of the bandwidth allocation algorithm in the BS, since this can affect jitter.

Second, there is the issue of the adaptive PHY. The same minimum reserved traffic rate for two different services, each to a different SS, may use substantially different resources. For instance, an SS operating at QPSK will use three times as much physical airlink resources for the same amount of data as a terminal operating at 64-QAM and the same FEC. Therefore, the CAC algorithm must be aware of the planned PHY mode of each SS and normalize the bandwidth requirements of services to common units to determine actual PHY resources consumed. The planned PHY mode is used because the CAC algorithm is concerned with guarantees, and it guarantees that contractual obligations can be met, but only within the bounds of system availability. The planned PHY modes dictate the bandwidth guaranteed available a sufficient fraction of the time to meet the operator's availability goals.

Since SSs quite often have environmental conditions allowing them to operate above their planned modulation (just not often enough to meet availability goals), there is quite often more airlink bandwidth available than CAC uses in admitting services. This bandwidth can be used for BE traffic. It can serve this purpose in two ways. For rtPS and nrtPS connections, this excess bandwidth is normally available to service demands in excess of the minimum reserved traffic rate, up to the maximum sustained traffic rate. Similarly, the excess bandwidth is available for BE services. The high probability of excess bandwidth in the system allows for substantial oversubscription of maximum sustained traffic rate. This can be especially useful for selling services that can peak at their port rate (for example, 100BaseT Ethernet) if bandwidth is available.

Adaptive CAC philosophies for adverse conditions

Once a function is determined for allowing services into the system based on the planned PHY modes of the terminals, there remains a need for a plan of action for when the system is faced with adverse conditions, e.g., when it rains hard enough for terminals to drop below their planned PHY modes to more robust but less efficient PHY modes.

First, it must be determined whether a terminal dropping to a more robust PHY mode actually causes oversubscription. The airlink on which the terminal operates may not be fully subscribed (in the sense defined in the previous section) with guaranteed traffic. Also, if the terminal in question has only BE services, there is no change in the ability to accommodate the guaranteed traffic within the system. Finally, especially in uncorrelated fades, some other terminals may be operating at more efficient PHY modes, offsetting any bandwidth availability problem.

Second, if it is determined that the actual PHY mode dropback causes an oversubscription problem, some action must be taken to retain some amount of services at adequate QoS. This allows the system to maximize the benefit of one of the major features of adaptive PHYs—the ability to still retain the link, albeit with reduced capacity, even in severe weather conditions.

Care must be taken in maintaining QoS and system stability on the temporarily oversubscribed airlink. Initially, it might be tempting to just let the system run with reduced capacity, increased delay, and reduced QoS because this condition may exist only a few minutes a year. In systems with no UGS or rtPS, this extremely simple approach may be very reasonable. Customers can be taught to accept slower response during extreme weather. But UGS, and to some extent rtPS, is very intolerant to change. If the system has a high enough ratio of UGSs, the T1/E1 TDM systems will start alarming due to insufficient bandwidth. A minor oversubscription situation could ultimately bring down all the phone lines in the sector. Obviously, this is not good.

Another simple approach is to temporarily suspend the services with guaranteed bandwidth for the terminal that is below its planned PHY mode (keeping in mind that, for VBR connections, suspending may be merely reducing the SCR). After all, the reason it is in this predicament is that it's raining hard enough to be in the 52 min a year (four 9's) when it's not guaranteed to be available anyway. This works and meets contractual obligations; in fact, the system may not need to suspend all of the services at that terminal, but just enough to resolve the oversubscription situation. That effectively gives the remaining services an availability slightly higher that the system goal.

This concept can be taken further and can lead to system availability greater than the goal, or it can be traded off for additional capacity by slightly increasing the size of the higher PHY mode regions within a sector. This is especially true in light of the fact that, while affecting only the services on the SS at lower modulation meets contractual guarantees, it's not really fair.

There are two main points that make this simple solution unfair. First, consider that the oversubscription situation may have come into being not because of this single SS, but because a sequence of SSs dropped below the planned modulation. This one just happens to push the system below the oversubscription threshold. Why should its customers be the ones to suffer? The second reason the solution is unfair is best explained pictorially.

In Figure 7–18, two SSs within the same sector have the same planned PHY mode. SS 1 is not quite close enough to the BS to have a more efficient PHY mode and still meet availability goals. SS 2 is in the opposite situation; it is barely close enough to the BS to meet availability with its planned PHY mode. In this situation, SS 2 will normally need to drop to a more robust PHY mode long before SS 1. In fact, SS 1 may never need to drop to a more robust PHY mode. With the previous simple solution to the temporary oversubscription problem, SS 1will be unfairly penalized for its relative geographical proximity to the BS.

There are numerous methods to make the situation fairer. In applying them, the CAC algorithm can rely on the fact that the availability of services has a minimum, as expressed by the availability goal, but that many terminals actually enjoy much greater availability. So the algorithm can "spread the pain," keeping the average availability the same while decreasing the worst-case availability. One method is to randomly, or in a round-robin fashion, suspend services with guaranteed bandwidth until the oversubscription

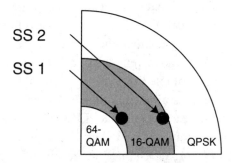

Figure 7–18:   Two terminals with same planned PHY mode

problem is alleviated. This is more fair than the previous approach, and it exceeds the availability goals of the system. However, it has the drawback that some of the suspended services may have been critical.

The problem of critical services can be handled by assigning services that have guaranteed bandwidth a precedence level relative to each other. Those with lower precedence are suspended first until the oversubscription problem is alleviated. Within a precedence level, the random or round-robin suspension algorithms may be used. The term *precedence* is used intentionally and should not be confused with CoS or QoS. It may not be acceptable, for instance, to suspend nrtPS first, followed by rtPS, with UGSs getting the axe last. Take, for instance, the business with both its phone lines and its point-of-sale (PoS) transactions going over the BWA system. While maintaining some phone lines during severe weather may be necessary, it is probably more acceptable to have reduced phone capability while retaining PoS capability than to retain all phone lines at the expense of PoS.

Multicast connections

The IEEE 802.16 standard supports multicast connections that allow a service on the DL to be transmitted to multiple SSs on the same airlink without duplicating the transmission to each subscriber. The transmission would be performed at the lowest common denominator of the PHY modes currently used on the DL for the set of subscribers receiving the service.

Multicast introduces an additional complication regarding security. IEEE Std 802.16 typically uses a pair of encryption keys per client so that one client's data cannot be read by another client on the same airlink. For multicast data, e.g., a TV program being broadcast to a number of subscribers simultaneously, this policy of one set of keys per client would effectively require the service to be transmitted individually to each client so it could be individually encrypted for each client. The IEEE 802.16 standard, however, allows per-service encryption keys shared by all recipients of the service, allowing the service to be broadcast once and received by many, but still by only those who subscribe. Clients that support this must be able to have multiple SAs. The target security association identifier (SAID) parameter associated with the service is used to determine which set of keys and which encryption method to use. The SSs' control and any unicast services must use the SA unique to the client. Any multicast or broadcast services are mapped to SAs unique to the service.

Multicast connections also have a unique interaction with ARQ. For multicast service, per-service MAC ARQ is not allowed. Further, since PHY HARQ is on a per-client basis, SSs receiving a multicast connection must be configured to not enable HARQ.

BW request/grant

The BW request/grant mechanism for the IEEE 802.16 standard was chosen to be efficient, low-latency, and flexible and to dovetail with QoS. Requests are made on a per-connection basis to ensure they can be properly used in fairness algorithms in the BS's UL scheduler. But grants are made to the SS, not to the connection. This increases efficiency; e.g., the MAPs are smaller due to an entry for the terminal, rather than each of the terminal's connections. Latency can also be reduced. The terminal can use the bandwidth for the best current need rather than for a reason dictated by the BS. For instance, if an SS asked for bandwidth for a BE connection, but an urgent need to go to a more robust DL PHY mode arises, the SS can use the bandwidth to send a DBPC-REQ (DL burst profile change request) message. This very important message is sent with lower latency than if the SS needed to wait for bandwidth explicitly granted for that purpose.

Also, there are no explicit acknowledgments (positive or negative) of requests. Either the SS gets a grant, or it does not. This saves the bandwidth that would have been used for the acknowledgments. Latency is reduced because the SS does not need to wait for an acknowledgment, just for a grant.

Giving grants to SSs rather than connections can result in inconsistent behavior. Even if the SS does not intentionally use the bandwidth differently from its requests, an SS with multiple services does not know exactly how much of a grant the BS intended for one service vs the other. Additionally, the lack of acknowledgments can cause the BS's perception of the SS's queue status to deviate from the actual situation.

The reality at the SS and the perception at the BS can get out of sync for a number of reasons:

- The BS does not hear a BW request (unrecoverable PHY error).
- The SS does not hear the allocation in the MAP (unrecoverable PHY error).
- The BS scheduler decides it does not have bandwidth right now for the particular service for which the SS requested bandwidth.
- The SS used bandwidth for a purpose different from that originally requested.

To rectify any discrepancy between the SS's reality and the BS's perception, the BW request/grant scheme is designed to be self-correcting. After a period, potentially based on the QoS of the service in question, if the SS still needs bandwidth for a service, it simply asks again. To avoid the BS's perception becoming further askew from reality by duplicate requests, the SS issues an *aggregate request*. An aggregate request tells the BS that this is the current state of the SS's queue for that service, allowing the BS to reset its perception of that service's needs.

Aggregate requests are necessary to guarantee the self-correcting nature of the request/grant scheme. However, there is a chance that a repeated aggregate request crosses the grant for that same bandwidth in the same frame. While this is very unlikely to happen with any regularity, it can, if the request is large, cause wasted allocations to the SS that would have been better served going to a different SS. Therefore, while aggregate requests are both necessary and sufficient for the request/grant mechanism to operate,[4] there is a potential waste that can be easily avoided by adding the concept of *incremental requests*. With an incremental request, the SS is simply saying "I need more bandwidth" for a particular service. The BS would add this BW request to its current perception of the bandwidth needs for that service rather than using the BW request to reset its perception of the services needs. In general, the airlink should be reliable. Therefore, most BW requests typically would be incremental, with only periodic aggregate requests to ensure the BS does not deviate too far from reality.

If a BS does happen to allocate bandwidth to an SS without a need, the SS can be polite and send an aggregate request for zero bandwidth. This deterministically informs the BS that the SS does not need bandwidth. Alternatively, the SS can just send fill bytes. The BS can key into the presence of fill bytes to modify its perception of the SS's bandwidth needs. Obviously, keying in on fill bytes can be dangerous if the SS implementation can generate fill bytes when the SS has data queued. But, then again, the request/grant scheme is self-correcting, and this situation will correct itself as well.

A few other BW request options exist as well. For UGS connections, which would typically be used for such services as legacy TDM voice, there is no need for BW requests since both parties know the recurring bandwidth needs of the service. However, clock skew can be a problem. To address this, if the SS sees UGS data starting to back up, it can set the SI bit in the GMSH, thereby requesting the BS to slightly increase the rate at which it automatically allocates bandwidth to the SS. This increase would continue until the SS stops setting the SI bit.

If an SS needs more bandwidth for a connection that is currently granted some over the airlink, the SS can piggyback an embedded incremental BW request in a GMSH of an MPDU for that connection.

Similarly, if the SS is receiving bandwidth that is sufficient only for its UGS connections, the SS can request to be polled, indicating it has a bandwidth

---

[4]   ETSI BRAN HiperACCESS uses the same request/grant scheme, but with only aggregate requests.

need on another connection, by setting the PM bit in the GMSH of an MPDU of a UGS connection. When the BS sees the PM bit set, it knows the SS needs to make a BW request and may choose to short-circuit its polling cycle for that SS and poll it immediately. In fact, if an SS has bandwidth needs for its UGSs sufficient large that it receives an unsolicited allocation every frame, the BS can refrain from polling the SS until it sets the PM bit, saving UL bandwidth.

Scheduling

Scheduling algorithms were intentionally left outside the scope of the IEEE 802.16 standard. The set of service offerings combined with scheduling algorithms is a major area for vendor differentiation. However, some concepts are common to all implementations, and some ideas were intended even while not explicitly made a part of the standard.

An IEEE 802.16 system has three main scheduling points. The most obvious one, and the one that most often comes to mind first, is the scheduling of the UL by the BS. The BS receives requests from SSs, processes those requests, and then creates and distributes UL MAP messages. This process must consider not only the outstanding BW requests but also polling, PHY maintenance needs, and the current UL PHY mode for each SS. The current UL PHY modes impact the amount of bandwidth, from a data rate perspective, available in the UL.

But, as previously mentioned, the UL MAP messages allocate bandwidth to SSs, not connections. This leads to the second scheduling point. Once the SS receives an allocation, it must decide which of its data to transmit during that allocation. To do this, it runs a scheduling algorithm analogous to that in the BS, but limited in scope to its own services, MAC management messages, and its individual allocation at its current UL PHY mode.

The third major scheduling point is the creation by the BS of the DL subframe. Like the BS scheduling of the UL, this task must take into consideration not only the DL data awaiting transmission but also the current DL PHY mode of each SS. DL MAC management messages must be included in this scheduling. The end product of this scheduling point is a DL subframe, complete with a DL MAP message indicating PHY mode transitions within the DL subframe, and the UL MAP message resulting from the BS UL scheduling.

The intent of IEEE Std 802.16 was that the DL MAP and UL MAP messages be transmitted using the most robust modulation currently in use on the airlink. This aids new SSs entering the network, and it guarantees that the MAP will be received with an low likelihood of error. Since the DL MAP is of variable length, it needs to be relevant to the current PHY frame; i.e., the frame containing it. If it instead mapped the PHY mode transitions of the next frame, it would need to know the length of the next DL MAP so it could map the first PHY transition. This is a nondeterministic problem. It requires an FEC structure for transmitting the MAPs that guarantees that the first DL PHY mode transition is contained in an FEC block that is not the last FEC block of the initial PHY mode. This guarantees at least one FEC block buffer for processing the first DL PHY transition. For most practical implementations, this will require the DL MAP to be processed in hardware or firmware rather than software.

For TDD systems, the UL MAP message can be for either the current frame or the next frame, as shown in Figure 4–7 and Figure 4–8 (in Chapter 4). The BS must choose one of these at the start and not change its choice. When the UL MAP is for the current frame, the latency of the request/grant cycle is reduced. This comes at an expense, however. The UL MAP cannot be for any earlier than the round-trip delay of the first SS scheduled in the UL plus the time required to be allotted for processing the MAP. This bounds the minimum DL subframe duration, reducing flexibility compared to when the UL MAP is for the following frame.

For FDD systems, the UL MAP can map allocations starting as early as the start of the current frame plus the time allotted for processing the MAP plus the round-trip delay. Alternatively, it can map allocations starting with the following frame. Anything within these two extremes is possible, as shown in Figure 4–9 and Figure 4–10 (in Chapter 4).

When an SS has UGSs, it does not need to request bandwidth for those services. In fact, it should not make BW requests for those services. The BS already knows the bandwidth need of the UGSs and allocates the bandwidth unsolicited, hence the name *unsolicited grant* services. Based on the frame duration and the maximum sustained traffic rate parameter, which should be the same as the minimum reserved traffic rate parameter for UGSs, the BS can determine how much periodic bandwidth the SS will need each frame. In granting the allocations, the BS must also take into account the ToS being offered and other parameters of the service.

For example, assume a WirelessMAN-SC backhaul for a cell site of a global system for mobile communications (GSM) network in the United States. The IEEE 802.16 system would most likely be backhauling TDM voice circuits and using a 1 ms PHY frame. A convenient way to backhaul such cell sites is with an ATM adaptation layer virtual circuit. There are various modes for such a circuit, but a popular mode generates 4192 ATM cells per second. Therefore, the circuit needs bandwidth equivalent to 4.192 ATM cells each PHY frame. Since the ATM CS does not allow ATM cells to be fragmented, the BS must, over the course of each second, allocate the SS at least 4 ATM cells in 808 frames and at least 5 ATM cells in the other 192 frames. How the BS distributes these allocations is implementation-dependent and outside the scope of the IEEE 802.16 standard, but it should be obvious that doing a better or worse job can have a big impact on jitter and on buffer requirements at the SS. In general, the BS should try to distribute the allocation of the fifth cell evenly across time.

Even with very good bandwidth allocation algorithms at the BS, extra care must be taken with UGS connections. For instance, the bandwidth allocation for a UGS connection will likely be grouped into a single burst for each UL subframe rather than distributed across the frame. There are a number of reasons for this:

- Reduced complexity of SSs that need to transmit only once per frame.
- Reduced overhead of a single burst rather than multiple smaller bursts (e.g., packing, burst preambles).
- The fact that, unless all you're doing is nailing up T1s or E1s, the scheduling algorithm will never be able to perfectly distribute the allocations within a frame if there is more than one SS in the system.
- The fact that, in TDD systems, the allocations must be compressed into a partial frame anyway.

The "clumping" that results is really the equivalent of adding one PHY frame duration of jitter into the connection. In our example, this is 1 ms. Jitter buffers at both ends of the airlink should be sufficient to remove at least one PHY frame duration of jitter. Be careful, however, because the way to eliminate jitter is to turn it into delay.

Since UGS connections are the preferred choice for transferring constant data rate services, it is important to provide mechanisms for each side to maintain synchronization. Potential problems may arise. The clock used by the BS for PHY frame generation may drift with respect to the clocks used by the network to clock the UGS data. This can lead to data in the UGS queues at the SS building up at a faster rate than the allocations to the SS can drain the queues. When this situation happens, the SS notices that its UGS queues are above an implementation-dependent threshold. The SS then sets the SI bit in a GMSH, indicating that, due to a clock mismatch, the BS should grant it slightly more bandwidth than agreed. Additionally, many types of UGSs need their clock at the SS to match the network clock used for the service on the backhaul. The SS has the airlink clock available, but this may not be locked to the network clock. Whenever the BS is providing UGSs, it also provides the CLK-CMP ([SS network] clock compare) message, which allows the SSs to derive the network clock, or multiple clocks, from the airlink timing. The method for deriving the network clock is described later in this chapter.

rtPS, nrtPS, and BE service all differ from UGSs in that they do not expect continuous, constant bandwidth allocations. They typically carry bursty traffic such as web browsing or VoIP. From the end user's point of view, they may be the aggregation of multiple bursty services. For instance, a single IEEE 802.16 rtPS may allow some number of simultaneous VoIP calls. For rtPS and nrtPS, the minimum reserved traffic rate should be statistically guaranteed. The parameters of such a guarantee are outside the scope of the IEEE 802.16 standard; but in general, if a service is requesting its minimum guarantee, it should receive it. On the other hand, if a service does not use its guaranteed minimum for some period of time, it should not build up an infinite credit. Care must be taken to strike a balance between guaranteeing a minimum rate on average for bursty traffic and building up too much credit, which could starve other services.

The main difference between rtPS and nrtPS is the expected response time. This is especially true once a service becomes active. For instance, a VoIP service may tolerate greater delay during initial setup of a call than during the life of the call. BE services, however, differ from rtPS and nrtPS in that they have no minimum reserved traffic rate. Depending on the CAC, fairness, and scheduling algorithms used, the guaranteed part of other CoSs could effectively starve BE services. If this condition is more than an infrequent transient phenomenon, the system is oversubscribed, and more channels should be made available if possible.

Along with satisfying minimum guarantees and distributing bandwidth in excess of guarantees, another aspect of scheduling is policing. In some networks, services may be sold under agreement guaranteeing some (or no) minimum rate, bursting to some peak rate. Network operators may not want to allow users to become accustomed to having access to all the system's bandwidth if no one else is using it. UGSs typically have a fixed data rate, but rtPS, nrtPS, and BE service can burst with a peak rate equivalent to the line rate of their input port to the SS. For instance, a service connected to the SS via a 100BaseT port can burst data upwards of 96 Mbit/s. At ingress points to the network, policing algorithms should use a function of the maximum sustained traffic rate and the maximum traffic burst parameters to allow instantaneous bursts at the line rate of the input port while statistically limiting the traffic over time to at most the maximum sustained traffic rate. Obviously, if it is desired to have no upper limit on the bandwidth a service can use when available, the maximum sustained traffic rate can be set greater than or equal to the line rate of the port or the maximum rate of the airlink.

Unicast polling

The first thing to understand about unicast polling in IEEE Std 802.16 is exactly what it means. The basic intention of unicast polling is to give the SS a contention-free opportunity to tell the BS that it needs bandwidth for one or more connections. Since the IEEE 802.16 bandwidth allocation scheme allows the SS to use any bandwidth allocation as it sees fit, the SS simply sees extra, unsolicited bandwidth. If the bandwidth is available, unicast polling is generally preferred to broadcast or multicast polling. It can be used to minimize delay and jitter as well as to provide more deterministic delay and jitter than broadcast and multicast polling, both of which have the potential for collision between two BW requests.

In keeping with the main purpose of the unicast poll, the BS has two primary options for how much additional bandwidth to give an SS. Which option a BS uses is a function of many things, including the SS's service mix, the connection activity, and the complexity of the BS's polling algorithms. If an SS has only a single, active connection, the BS may opt to give the SS only the additional 2 bytes necessary to piggyback a BW request in a GMSH. Alternatively, if the SS has no outstanding BW request or if it has multiple connections, the BS could opt to allocate the 6 bytes necessary for a BW request header. These two options represent the minimum necessary unsolicited allocations for unicast polling. The BS is free to grant more unsolicited bandwidth to an SS. This could allow the SS to transmit multiple BW requests for different services. There are additional reasons why the BS may want to vary the unsolicited grant size. These additional reasons will become more clear when discussing the circumstances that may stimulate the BS to grant unsolicited bandwidth.

In keeping with the main purpose of unicast polling, the BS would periodically grant unsolicited bandwidth at a rate it deems sufficient for the QoS requirements of the SS's services. The frequency can depend on many factors. The BS could merely grant unsolicited bandwidth on a periodic basis, such as every n frames. Alternatively, the BS could consider the CoS of the SS's services. For instance, an SS with an rtPS may be polled more frequently than one that has only nrtPS or BE service. Another factor that could be considered is activity. A BE service that has been idle for a few hours has much less likelihood of needing bandwidth soon than one that has transmitted data within the last second. The BS is free to take into consideration all these issues and any others the implementor thinks are important.

There are a few secondary purposes of unicast polling. First, for MAC management message exchanges on the basic or primary MAC management connections, the BS can predict the SS's need for response bandwidth. Such a predictive ability can greatly speed up MAC management protocols. Of course, the SS has the freedom to either use the bandwidth for its part of the protocol exchange or request bandwidth for its part of the exchange. The timing of these unsolicited allocations depends on the MAC management exchanges in which the BS and SS are engaged.

Additionally, since IEEE 802.16 systems have subscriber-level adaptive PHYs, the BS needs to periodically assess the UL quality from each SS individually. Periodic ranging is used for this purpose. Any unicast bandwidth allocation is considered a PHY maintenance opportunity for periodic ranging. Because the SS must transmit at least filler, even when it has no other data or requests to transmit, any unicast allocation gives the BS an opportunity to measure the receive signal quality of the SS's UL transmissions. Additionally, the SS may use any UL allocation to transmit a DBPC-REQ message for maintenance of the DL. Therefore, the BS must periodically send a unicast poll to the SS for the purpose of PHY maintenance. The frequency of this type of unicast poll is PHY-dependent. For example, to ensure proper PHY maintenance with 10 dB/s fading in a WirelessMAN-SC system, the BS should give the SS a PHY maintenance opportunity about once every 80 ms. If the SS's services are idle, this PHY maintenance opportunity takes the form of a unicast poll.

Finally, at the BS's discretion, any unused UL bandwidth in a frame may simply be allocated to various SSs, effectively polling them in a unicast fashion. This can be particularly effective as a semipredictive way of reducing the acknowledgment latency of TCP/IP transfers, resulting in better response time, such as faster File Transfer Protocol (FTP) transfer rates.

Broadcast polling

During times of congestion, it may be better for certain CoSs to reduce the UL overhead associated with unicast polling and poll SSs on a contention basis as groups. This is especially true of any SS that has only BE connections. Additionally, SSs whose services have been idle for some period would be good candidates for being polled as a group. When an SS enters the network, it is automatically part of the broadcast polling group. At the discretion of the BS scheduling algorithm, the BS may chose to provide contention-based BW request opportunities open to the entire set of SSs on the channel. These opportunities appear in the UL MAP as a special DIUC indicating a BW request opportunity, addressed to the Broadcast CID 0xFFFF.

The method of sending a BW request in such an allocation, common to all PHYs, is the transmission of the BW request header. However, for the OFDM and OFDMA PHYs, an additional anonymous BW request method exists, for which the allocation must be separately provided by the BS. Different techniques are used in OFDM and OFDMA to support anonymous BW requests. However, in both cases, in addition to the BW requests being anonymous, the techniques allow the transmission of simultaneous BW requests from multiple SSs in the same UL allocation.

For OFDM, this so-called *focused contention BW request* (as compared to the full BW request, which uses the BW request header) is achieved by sending a randomly selected 4-bit code on four out of all data subcarriers for the duration of two OFDM symbols, while sending no energy on all other subcarriers. A random selection of a group of four subcarriers by multiple SSs allows a low probability of collision within the same BW request opportunity. For OFDMA, this so-called *contention-based CDMA BW request* is achieved by modulating a randomly selected spreading sequence over the ranging subchannel in the traditional CDMA fashion.

Upon reception of these codes or sequences, the BS can provide UL allocations; however, since the originator is unknown, it indicates in the allocation the received code or sequence and the BW request opportunity in which it was received rather than the Basic CID of the requesting SS. One drawback of this method is that SSs may sometimes select the same code or sequence during the same BW request allocation, resulting in a collision within the resulting UL allocation. Another is that it is not possible to indicate the size of the UL allocation needed in the code or sequence. These drawbacks are offset by the efficiency of the requests themselves.

Multicast polling groups

Sometimes it may be desirable to used contention-based opportunities for groups of SSs, but the entire group may be large enough that the possibility of collision is too high. For situations such as this, SSs can be divided up into multicast polling groups. Multicast polling groups should not be confused with multicast connections. Multicast connections, as described in Chapter 6, are DL connections that are received by multiple SSs simultaneously, without duplicate transmissions on the same airlink channel. Multicast polling groups are groups of SSs that may respond to a BW request addressed to their multicast polling group address. Effectively, it is the same as the broadcast poll, but only selected SSs are members. This allows the BS to limit the number of SSs that may respond to a contention-based poll, which it may do for a variety of reasons.

- To reduce the possibility of collisions.
- To group SSs by like services, allowing certain groups to be polled more frequently than others, thus maintaining some determinism in QoS while still taking advantage of group polling.
- To provide the BS with a smaller set of SSs that it may decide to poll individually upon detection (assuming the BS has this capability) of a collision.

CLOCK COMPARISON

The clock compare function arose from the need to be able to support services that transferred data synchronously. In particular, synchronous T1s and E1s, while gradually being supplanted by asynchronous T1s and E1s, were relatively common when IEEE Std 802.16 was first approved and are still used today for certain voice backhaul applications. Also, the standard's developers noted that, since synchronous data transfer is generally faster, some future applications are likely to require synchronous transport. Such applications may include the streaming of audio to multiple devices. However, it is difficult to ensure that the clocks needed for synchronizing services across a network and the clocks used for generating the wireless transmission are the same or even synchronized with each other.

In TDD systems and in FDD systems that allow H-FDD devices, information is not provided continuously from a transmitter to the receiver, further aggravating the problem of synchronization of services. In IEEE Std 802.16, we add sophisticated PHY burst profile usage and MAC scheduling that can further disrupt the continuity desired by synchronous services.

Usually, but not always, the synchronous data are synchronized to the network clock that is also used to generate the backhaul. Theoretically, the symbol clock for the wireless link could be generated from the network clock, but this is typically not done for a variety of reasons:

- The airlink typically needs a stratum 1 reference. While the network clock is probably of high quality, there is no guarantee that the network clock is better than stratum 3 because it may have accumulated errors through the network.

- Especially in TDD systems, it is advantageous to synchronize BSs within a network. Depending on the network, the backhauls to and from each BS may not be synchronized among themselves.

- Rather than build a variety of backhaul synchronization means, it is easier for BS vendors to design airlink synchronization around a known, readily available stratum 1 clock source, such as a global positioning system (GPS).

This creates a situation such as shown in Figure 7–19. In this example, the BS is synchronized to the 10 MHz reference typically provided by a GPS receiver collocated with the BS. It may further use the 1 pps (pulse per second) reference typical of a GPS receiver to synchronize PHY frame boundaries, but this is immaterial to the clock compare discussion. The BS generates the symbol clock for the airlink synchronized to the 10 MHz reference. The SS synchronizes its symbol clock to the airlink, effectively synchronizing to the 10 MHz reference from GPS.

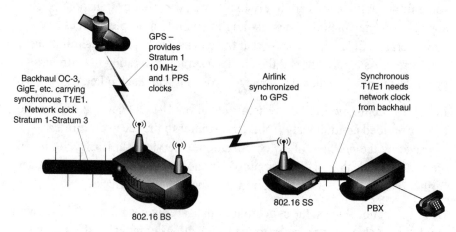

Figure 7–19:   Clock comparison

However, any network clocks derivable from the backhaul or services on the backhaul are not guaranteed to be synchronized with GPS. They could be synchronized to a stratum 1 rubidium oscillator clock generator somewhere in the network or to some other clock generator of similar or worse quality, or they could be synchronized to GPS, but with accumulated errors. But, regardless of its quality or its relationship to GPS, one of these network clocks is the one to which the T1/E1 recreated by the SS must be synchronized.

The SS must regenerate the network clock for the service from the airlink clock. By itself, there is insufficient information to do this with accuracy sufficient to prevent bit slips on a synchronous T1 or E1. Therefore, the BS helps out. Every 50 ms, the BS notes the difference between the 10 MHz reference used to generate the airlink and an 8.192 MHz clock generated from the desired network clock. In the clock compare algorithm, 10 MHz was chosen due to the predominance of GPS as the choice for BS reference, and 8.192 MHz was chosen since it is a commonly occurring divisor of most network clocks.

Now the SS can quantify the difference between the two clocks at known intervals. The SS then adjusts the network clock it is regenerating so that the difference between the 8.192 MHz reference it creates for the network clock and the 10 MHz reference it generates from the airlink symbol clock tracks the difference between the two source clocks at the BS. This can be done with sufficient accuracy to prevent bit slips on a synchronous T1 or E1. It has applicability to any synchronously clocked services carried over the airlink.

ARQ OPERATION

ARQ parameters and timers

This section describes the set of parameters and timers required for ARQ operation. Table 7–2 provides the basic IEEE 802.16 ARQ parameters and their usage. An ARQ implementation may also define additional parameters. For example, an ARQ block may have a maximum retry limitation in addition to its lifetime. The acknowledgment generation at the receiver may also be controlled by an acknowledgment frequency parameter (specifying how often to generate ARQ feedback) and/or a maximum ARQ feedback size.

Table 7–2: ARQ parameters and timers

Parameter	Description
ARQ_BSN_ MODULUS	The maximum number of unique BSN values ($= 2^{11}$)
ARQ_ WINDOW_ SIZE	The maximum number of consecutive ARQ blocks in the sliding window maintained by the transmitter and receiver. In order for the SR ARQ algorithm to work correctly, this window size cannot be greater than half of the ARQ_BSN_MODULUS. The transmitter and receiver state machines use the following additional variables to manage the ARQ blocks within their respective sliding windows: • ARQ_TX_WINDOW_START: transmitter variable—All ARQ blocks with BSNs up to one less than this value have been acknowledged by the receiver. Therefore, this is the BSN of the first (unacknowledged or not yet transmitted or not yet available for transmission) ARQ block in its sliding window. • ARQ_TX_NEXT_BSN: transmitter variable—The BSN of the next ARQ block to send. Therefore, ARQ_TX_WINDOW_START <= ARQ_TX_NEXT_BSN <= (ARQ_TX_WINDOW_START + ARQ_WINDOW_SIZE) • ARQ_RX_WINDOW_START: receiver variable—All ARQ blocks with BSNs one less than this value have been successfully received. • ARQ_RX_HIGHEST_BSN: receiver variable—The BSN of the highest ARQ block successfully received, plus one. This variable is related to the window size as follows: ARQ_RX_WINDOW_START <= ARQ_RX_HIGHEST_BSN <= (ARQ_RX_WINDOW_START + ARQ_WINDOW_SIZE)
ARQ_ BLOCK_ SIZE	The ARQ block size negotiated during connection setup.

Table 7–2: ARQ parameters and timers (Continued)

Parameter	Description
ARQ_ BLOCK_ LIFETIME	The maximum amount of time an ARQ block will be held by the transmitter after the initial transmission of the block. A timer with this value is typically set for every ARQ block after the first transmission. If the block is successfully transmitted before its lifetime expires (i.e., upon reception of an acknowledgment), this timer is cancelled. If the transmitter is unable to successfully transmit the block within its lifetime subject to retransmissions, the ARQ block is discarded.
ARQ_ RETRY_ TIMEOUT	The minimum time between a transmission (or retransmission) and a subsequent retransmission of an ARQ block. A block is retransmitted if no acknowledgment is received within this timeout and the ARQ block is still eligible for retransmission.
ARQ_ SYNC_ LOSS_ TIMEOUT	The maximum time the transmitter or receiver keeps its respective TX/RX_WINDOW_START variables at the same value (i.e., sliding window remains fixed) before declaring a loss of synchronization between the transmitter and receiver. This timeout is meaningful only if the connection is actively transferring data and the TX/RX_WINDOW_START has not moved. The transmitter and receiver have their own way of determining whether a connection is actively transferring data.
ARQ_RX_ PURGE_ TIMEOUT	The maximum time interval the receiver waits after successfully receiving an ARQ block that did not advance the ARQ_RX_WINDOW_START. After this timeout, the ARQ_RX_WINDOW_START is advanced by the receiver.
ARQ_ DELIVER_ IN_ORDER	If enabled, the ARQ receiver must deliver the MSDUs to the upper layers in order. If disabled, MSDUs may be delivered to the upper layer out of order.

ARQ protocol messages

Apart from a stand-alone message to transport the ARQ feedback, as described in Chapter 6, a few additional MAC management messages are necessary to exchange state information between the transmitter and receiver and keep their ARQ states in sync. These messages are transported in a management connection using the TLV format.

ARQ Feedback management message

The ARQ Feedback management message (Type 33) may carry any size of ARQ feedback up to the maximum allowed by the connection. It is used by the receiver when no other piggybacking opportunities are available to transport the ARQ feedback payload.

ARQ Discard management message

The ARQ Discard management message (Type 34) is used by the transmitter to instruct the receiver to skip a certain number of ARQ blocks. This message, of fixed size, carries the CID and BSN, where the BSN indicates the last ARQ block that the transmitter wants to discard. The receiver should skip up to the indicated BSN and update its sliding window variables accordingly. For an ARQ receiver, the reception of an ARQ Discard management message is equivalent to the successful reception of the ARQ blocks up to the specified BSN. Therefore, in response, the receiver generates an ARQ Feedback IE with the appropriate acknowledgment (i.e., cumulative acknowledgment with the BSN from the discard message).

ARQ Reset management message

The ARQ Reset management message may be sent by the transmitter or the receiver to reset the transmitter and receiver state machines. This message, which resets all variables in the transmitter and receiver to zero, is used when all other attempts to synchronize the state machines fail. The reset message carries the CID, a two-bit Type field, and a two-bit Direction field. The Type field indicates whether the reset message is the original reset message from the initiator, the acknowledgment from the responder, or the confirmation from initiator. A three-way handshake is necessary to complete the reset operation. The Direction field indicates whether the CID in the message refers to an UL or DL CID. For transport connections, the Direction field is reserved. However, for the secondary management connection, the same CID is used in both the UL and DL for a specific SS. Therefore, if ARQ is enabled for the secondary management connection, the Direction field is used to identify the source of the reset message.

BSN comparison

Both the transmitter and receiver ARQ state machines need to take actions based on comparison of BSNs (e.g., comparing the BSN from a message or received MPDU with a locally maintained value). Since a simple numerical comparison is inaccurate, due to the size of the BSN space and sliding window and the rolling over of BSN values, the BSN values need to be normalized prior to comparison. The normalized BSN, BSN', is defined as BSN' = [(BSN − BSN_Base) *mod* ARQ-BSN-MODULUS], where BSN_base is ARQ_TX_WINDOW_START for the transmitter and ARQ_RX_WINDOW_START for the receiver.

ARQ transmitter

The primary functions of the ARQ transmitter are the following:

1. Determining whether one or more ARQ blocks are qualified for transmission and whether any ARQ blocks need to be retransmitted.
2. Processing ARQ feedback, extracting acknowledgment information, and updating the local state variables, including the ARQ_TX_WINDOW_START, if necessary.
3. Generating a discard message when an ARQ block's lifetime is expired without a successful transmission.
4. Generating a reset message whenever the ARQ_SYNC_LOSS_TIMEOUT is expired.
5. Processing reset messages from the receiver.

The first three functions of the transmitter state machine can be best illustrated using the state transition of ARQ blocks. Figure 7–20 shows the ARQ block states at the transmitter.

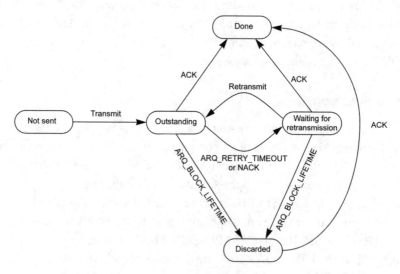

Figure 7–20: ARQ block states at the transmitter

As described in Chapter 6, an MSDU received from the CS is divided into one or more ARQ blocks based on the block size negotiated during the connection setup. An ARQ block starts at the *not-sent* state. An entity outside of the ARQ transmitter, typically the scheduler, is responsible for determining transmit opportunities for a specific connection, based on the QoS requirements and the number of queued ARQ blocks. The exact interface between the scheduler and ARQ is not defined in the IEEE 802.16 standard.

Based on the maximum number of ARQ blocks that can be transmitted for a given opportunity, the ARQ transmitter builds an MPDU of ARQ blocks comprising first transmissions and/or retransmissions. The ARQ transmitter has to follow the packing and fragmentation rules defined for this connection in constructing the MPDU. In order to qualify an ARQ block in not-sent state for transmission, the ARQ transmitter must make sure that the block falls within its window size. Once a block is qualified, it is assigned the ARQ_TX_NEXT_BSN as described in Table 7–2 and the ARQ_TX_NEXT_ BSN is incremented by one.

Once an MPDU is constructed and sent for transmission, the ARQ blocks carried in that MPDU become outstanding, i.e., enter the *outstanding* state. Once in the outstanding state, the ARQ block remains in that state until one of the following four events occurs, whichever happens first: (1) Positive acknowledgment (ACK) is received, (2) Negative acknowledgment (NACK) is received, (3) ARQ_RETRY_TIMEOUT is exceeded, 4) ARQ_BLOCK_ LIFETIME is exceeded. If a negative acknowledgment is received or an ARQ_RETRY_TIMEOUT occurs, the ARQ block is placed in *waiting-for-retransmission* state. If the ARQ_BLOCK_LIFETIME expires, the ARQ block is moved to the *discarded* state, irrespective of the current state (i.e., outstanding or waiting-for-retransmission).

Similarly, if an acknowledgment is received, irrespective of the current state, the ARQ block is considered successfully delivered. If the BSN of this block is equal to the ARQ_TX_WINDOW_START, then the ARQ_TX_WINDOW_START is advanced by one. Whenever the ARQ transmitter has the opportunity to build a new MPDU, it must give preference for retransmissions, i.e., blocks in the waiting-for-retransmission state must be transmitted before attempting to transmit any blocks in the not-sent state. If there is more than one ARQ block in the waiting-for-retransmission state, the block with the lowest BSN must be transmitted first.

Whenever an ARQ block is marked as discarded (i.e., ARQ_BLOCK_ LIFETIME expires), the transmitter sends an ARQ Discard management message to the receiver. This message can be cumulative; therefore, multiple consecutive discard orders can be aggregated into a single discard message that includes the highest BSN to discard. Since it is still possible to receive an acknowledgment while in discarded state and the receiver waits up to ARQ_RX_ PURGE_TIMEOUT before advancing a stagnant ARQ_RX_WINDOW_ START (i.e., deciding to discard on its own), a discard message may be delayed up to (ARQ_RX_ PURGE_TIMEOUT + ARQ_RETRY_ TIMEOUT). In order to quickly recover from possible out-of-sync or other disconnects between the transmitter and receiver, the transmitter repeats discard orders at ARQ_ RETRY_TIMEOUT intervals until an acknowledgment is received for the discarded BSN.

ARQ feedback processing

Whenever an ARQ feedback (ACK or NACK) is received for a specific BSN, the transmitter checks to see whether the BSN falls within its window and whether it is less than ARQ_TX_NEXT_BSN. If either of those conditions is not met, the positive or negative acknowledgment is ignored as invalid.

All the acknowledgment types supported by the IEEE 802.16 standard can be decoded into a cumulative and/or selective acknowledgment. For example, the selective with cumulative acknowledgment and the sequence acknowledgment messages can be decoded into two separate acknowledgments: a cumulative and a selective.

Whenever a valid cumulative acknowledgment is received, the ARQ_TX_WINDOW_ START is set to BSN + 1 of the cumulative acknowledgment. For a selective acknowledgment bitmap, the ARQ_TX_WINDOW_START is advanced for every bit with a positive acknowledgment, whose BSN is equal to ARQ_TX_WINDOW_START.

There may be a delay between the reception of ARQ blocks at the receiver and the actual transmission of the ARQ feedback to the transmitter. This may create a situation in which the transmitter receives a negative acknowledgment for an ARQ block just after it was transmitted and successfully received by the receiver. In order to address this problem, an optional parameter RECEIVER_ARQ_ACK_ PROCESSING_TIME (specified in milliseconds) is defined. This parameter may be provided by the receiver (BS or SS) to the transmitter during connection setup or change. The transmitter may use this information to intelligently interpret ARQ feedback and make retransmission decisions.

ARQ receiver

The primary functions of the ARQ receiver are as follows:

1. Receiving ARQ blocks within its receive window and updating the appropriate state machine variables and timeouts.

2. Generating ARQ feedback (ACKs and NACKs) whenever necessary.

3. Generating a reset message whenever the ARQ_SYNC_LOSS_ TIMEOUT is expired.

4. Processing discard messages and generating acknowledgment (ARQ feedback) in response to indicate to the transmitter the reception of the discard message.

5. Processing reset messages from the transmitter.

Figure 7–21 shows the ARQ block reception procedure. As describer earlier, the CRC-32 is mandatory for ARQ connections. If the CRC check fails, all the ARQ blocks in that MPDU are considered invalid. The CRC check is typically performed outside of the ARQ receiver, and only ARQ blocks that passed the CRC check are handed over to the ARQ receiver. If the MPDU is packed, it needs to be unpacked prior to the examination of ARQ blocks. The reassembly occurs after all the blocks for a specific MSDU are received correctly.

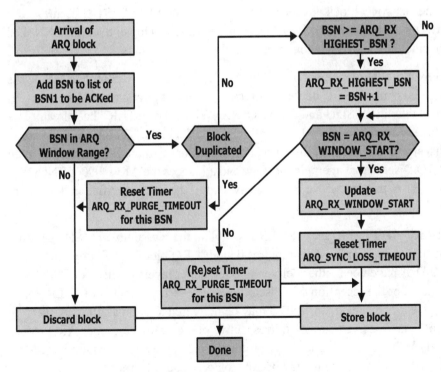

Figure 7–21:   ARQ block reception

The first step after the reception of an ARQ block is to add the BSN to the list of BSNs to be acknowledged and to determine whether its BSN is within the ARQ receive window. The ARQ receiver also checks to see whether this block is a duplicate block and discards duplicate blocks after resetting the ARQ_RX_ PURGE_TIMEOUT timer corresponding to that BSN. If this is not a duplicate block, then the BSN is checked against the ARQ_RX_ HIGHEST_BSN. If the BSN is greater than ARQ_RX_HIGHEST_BSN, then ARQ_RX_ HIGHEST_BSN is set to BSN + 1. Then the receiver checks to see whether the BSN advances the ARQ_RX_WINDOW_START. If it is equal to the ARQ_RX_WINDOW_START, then the value is updated and ARQ_SYNC_ LOSS_TIMEOUT timer is reset.

Whenever an ARQ block that did not advance ARQ_RX_WINDOW_START is received, an ARQ_RX_PURGE_TIMEOUT is started for that block. Every time an ARQ_RX_PURGE_TIMEOUT expires, the ARQ_RX_WINDOW_ START is set to the BSN of the next block that has not yet been received. When an ARQ block advances the ARQ_RX_WINDOW_START, all ARQ_RX_PURGE_TIMEOUT timers set for BSNs lower than this BSN (i.e., outside the receive window) are cancelled.

Only if the ARQ receiver has all blocks corresponding to complete MSDUs is the reassembled SDU delivered to upper layers. If only partial blocks are available for an SDU and the ARQ_RX_WINDOW_START has already moved past the required BSNs, all ARQ blocks in the incomplete SDU are discarded. The receiver also discards all blocks up to the BSN specified in an ARQ Discard management message and advances the ARQ_RX_WINDOW_ START to the first BSN not yet received after the BSN specified in the discard message.

If ARQ_DELIVER_IN_ORDER is enabled, the reassembled MSDUs must be delivered in order. This requires that all ARQ blocks of all previous MSDUs have been either delivered to the upper layers or discarded due to a timeout or the reception of a discard message. When ARQ_DELIVER_IN_ ORDER is disabled, an MSDU may be delivered as soon as all its ARQ blocks are received correctly, irrespective of the state of previous ARQ blocks and MSDUs. Most higher layer protocols and applications require in-order delivery, although some may be able to accept out-of-order delivery.

ARQ feedback generation

The frequency of acknowledgment generation is not defined in the IEEE 802.16 standard. It is up to implementations to define a frequency or other innovative algorithms to decide when to generate an acknowledgment. However, all received ARQ blocks must be acknowledged.

All acknowledgments outside the receive window must be cumulative, irrespective of whether those blocks were actually received correctly. As described in earlier subsections, the ARQ_RX_WINDOW_START may be advanced for a variety of reasons: (1) the reception of an ARQ block with BSN equal to ARQ_RX_WINDOW_START, (2) the reception of a discard message, or (3) a purge timeout. In all of these cases, the ARQ receiver generates an ARQ Feedback IE with cumulative acknowledgment at least once to indicate the reception of all ARQ blocks less than ARQ_RX_WINDOW_START. For all other ARQ blocks received that fall within the receiver window, the generated acknowledgments may be cumulative (if applicable), or selective, or a combination of both.

ARQ state machine reset and resynchronization

The ARQ reset mechanism is used to recover from an out-of-sync state, with the transmitter and receiver having lost synchronization due to a sequence of lost messages or an incorrect implementation. The reset operation may be initiated by the transmitter or the receiver.

- *Transmitter-initiated reset:* A reset is initiated by the transmitter when the ARQ_TX_WINDOW_START has not moved for ARQ_SYNC_LOSS_TIMEOUT while the ARQ connection is active. The process of transmitter-initiated reset, along with the actions taken by the receiver, is given in Figure 7–22. The timer T22 used in Figure 7–22 is defined in the IEEE 802.16 standard as the *"Wait for ARQ-reset"* timer, with the default value of 0.5 s.

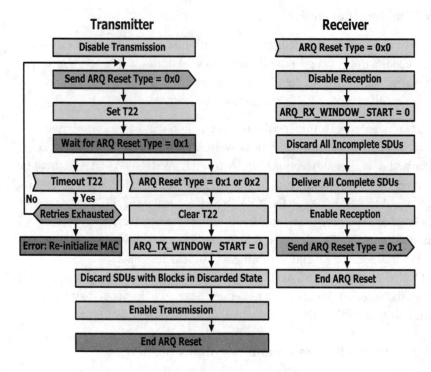

Figure 7–22:   Transmitter-initiated reset

- *Receiver-initiated reset:* The receiver, like the transmitter, can initiate an ARQ reset. It does so when the ARQ_RX_WINDOW_START has not moved for ARQ_SYNC_LOSS_ TIMEOUT while the ARQ connection is active. Figure 7–23 shows the process of receiver-initiated reset. The timer T22 used here is the same timer used in transmitter-initiated reset.

Both the reset procedures take into account the case where both the transmitter and receiver independently initiate a reset. In the case of an ARQ reset error, the SS should reinitialize its MAC. The behavior of the BS is undefined, and it is left to the implementation to take the appropriate action when an ARQ reset error occurs.

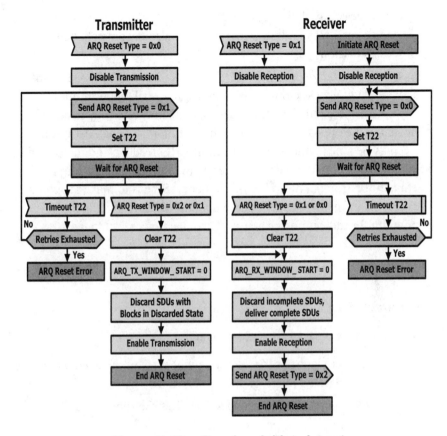

Figure 7–23: Receiver-initiated reset

Interaction with scheduler

The interactions between a scheduler and ARQ are not defined in the
IEEE 802.16 standard. It is up to the implantation to define the exact
behavior, including the following:

- *How to count retransmissions?* The ARQ protocol requires that the
  retransmissions be given preference over first transmissions. However, a
  MAC-level QoS scheduler has to decide whether to fully or partially count
  the retransmissions or count only the first transmissions toward the
  connection's specified limit. For a higher layer protocol or application,

only the successfully delivered MSDUs count toward its perception of QoS. Moreover, delay and jitter management also become challenging with retransmissions.

- *How to allocate bandwidth for ARQ feedback?* Although ARQ Feedback IEs may be piggybacked on another connection, such opportunities are not always available or sufficient. In such cases, bandwidth may need to be allocated in the opposite direction exclusively for carrying ARQ feedback. A BS may be able to allocate and transmit ARQ feedback going toward an SS on demand from the ARQ receiver. However, for transporting ARQ Feedback IEs from an SS to the BS, either an SS has to make an explicit BW request or the BS has to proactively and periodically allocate bandwidth for this purpose when ARQ connections are active.

HARQ OPERATION

HARQ parameters

The following HARQ parameters are defined:

- **SPID:** Subpacket identifier for the four subpackets of the encoder packet, which has a predefined range of 0b00 to 0x11.
- **ACID:** ARQ channel identifier, which specifies the subchannel for the subpacket transmission.
- **AISN:** ARQ identifier sequence number, which is toggled between 0 and 1 on successful transmission of an encoder packet.

The following subsections describe the usage of these parameters by HARQ protocol. The HARQ Control IE in the DL MAP or UL MAP is used to signal these parameters for a specific DL or UL HARQ burst.

Subpacket transmission and acknowledgment generation

When HARQ is enabled, the MAC delivers the constructed MPDUs to the HARQ PHY module responsible for concatenation, encoding, and subpacket generation. Since HARQ operates at the FEC level, the FEC encoder is responsible for generating the subpackets. Once generated, the subpacket with SPID 0b00 is always transmitted first. Subsequently, any one of the subpackets with SPID 0b00, 0b01, 0b10, or 0b11 may be transmitted based on the positive or negative acknowledgment feedback from the receiver. A subpacket may be transmitted more than once, and any subpacket may be skipped, with the exception of the subpacket with SPID 0b00.

Upon receiving a subpacket, the receiver's FEC decoder attempts to decode the original encoder packet. If the original is successfully decoded, a positive acknowledgment is sent to the transmitter; otherwise, a negative acknowledgment is sent. The transmitter stops sending additional subpackets from the same encoder packet if a positive acknowledgment is received. The transmitter may transmit one of the other subpackets if a negative acknowledgment is received, and this process may continue until the receiver successfully decodes the original encoder packet. A 1-bit HARQ AISN, toggled between 0 and 1, is used to indicate a new encoder packet transmission on the same HARQ channel. Therefore, upon successful decoding of an encoder packet, the transmitter toggles this bit to indicate this fact to the receiver of subpackets belonging to the next encoder packet.

As described in Chapter 6, the synchronous HARQ positive and negative acknowledgment signaling makes use of a dedicated subchannel in the UL and a compact DL MAP with acknowledgment bitmap in the DL. However, the BS has full control over both UL and DL and may decide not to send an acknowledgment bitmap at a fixed frame offset. If no acknowledgment bitmap is present in the expected frame, the SS retains the HARQ burst and the corresponding subpackets. The BS may explicitly request a retransmission later, using the HARQ Control IE in the Compact MAP IE.

The maximum number of HARQ retransmissions in the UL and DL is advertised in UCD and DCD messages, respectively. The default maximum number of HARQ retransmissions in both DL and UL is set to four.

Performance and QoS implications

As described in Chapter 6, the stop-and-wait HARQ protocol uses synchronous acknowledgments[5] that are sent after a fixed delay specified in

---

[5] The BS may optionally decide not to send a synchronous acknowledgment and request a retransmission using an HARQ Control IE.

frame offset in both UL and DL. However, the number of subpacket transmissions required for successful decoding may vary, based on channel conditions. Moreover, the transmission/retransmission of the subpackets is not synchronous. The ability to support multiple HARQ channels per terminal does offer some additional performance, as multiple encoder packets may be pending at any instant. The ACID, carried in the MAPs, uniquely identifies the channels.

While HARQ may offer some performance improvements due to SNR gain and coding gain by incremental redundancy, the lack of control from the MAC and the inherent limitations of stop-and-wait ARQ may affect the type of QoS that can be guaranteed for an HARQ–enabled SS.

Chapter 8 Security

PKM protocol and cryptographic methods

The IEEE 802.16 standard defines security mechanisms affording privacy to subscribers and protecting operators from theft of service. These objectives are achieved by encrypting the data between the BS and SS, by utilizing digital certificate-based authentication of the SS, and by employing an authenticated client-server key management protocol called *privacy key management* (PKM). The PKM protocol allows the BS, as server, to control the distribution of keying material to the SSs. The security protocols of IEEE Std 802.16-2004 [B20] do not offer authentication of the BS. This leaves open the possibility of an SS associating with a malicious BS. However, for a fixed wireless system operating in licensed bands, this should not be a grave concern since the BS equipment may not be readily available; furthermore, operation of a malicious BS can be easily detected by a legitimate operator. Nevertheless, BS authentication has been specified in IEEE Std 802.16e.

SECURITY ASSOCIATIONS (SAS) AND CRYPTOGRAPHIC SUITES

The packet data encryption function in IEEE 802.16 resides in the security sublayer between the MAC CPS and PHY, as shown in Figure 4–1 (in Chapter 4). In the transmitter, the MPDUs are mapped to an SA, which dictates the processing to be applied to a given plaintext MPDU. The receiver uses the CID to determine the correct SA and applies the corresponding processing to the received MPDU.

The SA is a set of security-related information that is shared between the transmitter and receiver, with help from the PKM protocol. Each SA contains information on the cryptographic suite used for that SA. The SA may also contain keys, such as the traffic encryption keys (TEKs), along with their lifetimes and other associated state information. The exact content of the SA depends on the algorithm for encrypting and authenticating the MPDUs and on the algorithm used for TEK exchange specified in the cryptographic suite. IEEE Std 802.16 defines two main cryptographic suites, as described in more detail below.

An SS entering a network will set up at least one SA, called the *primary SA*, with the BS. The BS may also set up a number of static and dynamic SAs. Static SAs are provisioned at the BS, while dynamic SAs are set up and torn down along with opening and closing service flows. In contrast to the primary SA, the static and dynamic SAs can be shared by the BS and multiple SSs.

Encrypted MPDUs

Only the payload part of an MPDU (which, as defined in the IEEE 802.16 standard, includes the MAC subheaders) is encrypted. The generic MAC header is always sent in cleartext. The cleartext header contains the two EKS bits indicating the LSBs of the TEK sequence number of the key that was used to encrypt the PDU. This allows smooth key transitions in both directions.

Data encryption with DES in CBC mode

One of the two main cryptographic suites defined in the standard specifies data encryption with DES in CBC mode. The MPDU data authenticity is not checked when using this method. The exchange of the 64-bit TEK is performed using 3DES in code book mode. During the TEK exchange, the BS delivers parameter CBC-IV, which acts as the basis for calculating the IV required for encryption processing. The encryption process involves the following steps:

1. The IV is derived by calculating the XOR of the CBC-IV and the frame number in which the MPDU transmission takes place. For UL transmissions, the XOR is taken between the frame number in which the UL MAP allocating the transmission was received and the CBC-IV.

2. The IV is XORed with the first 8 bytes of plaintext MPDU payload, and the result is then DES-encrypted to get the first 8 bytes of ciphered payload.

3. The ciphertext from the previous step is XORed with the next 8 bytes of plaintext payload and DES-encrypted to get the next 8 bytes of ciphered payload.

4. Step 3 above is repeated until less than 64 bits of payload remain unciphered. To cipher the n residual bits, the previously attained ciphertext is DES-encrypted once more. The XOR between the n MSBs of this encrypted result and the n plaintext bits is then computed. The n last bits of ciphered payload are made up of the result of this calculation.

In the receiver, the plaintext is recovered by performing a DES decryption of the ciphertext block and XORing it with the previous ciphertext block (the IV in case of the first 8 bytes). To recover the residual bits, the last full ciphertext block is encrypted once more, and the relevant part of result (the n MSBs) is XORed with the ciphered residual to get the plaintext.

The level of security provided by this suite is not very strong. DES uses keys that are considered by security experts to be too short to be secure. Also, there are some issues with the predictable way in which the cipher chain is initialized. Furthermore, the lack of MPDU authentication can be considered a weakness of the protocol. However, this suite has the benefit of not introducing any data expansion. If the weaker security of this mode is a concern, then the AES-CCM suite described below should be employed instead.

AES-CCM

The advanced encryption standard (AES) is the replacement for DES and considered to be state of the art. The AES is a block cipher with a block length of 128 bits. The keys can be 128 or 256 bits long. IEEE Std 802.16 defines a cryptographic suite that utilizes the AES algorithm in the counter with CBC-MAC (CCM) (see [B40]) mode with 128-bit keys. In addition to data privacy, the suite provides integrity protection for the MPDUs. The encryption method is the same as the one used in IEEE Std 802.11i[6]. The TEKs are exchanged

---

[6] IEEE Std 802.11 and IEEE Std 802.16 construct data fields differently. Also, IEEE Std 802.11i defines its own protocol for key exchange.

using AES in code book mode. The confidentiality and data integrity protection provided by the AES-CCM is considered very strong.

The AES-CCM processing adds two fields to the MPDU payload: a 4-byte packet number (PN) and an 8-byte message authentication code. The MPDU payload before and after processing is shown in Figure 8–1.

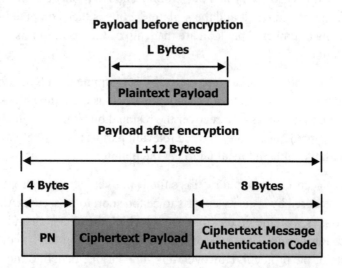

Figure 8–1: Encrypted payload format in AES-CCM mode

The PN space is split in two—the range 0–7FFFFFFF is used for DL and the remainder for UL—because only one MPDU payload should be encrypted with a given key and PN.

The message authentication code is calculated using AES in CBC mode, with counter mode encryption. The steps of the AES encryption process are described below:

1. A set of blocks $B_0,...,B_r$ are formed from the MPDU.

2. A temporary CBC checksum T is calculated over these blocks.

3. Using the nonce (see Figure 8–2) and a counter incremented by one for each block, a set of counter blocks (see Figure 8–3) $S_0,...,S_m$ is formed.

4. The first counter block S_0 is encrypted and stored.

5. Block S_1 is encrypted, and the XOR with the first 128 bits of the MPDU payload is calculated to gain the first 128 bits of encrypted MAC payload. The same is done for the subsequent blocks, if any. If the length MPDU payload is not a multiple of 16 bytes, the last block is shortened.

6. The message authentication code is derived by calculating T XOR S_0 and keeping only the eight most significant bytes.

7. The PN, the encrypted payload, and the message authentication code are concatenated to form the encrypted MPDU payload.

The initial block B_0 used in the CBC calculation is shown in Figure 8–2. By including the generic MAC header (save the redundant HCS) in the nonce, the header also will be afforded integrity protection. The remaining blocks are formed by splitting the plaintext MPDU payload in 16-byte blocks (padding with zeros to the next 16-byte boundary, if necessary).

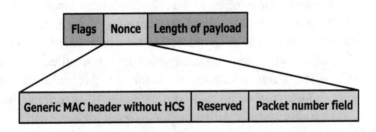

Figure 8–2: Initial CBC block and nonce

The counter blocks are shown in Figure 8–3, where the nonce is the same as above.

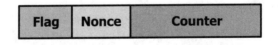

Figure 8–3: Counter blocks

The processing in the receiver is similar to that in the transmitter, but the order of operations is different. The receiver starts by generating the encrypted counter blocks and decrypts the payload. After decryption, it calculates the

CBC checksum and compares this against the message authentication code. If the check fails, the receiver discards the MPDU.

KEY MANAGEMENT

Key management is handled by the authorization and TEK state machines in the SS. The authorization state machine, initialized on SS power up, handles the authorization state and AK management. In the Authorization Reply message, the SS receives the AK and the list of SAs from the BS. For each SA, the SS starts an instance of the TEK state machine that handles the key management of that SA. A TEK state machine is also instantiated upon reception of the SA Add message that creates a dynamic SA.

AK management

The AK is always accompanied by its associated lifetime. Upon reception of an AK, the SS starts a timer to measure its remaining lifetime. At the instant named "AK grace time," which occurs before the expiration of the AK lifetime, the SS initiates a reauthorization procedure to acquire a new AK. The reauthorization is almost identical to the initial authorization, but the Authorization Information message is omitted. In the BS, assuming the SS continues to be authorized, the Authorization Request message will result in a new AK being activated with a lifetime that extends beyond the lifetime of the AK that is about to expire. In this transition period, the BS will have two active AKs for an SS. The transition period and key usage are illustrated in Figure 8–4. The AK is used to derive keys for MAC management message authentication and TEK exchange. As soon as the new AK is activated, the BS starts using the keys derived from the new AK.

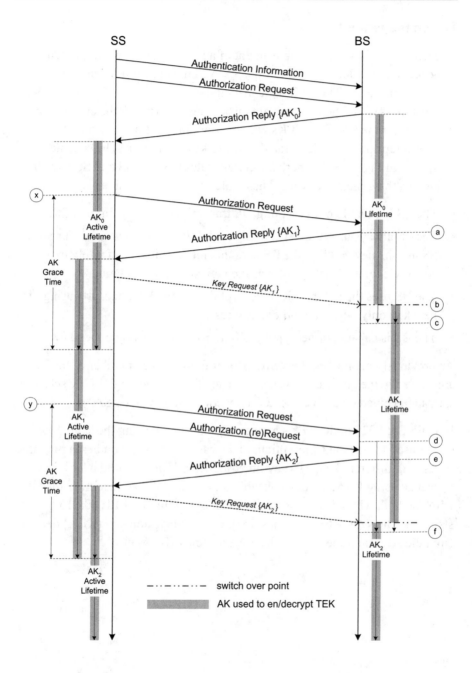

Figure 8–4:   AK management in BS and SS

TEK management

As mentioned above, the establishment of a new SA either during initial authorization or after reception of the SA Add message will instantiate a new TEK state machine. The TEK state machine will trigger the SS to send a Key Request message to the BS to obtain the keying material for the cryptographic suite used for the SA. Two TEKs with overlapping lifetimes are maintained to prevent interruptions in data transfer due to key expiration. The scheme is illustrated in Figure 8–5. There are certain rules that the BS and SS have to follow for this scheme to work. These rules are summarized below:

- The SS is capable of decrypting DL transmissions using either TEK.
- The SS always uses the newer TEK for UL transmissions. Once it sees the SS use the newer TEK, the BS uses this information as an implicit acknowledgment that the SS has the new keying information.
- The BS uses the older TEK to encrypt data in the DL and starts using the new key only when the old one expires.
- The BS is capable of decrypting UL transmissions using either TEK.

As previously mentioned, the two LSBs of the sequence number of the TEK are carried in the generic MAC header to aid the receiver in the TEK selection for data received on a given MAC connection using a given SA.

In IEEE Std 802.16, the TEKs are usually exchanged using the same block cipher used for the data encryption; in the case of DES, the cipher is run three times with different keys. The symmetric methods are favored over key exchange using RSA, as they consume less computational resources than the RSA method, which is also specified as an option in the IEEE 802.16 standard. The symmetric methods use the key encryption key (KEK) derived from the AK using the Secure Hash Algorithm 1 (SHA-1).

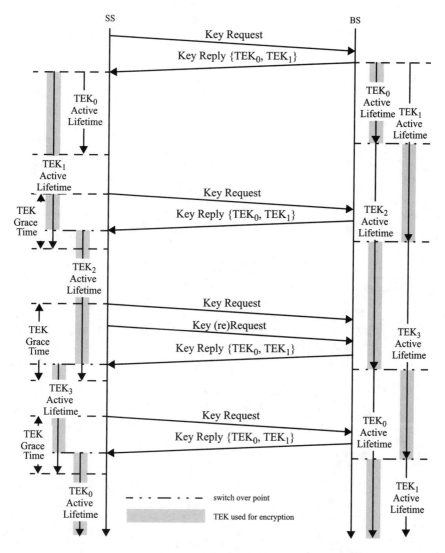

Figure 8–5: TEK management in BS and SS

Chapter 9 Mesh

MAC and PHY extensions for mesh

INTRODUCTION

The concept of mesh radio systems, as an alternative to PMP, has for the last decade been a hot topic for commercial communications, evolving from earlier military work done by the Defense Advanced Research Projects Agency (DARPA) in the United States. Beyond the fixed mesh specification already available in IEEE Std 802.16-2004 [B20], the IEEE 802.16 Working Group proposed in January 2005 the development of an amendment to facilitate mobile multihop relays for fixed and mobile infrastructure. The difference between mesh and infrastructure relay is that a mesh network allows connectivity between any set of end-user and infrastructure devices without manual intervention, where the number and location of end-user devices will fluctuate. An infrastructure relay, on the other hand, is topologically limited to a tree structure with a BS at its root and infrastructure relays as branches.

Mesh is of interest not only to the IEEE 802.16 community but also to the IEEE 802.11 and IEEE 802.15 communities. In particular, the IEEE 802.11s Task Group is currently developing a mesh amendment for WLANs, and the IEEE 802.15 Task Group 5 is currently developing a mesh amendment for WPANs.

The typical reasons for designing or deploying mesh include the following (see [B52]):

- *Communication within a local community:* This is where mesh performs best, but it is not a very typical scenario.
- *Capacity increase:* When the ingress and egress of traffic is distributed throughout a mesh network, the capacity increase is easy to comprehend as traffic can be sent simultaneously over different links due to the spatially

distributed nature of a mesh network. However, when most of the traffic within the mesh is being aggregated to a very limited number of egress points (such as the core network connection), then increased capacity can be difficult to achieve. Compared to the PMP, the capacity of a mesh network represents a balance of increased overhead to schedule the mesh traffic and forwarding of data over multiple links, both of which reduce capacity, vs the substantially increased capacity of each individual link with the mesh network resulting from shorter links and the ability to route around links severely degraded by obstacles in the transmission path.

- *Alternative paths:* Mesh systems that can provide alternative paths add resilience against link failures in the network.

- *Cost-effective range extension:* Because the nodes of a mesh are typically at least an order of magnitude less expensive than a BS and comparable in cost to PMP SS devices, mesh can be a cost-effective method of range extension of a traditional PMP network when the user population is too sparse to justify the extra BSs. Keep in mind that, if the node density decreases beyond a certain point, the mesh becomes ineffective. This is shown in Figure 9–1 and Figure 9–2. The typical mesh deployment hence requires so-called "seeding" to ensure an initial minimum coverage of the area.

- *Cost-effective deployment:* In fixed deployments, coverage is typically inadequate to allow a PMP SS to be installed in an arbitrary location. In some cases, a directional antenna must be installed and aligned, at substantial cost. For logical mesh networks, coverage is omnidirectional, and user self-installation may be enabled.

These tradeoffs can determine whether PMP, a true mesh, or something like a multihop relay is most appropriate.

Two general categories of mesh systems, *logical* and *physical,* are considered here. Logical mesh uses broad-beam or omnidirectional antennas to form logical links to neighbors. These links are considered logical because the hardware configuration does not change for different links. Physical mesh, by contrast, uses substantially directional antennas to create physical links. Links are considered physical because the hardware configuration (via beam forming or via switching, for example) changes to form different links with different neighbors. Both varieties have challenges and benefits.

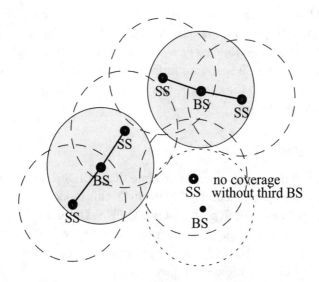

Figure 9–1:   Three BSs required for full coverage

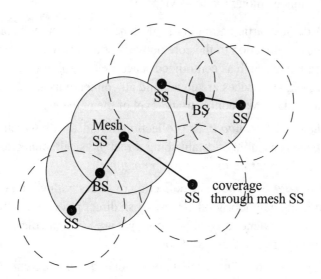

Figure 9–2:   Two BSs required with mesh SS

IEEE Std 802.16 includes both concepts to some extent. The logical mesh is completely specified, whereas hooks are in place for the physical mesh, known in the standard as *directed mesh*. Although it was never fully specified in the standard, it is worth spending further discussion here on the topic, especially since some of the concepts may be reused in the proposed mobile multihop relay amendment.

LOGICAL MESH

Since the logical mesh was incorporated into the IEEE 802.16 standard primarily by the proponents of the OFDM PHY, this mode was defined to function only with the WirelessMAN-OFDM PHY. Despite the fact that the mesh MAC uses many of the same messages used by the PMP MAC (there are only five mesh-specific MAC message formats), specifically in the upper MAC, the two architectures are not interoperable.

The logical mesh architecture is based on the concept of neighborhood, where distinction is made among the following:

- *Node:* A device participating in the mesh network. Although some devices may have a connection to the core network and some may be assigned special functions such as centralized schedule controller (Mesh BS) or master clock, all devices are equal, and all connectivity is peer to peer at the most fundamental connectivity level of the network.

- *Immediate neighborhood:* The collection of nodes with which the station of reference is capable of establishing direct physical connection. These nodes are also referred to as *one-hop neighbors*.

- *Extended neighborhood:* The collection of nodes with which the nodes in the immediate neighborhood can establish direct physical connection. Stations in the extended neighborhood that are not immediate neighbors are referred to as *two-hop neighbors*.

- *Cell:* The collection of nodes that are associated with the same coordinated (distributed and/or centralized) scheduler. Between cells, only uncoordinated distributed scheduling can take place. A node that has an interface to the core network or controls the centralized scheduler is termed *Mesh BS*. Typically these nodes include both functionalities. All other nodes are also termed SSs for the sake of simplicity.

- *Network:* A collection of cells where nodes may arbitrarily associate with the cell of choice and where synchronization takes place between the cells.

This concept is graphically depicted in Figure 9–3.

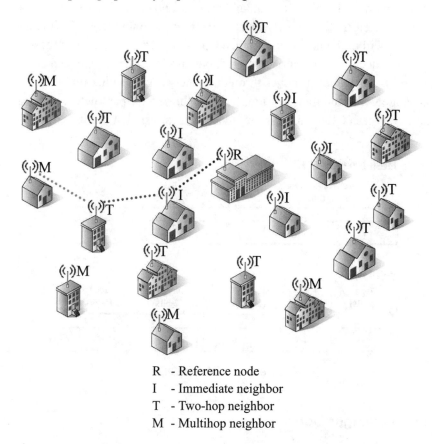

R   - Reference node
I   - Immediate neighbor
T   - Two-hop neighbor
M - Multihop neighbor

Figure 9–3:   Neighborhood definitions

The definition of the mesh architecture consists of essentially four aspects that are different from PMP: PHY frame format, network entry, network configuration/maintenance, and scheduling of data exchanges. The first aspect is dealt with in Chapter 11; the others are explored below.

Mesh connections and addressing

Beyond these aspects, the single most substantial difference between logical mesh and PMP is the usage of the concept of CIDs. In PMP, each 16-bit CID represents a unique, logically numbered connection to an SS with a defined set of QoS parameters. In a logical mesh network, the CID is used to convey a Link ID, unique only in the local neighborhood, and a set of broadcast/unicast service parameters. To uniquely identify the destination node, a 16-bit Node ID, unique within the network, is appended to the MAC header. The mesh system is hence not connection-oriented, but applies priority on a packet-by-packet basis. As a packet moves through the network, the Link ID in the CID will change with each transmission, whereas the service parameters typically do not. The mesh CID and mesh Broadcast CID are shown in Figure 9–4.

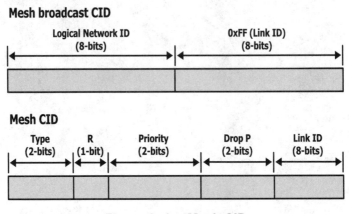

Figure 9–4:   Mesh CIDs

Network configuration

Network configuration is achieved by means of broadcasting the MSH-NCFG (mesh network configuration) message. It contains both physical information of the device itself, such as the transmit power as well as network time-synchronization information, nearness to known Mesh BSs, a list of all known neighbors with which connections exist, and information about the next transmission of the MSH-NCFG from the node. MSH-NCFG is also used to assist with network entry, which is explained later.

The time-synchronization information consists of frame number, the control slot number, and the synchronization hop count. Because control slots have a fixed length, nodes are able to determine from this number when the frame must have started. Fine time synchronization is not done with this method; for this, the PHY is responsible. The synchronization hop count provides the number of hops by which the node is removed from the master synchronization node, which is typically a Mesh BS. Nodes will defer in their synchronization timing to nodes with a lower synchronization hop count. In a system that consists of multiple cells, it is possible to have only one master synchronization node so that all cells are synchronized.

Nearness to a Mesh BS is computed with two metrics. The first is a straightforward hop-count, which for any node is the lowest value reported by any of its physical neighbors, plus one. The second metric is the required energy per bit across all hops to the BS, which is computed by all nodes as the minimum of the value reported by its physical neighbors plus the ratio of the transmit power and achievable physical data rate to that neighbor. This second metric allows a better capacity tradeoff as it balances the number of hops vs the data rates over those hops.

Connections are split into two categories: *physical* and *logical*. A physical connection exists merely by means of being able to decode the MSH-NCFG message of the other nodes. Neighbors with which only a physical connection exists cannot be addressed for data transmissions. Logical connections require an underlying physical connection and are addressable for data exchanges.

When listing its physical neighbors with the MSH-NCFG message, a node indicates, for each neighbor, whether a logical link is present. It can also flag that, for this link, a logical link is desired, or, when a neighbor put up such a request flag, that the request is accepted. The physical connection information also contains schedule information for the next transmission opportunity as reported by the neighbor. For logical connections, the list also contains information such as link quality, used burst profile, and transmit power.

Transmissions of MSH-NCFG are scheduled in a distributed fashion. A substantial computational burden would result if, in a single control slot, all nodes compete to transmit a MSH-NCFG. Therefore, a method is provided to limit the control slots in which a node can compete as well as the number of nodes that can compete in any particular slot. This method depends on the broadcasting by each node of two values, *Xmt Holdoff Exponent* (XHE) and *Next Xmt Mx* (NXM), and on reporting these values to each neighbor within the physical neighborhood so that the entire two-hop neighborhood of a node has access to the values for this node.

Xmt Holdoff Exponent is used to compute the number of control slots following the next MSH-NCFG transmission during which the node will not be eligible to transmit another MSH-NCFG as being $2^{XHE + 4}$. Both *Xmt Holdoff Exponent* and *Next Xmt Mx* are used to compute the eligibility interval for the next MSH-NCFG as

$$2^{XHE}(NXM) < \text{Next Xmt Time} \leq 2^{XHE}(NXM + 1)$$

For any control slot that falls within an eligibility interval, a node computes the list of nodes in its two-hop neighborhood whose eligibility intervals also contain this control slot. Each node then applies an algorithm to see whether it wins the election among this list of nodes. This guarantees that nodes transmitting MSH-NCFG simultaneously are at least three hops apart, which is generally enough to avoid mutual interference.

Network entry

Network entry for mesh devices consists of five steps:

- Scanning for active networks and coarse synchronization.
- Obtaining network parameters.
- Opening sponsor channel.
- Negotiating basic capabilities.
- Performing node authorization, registration, IP connectivity establishment, time-of-day establishment, and operational parameters transfer.

The first three are PHY and lower MAC functions that are substantially different from PMP behavior and are described below. The capabilities negotiation is identical to that performed under the PMP architecture, with the new node sending the SBC-REQ to the sponsor node (the node assisting the new node with network entrance). This capability exchange will be performed with any node with which a new logical connection is established (contrary to the entire network entry process, which is performed only once). The last step is similar to behavior under the PMP architecture, with the main difference being that the new node tunnels the message exchanges over UDP between itself and the mesh BS through the sponsor node. Because this process is transparent to the sponsor node, the sponsor node cannot determine when the upper MAC initialization is successful; therefore, it is the new node that, as shown in Figure 9-5, indicates completion of this phase to its sponsor node.

Scanning for active networks and coarse synchronization

Scanning is a lower PHY functionality and hence architecture independent. A node will listen to the medium cycling through all available channels until it detects energy and a strong correlation between the known preamble sequence and the received signal. This correlation allows the PHY to synchronize itself to the timing of the start of that burst. The new node will then continue listening to subsequent bursts until its MAC detects MSH-NCFG messages. This synchronization process is termed *coarse synchronization* and should not be confused with the coarse and fine timing synchronizing the lower PHY performs when synchronizing the PHY to the burst itself.

Obtaining network parameters

A new node is required to continue monitoring the channel for MSH-NCFG messages until it hears the same node at least twice, ensuring that the new node has collected a reasonable subset of potential neighbors. It also will have to wait until it receives at least one MSH-NCFG with a Network Descriptor subfield, which is one of the five defined subfields to the MSH-NCFG. The network descriptor contains such parameters as the number of control slots, the scheduling methods used, information on used channels, and criteria for the use of burst profiles as well as the Operator ID; and it is only occasionally attached to the MSH-NCFG by each node. The entire process of obtaining network parameters may take seconds, but in a fixed system, it is required only when the system is turned on (the neighbor list and network parameters can be maintained across losses of synchronization that make reentry necessary).

Opening sponsor channel

Once the collection of network parameters is completed, the node starts its request to join the network, as outlined in Figure 9–5.

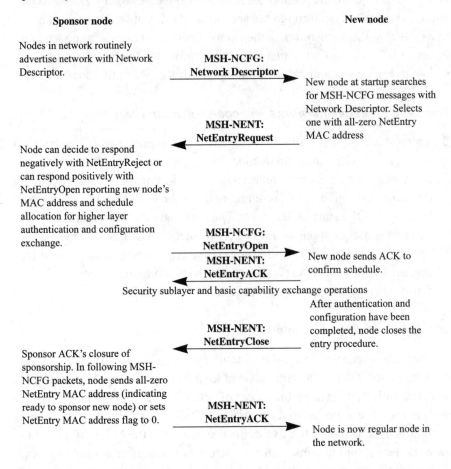

Figure 9–5: Mesh network entry: Opening a sponsor channel

For the purpose of joining the network, the node selects any of its neighboring nodes that are not currently assisting another node in network entry and transmits a MSH-NENT (mesh network entry) message with the NetEntryRequest subframe in the control slot reserved for network entry (see Chapter 11). The solicited node indicates its acceptance of the request by setting the new node's MAC address in the next MSH-NCFG and publishing an arranged schedule for data transfer in the data portion of the frame to this new node; the solicited node then takes on the role of sponsor. (A solicited node that is currently assisting another candidate node in network entry will signal this fact by setting that candidate's MAC address in its MSH-NCFG messages until its role as sponsor is terminated.) Upon reception of the MSH-NCFG and schedule, the new node confirms the schedule with a MSH-NENT message with the NetEntryAck subfield and proceeds with the fifth step of network entry by use of the facilitated schedule. Upon completion of this fifth step, an exchange of MSH-NENT messages with NetEntryAck, initiated by the new node, terminates the sponsor's role and adds the new node into the network. Due to the tunneling, this last exchange is initiated by the new node.

Distributed scheduling

Distributed scheduling is also based on the assumption that if a node is not part of the extended neighborhood, the probability of interference is sufficiently low to be irrelevant. The distributed scheduling mechanism, therefore, ensures that the entire two-hop neighborhood is aware of the pending data transmission to ensure a near-collision-free transmission. The distributed scheduling protocol uses only the MSH-DSCH (mesh distributed scheduling) message format to execute a three-way handshake for exchanging: requests (in which data slots in which the medium is free to fill those requests are indicated), grants, and grant confirmations. This handshake occurs in the control portion of the frame, which is collision-free using the distributed algorithm also used for MSH-NCFG, with the two parameters (as for MSH-NCFG) sent with every coordinated MSH-DSCH message. The transmitting party sends out a request, consisting of a demand level and a persistence ranging from 1 to 128 frames, and the status of the medium of which it is aware. The status consists of "busy," "free for transmission and reception," "free for transmission," or "free for reception." This effectively notifies its immediate neighbors that the medium will be busy during the time indicated as free for the indicated direction of the transmission and during the time indicated as busy due to other known scheduled transmissions among the one-hop neighbors. The receiving party responds with a grant in some or all of the requester's indicated free time, effectively notifying its immediate neighbors that the medium is reserved for the particular transmission during that period, which those neighbors will then mark as busy in their own schedules. The requester then confirms the indicated grant. The free medium time is computed as the period that none of the immediate neighbors has indicated as busy.

The only difference between coordinated and uncoordinated distributed scheduling is that for uncoordinated distributed scheduling, the handshake takes place in the data portion of the frame, making it susceptible to collisions due to attempts by other stations to set up an uncoordinated schedule.

The method is depicted in Figure 9–6, where node 1 is the node requesting the schedule, and node 2 is the node at which the request is directed. Node 3 is an immediate neighbor of node 1, but not of node 2; node 4 is an immediate neighbor of node 2, but not of node 1. Node 5 is a two-hop neighbor (i.e., within the extended neighborhood) of node 1.

When node 1 issues the request including an indication of potential free slots for the allocation to meet the request, it is received both by node 2 and node 3. Until it receives the grant confirmation sent by node 1, node 3 will assume the request will be granted in the explicitly indicated free slots and mark those temporarily as busy in its schedule. Node 2 will reply to node 1 with a grant in some or all of the indicated free slots, which also notifies node 4 of the allocation, and node 4 will mark those free slots as busy in its own schedule. The grant confirmation sent by node 1 finalizes the handshake between node 1 and node 2 and also serves to notify node 3 of the exact slots being used, as a result of which node 3 may mark the unused slots as free again. Because node 3 will advertise its availability before node 1 sends its grant confirmation (because of the cyclic way the control schedule is designed), node 5 (and any other node in the two-hop neighborhood) will effectively avoid any transmissions within the two-hop neighborhood during the same slots because they know from the free slots advertised by the one-hop nodes that those slots are being used by one of their two-hop neighbors.

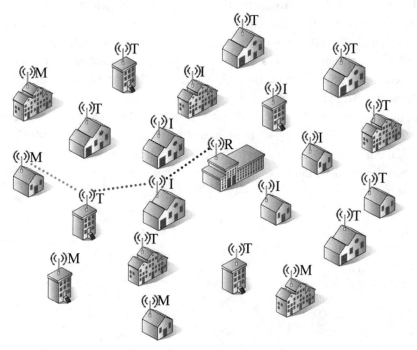

I - Immediate/one-hop neighbors of either transmit or receive node
T - Two-hop neighbor
M - Multihop neighbor

Figure 9–6: Distributed scheduling

Centralized scheduling

The concept of centralized mesh is based on the notion that most traffic passes
the gateway into the core network and that the node immediately connected to
this core network, the Mesh BS, determines the flow of traffic throughout the
entire cell. To achieve this, the BS dynamically splits the mesh-connected
nodes up into a tree structure, which, by means of a schedule decided by the
BS, is then used to arrange upstream and downstream traffic allocations. In a
pure OSI model, knowledge of the routing tree would exist only at the routing
layer. However, the IEEE 802.16 mesh specification defines knowledge of the
tree at the MAC, which enables higher MAC efficiency.

To be able to decide how to update the tree structure (which at network initiation consists only of the BS itself), the current tree structure is used for a fixed schedule to pass down the tree MSH-CSCF (mesh centralized scheduling configuration) messages, which contain the number of nodes attached to the BS and for each of those nodes its ID and the IDs of all its child nodes. An example of a MSH-CSCF list is shown on the right-hand side in Figure 9–7, where the ID of each node is indicated with a single character, and the burst profile indicators in both directions are omitted. The order in which the nodes are listed in the MSH-CSCF is irrelevant, but it is used to indicate the child-parent relationship from which each node can reconstruct the entire tree.

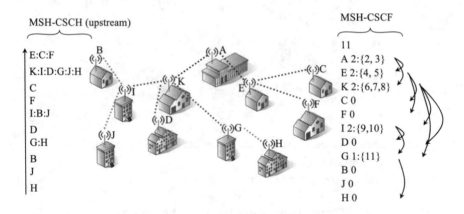

Figure 9–7: Centralized scheduling

The transmission rule for MSH-CSCF is that the BS transmits it first, followed by all its immediate children in list order, followed by the children of the first listed child in list order, etc. In the example in Figure 9–7, the transmission order is, therefore, identical to the list order; note that nodes without children do not rebroadcast the schedule. Due to its overhead burden, MSH-CSCF is used only occasionally, depending on the volatility of the network, whereas incremental changes can be distributed through the frequently sent MSH-CSCH (mesh centralized scheduling) message.

The MSH-CSCH messages serves three purposes:

- To distribute BW requests and grants up and down the tree, respectively. The requested and granted bandwidth is indicated as a function of instantaneous bits per second. The BS and all other nodes use their knowledge of the burst profile on each branch of the tree to translate this into an allocation of time. To facilitate the requests, each node uses a table the size of its branch in the MSH-CSCF list and fills, in MSH-CSCF list order, the BW requests for itself and all nodes down its branch. For example, in Figure 9–7, node K transmits a table with six entries, containing the BW requests of itself and the five nodes (I, D, G, J, and H) in its branch. The list order within the table is dictated by the list order in the MSH-CSCF, shown also in Figure 9–7, which all nodes including node K received from the BS. The transmission order for the requests, shown on the left side of Figure 9–7 starting with node H, is the reverse of the MSH-CSCF message, resulting in the table growing in size until the BS finally combines the pieces of the table into the full table. In their own request (for example, the entry for node K in the request table sent by node K), nodes indicate only the bandwidth needed for traffic that terminates in the node itself, not the cumulative traffic from the nodes downstream. In Figure 9–7, the upstream order of transmissions for MSH-CSCH with the request content of each transmission is shown on the left-hand side.

- To allow a node to request the BS to facilitate bandwidth to add a new node to the cell. The node initiating the request is termed a *sponsor node*.

- To enable the BS to notify the nodes of changes to the tree structure or changes in burst profile, which may be necessitated by changes in the propagation conditions. In this case, nodes and their new parent (or old parent, if only the burst profile changes) are identified with their respective list number in the last MSH-CSCF, which are 1-byte IDs.

Note that MSH-CSCH currently does not allow a node to send up a request for a downstream flow starting in the node, as the indication for this is limited to grants (which are issued by the Mesh BS). Hence, this requires that all downstream traffic be routed through the Mesh BS. As a result, the Mesh BS would need to inspect traffic to predict the required downstream flow grants, and, although this need does not break the specification, it is impractical. This should, therefore, be considered an error in the specification, which is correctable by the removal of a single "if" statement in the MSH-CSCH frame format.

Note also that there is no facility for nodes to pass up changes to the tree except for the facilitation to add new nodes. This is done because deciding the tree structure is primarily a routing layer function, and hence the indication of the parameter set used by the Mesh BS to decide the tree is left to this layer. In other words, in practice, if a node loses the physical connection with its parent node in the tree, it would have to use the distributed scheduling mechanism using MSH-DSCH to achieve an update to the tree through another neighbor.

DIRECTED MESH AND POINT-TO-POINT (PTP)

Directed mesh in IEEE Std 802.16 can be thought of to some extent as transient PtP links from devices that could act as repeaters for the BS. The hooks for both directed mesh and PtP have been added for systems that use the WirelessMAN-SC PHY, but neither has been completely developed in the standard.

While a BS talking to a single SS in PMP mode is in principle a PtP link, it carries a significant amount of MAC overhead that is essential in the PMP case, but simply adds inefficiency to PtP operation. Furthermore, PtP range can be quite long through the use of highly directional antennas. This concept is shown in Figure 9–8.

The basic PtP link illustrated is a single-path link between the BS and an SS. With directed mesh, the concept is expanded; the SS has some ability to repeat all or some of the BSs transmission on another beam, in a different direction, and possibly on a different frequency. Such a system might appear as shown in Figure 9–9.

Physical or directed mesh can have many benefits. The reduced beam width can significantly improve the carrier to interference ratio (C/I) and improve frequency reuse. This gives high spectral efficiency. Conversely, network entry is greatly complicated, and the antenna systems are more expensive than with a standard PtP SS.

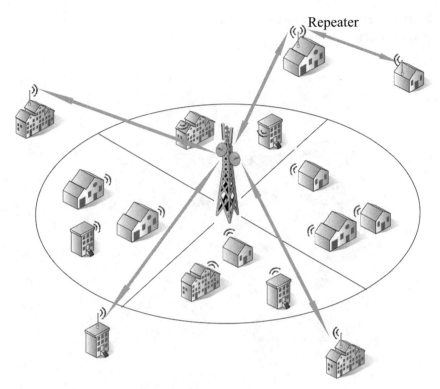

Repeater

Figure 9–8: Addition of PtP

Interference impact on reuse reduced due
to a smaller arc of impact

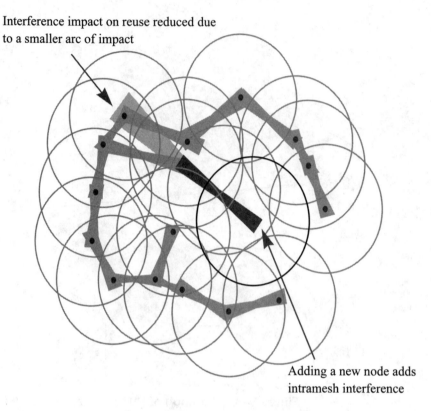

Adding a new node adds
intramesh interference

Figure 9–9: Physical or directed mesh

Chapter 10 PHY: WirelessMAN-SC

Single-carrier PHY for 10–66 GHz

The WirelessMAN-SC PHY specification is intended for operation in the 10–66 GHz frequency range. At these frequencies, LOS operation is a practical necessity due to the propagation characteristics. This makes mobility infeasible. While radios operating at these frequencies are typically expensive, highly directional antennas of practical size are available; therefore, given a LOS, multipath propagation can be negligible. This allows for a low-complexity, single-carrier PHY with good performance, even over a broadband channel. This is compatible with the large spectrum allocations in these frequency bands and the user data rates necessary to justify the radio costs with today's technologies.

As the original IEEE 802.16 PHY specification, WirelessMAN-SC set some precedents. For instance, both TDD and FDD configurations are supported in order to allow operation in worldwide spectrum allocations. Care was taken to make the FDD and TDD systems as similar as possible to minimize the impact of implementing both variants. Both have the same basic frame structure supporting adaptive burst profiling. The main difference is whether the DL subframe and UL subframe are sequential or simultaneous in time. The FDD case looks very much like the TDD case, but with the DL and UL subframes "stretched" toward each other so that they overlap. The FDD configuration also has special features to support half-duplex SSs.

The allocation of channel time (generally known in the IEEE 802.16 standard as *bandwidth allocation*) at the PHY is based on a combination of time division multiple access (TDMA) and demand-assigned multiple access (DAMA). In the UL, the channel is divided into a number of time slots (i.e., PSs) that are assigned to various purposes on a frame-by-frame basis by the BS MAC, based on demand. Time slots are available for ranging, contention-based BW requests, and user traffic. In the DL, the data bound for each SS are multiplexed in a TDM fashion into a single stream of data and available for reception by all SSs within the same sector. This stream may change its PHY burst profile on QAM symbol boundaries within the frame to allow a tradeoff of robustness for efficiency. To support H-FDD SSs, provisions are made for resynchronization to the DL in midframe.

FRAME STRUCTURE

The WirelessMAN-SC PHY, like the other IEEE 802.16 PHYs, operates in a framed format. The frame structure eases the task of PHY synchronization. Within each frame, in the TDD case, are a DL subframe and an UL subframe. The DL subframe begins with information necessary for frame synchronization and control. While the DL data PDUs are addressed to individual SSs or to a broadcast address, the DL MAPs describing the DL merely indicate the occurrences of DL PHY burst profile changes; they do not indicate to which SS a series of time slots is directed. So each SS attempts to receive all portions of the DL except for the bursts whose profiles either are not implemented by the SS or are less robust than its current operational DL burst profile. The SS filters the data by CID, retaining only the data intended for it. Encryption ensures it cannot make use of data destined for other SSs anyway.

The WirelessMAN-SC PHY allows three different frame durations: 0.5 ms, 1 ms, and 2 ms. While not strictly required, the different frame durations are typically associated with different channel bandwidths, allowing a tradeoff of overhead vs protocol latency, with 1 ms frames being the most typical. Frame durations of 0.5 ms might typically be used for large channel bandwidths of 50 MHz or 56 MHz, while 2 ms frame durations might be preferred for channel bandwidth less than 20 MHz.

To keep the MAPs compact and to ensure that allocations are on byte boundaries, the building blocks for allocation are PSs of four QAM symbols each, yielding 1, 2, or 3 bytes, respectively, for QPSK, 16-QAM, and 64-QAM modulations. In the DL, the allocations are at this PS granularity. Because the number of QAM symbols in a 2 ms frame on a 56 MHz wide channel is greater than can be represented by the fields in the UL MAP (a 2 ms frame on a 56 MHz channel is allowed, but not recommended), the granularity of the UL allocations are 2^m PSs (m ranging from 0 to 7). The typical value for m is 0 (1 PS granularity).

The basic structure for the DL subframe is taken from the TDD configuration of the system and is shown in Figure 10–1. The DL subframe starts with a frame start preamble used by the PHY for synchronization and equalization. This is followed by the frame control section, consisting of the DL MAP and UL MAP, broadcast to all SSs indicating the PSs at which the various bursts begin. The frame control section is followed by a TDM section carrying data and organized into burst profiles ordered by decreasing burst robustness. For example, assuming the FEC parameters do not change, data transmission begins with QPSK modulation, followed by 16-QAM, followed by 64-QAM. In the case of TDD, a TTG follows the DL subframe, allowing time for the BS to switch from transmit to receive and for the first SS of the UL to switch from receive to transmit mode.

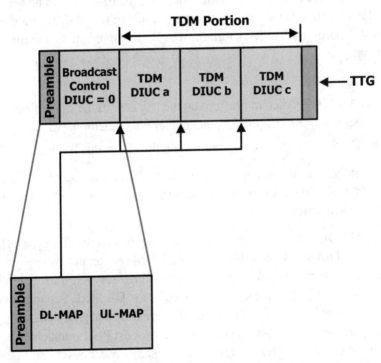

Figure 10–1:   TDD DL subframe structure

The structure of the DL subframe for the FDD case can be quite similar to that of TDD operation, but becomes more complex with the introduction of H-FDD SSs. In the FDD case, the BS is full-duplex, but both full-duplex and half-duplex SSs are allowed. Although a full-duplex SS is capable of functioning as a half-duplex device, the IEEE 802.16 standard assumes that H-FDD SSs differ in being incapable of receiving while transmitting. This assumption allows for simplicity and cost savings in designing the H-FDD SS. When only full-duplex SSs are used, the FDD DL subframe is identical in structure to the TDD DL subframe except for the lack of a TTG. H-FDD SSs complicate matters, however, because whenever they are directed to transmit in the UL, they potentially lose synchronization (in particular, phase synchronization) with the DL. This makes it difficult or impossible for them to receive more DL data until resynchronized. As discussed below, a special preamble is specified for this purpose.

The FDD DL subframe structure is shown in Figure 10–2. As can be seen, the initial TDM portion is identical in structure to that of the TDD DL subframe. During this portion of the DL subframe, the BS can transmit data to the following:

- Full-duplex SSs.
- H-FDD SSs scheduled to transmit on the UL later in the frame than when they receive in the DL (the SSs that do not lose synchronization).
- H-FDD SSs not scheduled to transmit on the UL in this frame.

Unfortunately, in a system that has primarily H-FDD SSs, use of only the TDM portion could waste upwards of half the channel bandwidth, due to SS loss of synchronization.

To allow H-FDD SSs to transmit earlier than they receive in a frame, a TDMA section was introduced that differs from the TDM section in two important respects. First, extra preambles are inserted to allow resynchronization with the DL. The positions of these are described in the DL MAP. Second, given the extra preambles for resynchronization, there is no longer a need to smoothly transition through progressively less robust PHY burst profiles; therefore, the burst profiles used may occur in any order convenient to the BS's scheduling algorithm.

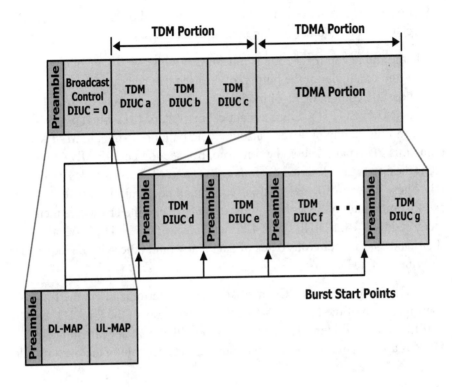

Figure 10–2:   FDD DL subframe structure

Since the H-FDD SSs are nearly synchronized with the BS, the resynchronization preambles are shorter than the frame start preamble. The frame start preamble is 32 bits, created from the repetition of a 16-bit constant amplitude zero autocorrelation (CAZAC) sequence. The resynchronization preambles are 16 bits, created from the repetition of an 8-bit CAZAC sequence.

For WirelessMAN-SC, the frame control section of the DL always contains a DL MAP and a UL MAP, encoded with an especially robust well-known modulation/FEC combination (using QPSK). The DL MAP and UL MAP are structured so that the frame control section is always at least two FEC blocks in duration, but the first DL PHY transition will always be identified in the first FEC block of the frame, even if the DL MAP message does not terminate in that block. This ensures that the SS will be warned of the upcoming PHY burst change at least one FEC block in advance.

While the frame control section is transmitted with a well-known burst profile, the remainder of the burst profiles used can be any of the over 4000 possible combinations of modulation and FEC allowed by the WirelessMAN-SC PHY. Since a practical number of PHY burst profiles is closer to four than to 4000, and also to avoid ballooning the DL MAP with numerous PHY parameters, the frame control section periodically contains DCD messages that describe the four or so (up to 13) burst profiles used in the DL. These burst profile descriptions contain all necessary information, such as FEC, modulation, SNR thresholds, etc. These parameter sets are then identified by a 4-bit DIUC. This 4-bit value is used in the DL MAP as a shorthand for the numerous bytes of data that would otherwise be required to describe each PHY burst.

Similarly, for the UL, the UCD message appears periodically in the frame control section of the DL. It describes the parameters of the UL PHY burst profiles in use and identifies them by a 4-bit UIUC that appears in the UL MAP as a shorthand for the actual parameter sets describing the burst profiles.

DL CHANNEL ENCODING

The DL data sections are used to transmit data and control messages to specific SSs. The data are always FEC coded and are transmitted at the current DL operating modulation of each SS. In the TDM portion of the DL subframe, the data are transmitted in decreasing order of burst profile robustness. Within a particular burst, the data transmitted at a given burst profile may contain data for multiple SSs multiplexed together. It is the responsibility of the SS to extract its data from data destined for other SSs sharing a common operational DL burst profile. In the TDMA section, the data are grouped in separate burst profiles that need not be in order of robustness. The BS may, but is not required to, segregate the data by SS as well as by burst profile. However, the DL MAP generally does not indicate the target SS; therefore, SSs listen to all bursts for which they are capable.

While a new DL burst always starts on a PS boundary, the choice of FEC parameters and modulation may cause FEC blocks to end in the middle of a PS or even in the middle of a QAM symbol. However, they always end on bit boundaries. The next FEC block starts on the next available bit, without padding to the next symbol or PS boundary. The last data for a burst may not be sufficient to fully fill an FEC block. In this case, the last FEC block of the burst may be shortened. Unused bits in an unshortened FEC block or the bits used to allow realignment of the next DL burst to a PS boundary are padded with all ones. Figure 10–3 illustrates the allocation of data to a DL burst. If the burst contains a preamble, as is the case for the frame control section and TDMA bursts, the preamble is considered part of the burst and precedes the first FEC block of the burst.

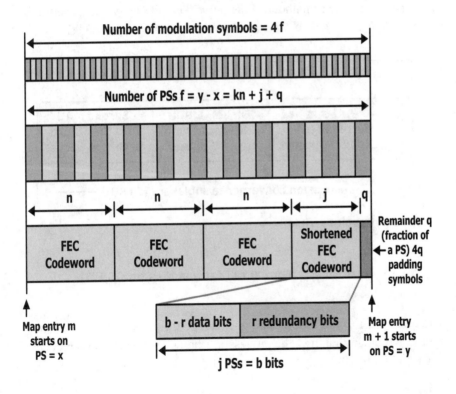

Figure 10–3:   DL MAP usage with shortened FEC blocks—TDM case

Within each FEC block, additional information is added to help resynchronize the MAC to the next MPDU within the bit stream in the case of an unrecoverable error in an FEC block. This capability is handled by the transmission CS. Since MPDUs are variable in length, while the FEC blocks of a burst are of fixed length, it is common for MPDUs to begin or end in the middle of an FEC block. The MAC header contains the length of the MPDU, which is required for extracting the MPDUs from the received bit stream. Unfortunately, if an FEC block has unrecoverable errors, this length information may be lost or erroneous, causing difficulty in finding the start of the next MPDU. To avoid the risk of the SS losing all data until the end of the PHY burst, each FEC block uses the first byte of its payload not for data but for a field indicating where, if anywhere, within the FEC block the next MPDU starts. This allows the SS to quickly resynchronize to the next MPDU rather than losing the remainder of the burst. This FEC payload structure is shown in Figure 10–4. The same mechanism is employed in the UL.

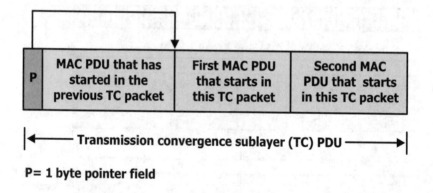

P= 1 byte pointer field

Figure 10–4:   Format of DL transmission CS PDU

In addition to FEC and modulation, the data go through a randomization process and pulse shaping. The processing chains for transmit and receive are shown in Figure 10–5.

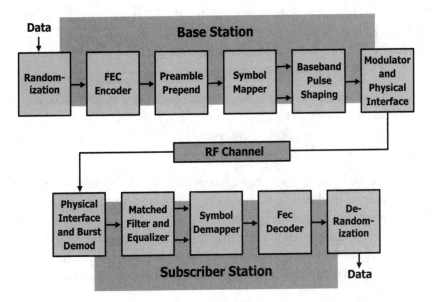

Figure 10–5: Conceptual block diagram of DL PHY

There are four basic types of FEC used in WirelessMAN-SC. These can be further parameterized to give many possible combinations.

The first type, which is specified as mandatory, is a Reed-Solomon (RS) outer code over Galois field (GF) 256 with no inner code. It is generally used for user data except when, even with QPSK modulation, a stronger coding is desired. It can be parameterized with the information block length K variable from 6 to 255 and with error correction capability to correct errors from $T = 0$ to 16 bytes, depending on the number of parity bytes $R = 2T$.

The second type of FEC, also mandatory, is used for the frame control section and for transmissions that need to be very robust. The RS is used with an inner (24, 16) block convolutional code (BCC) derived from a four-state, nonsystematic punctured (7, 5) convolutional code. In this case, the number of information bytes K for the outer RS code must be an even number so that the total size $(K + R)$ is also even, since the inner BCC operates on pairs of bytes. In the frame control section, the parameters are set to $K = 26$ and $R = 20$.

The third FEC type is optional and combines the RS outer code with a (9, 8) parity check code. In this case, a parity bit is calculated for each RS symbol individually and inserted as the LSB of the resulting 9-bit word. The parity bit is the XOR of all 8 bits in the symbol.

The fourth and final FEC type is an optional block turbo code (BTC). If implemented, both (39, 32) × (39, 32) and (53, 46) × (51, 44) codes are required. Shortening can apply to the last FEC block of a BTC-encoded burst, as in the RS case.

To maximize the spectral efficiency of the airlink, the WirelessMAN-SC PHY uses a multilevel modulation scheme. The modulation is optimized for each SS based on the quality of the RF channel. If link conditions permit, a more efficient modulation is used to maximize the tradeoff between bandwidth and robustness. If the RF quality degrades, the BS changes to a more robust modulation for that subscriber. If it improves, it changes to a more efficient modulation.

The WirelessMAN-SC PHY supports QPSK, 16-QAM, and 64-QAM providing 2 bits, 4 bits, and 6 bits per QAM symbol, respectively. Two power adjustment rules can be used when transitioning from one modulation to another: *constant peak power* and *constant average power*. In the constant peak power scheme, corner points are transmitted at equal power levels regardless of modulation. In the constant average power scheme, the signal is transmitted at constant average power regardless of modulation. The power adjustment rule is communicated via the DCD message. If a burst at a specific modulation does not end exactly on a PS boundary, the remainder of the burst is filled with zero bits.

Prior to modulation, the I and Q signals are filtered by square-root raised cosine filters with the excess bandwidth factor α set to 0.25.

UL CHANNEL ENCODING

In the WirelessMAN-SC PHY, three types of bursts may be transmitted in the UL subframe:

• Contention-based initial ranging bursts.

- Contention-based multicast or broadcast BW requests.

- Dedicated data bursts in response to unicast bandwidth grants allocated to individual SSs.

In a given UL subframe, any of these burst types may be present (or absent) and may occur in any order and quantity. The scheduler at the BS indicates which bursts are present via the UL MAP message. The bursts are separated by subscriber station transition gaps (SSTGs), allowing for the ramping down of the previous burst and the ramping up and preamble of the next burst. The preamble allows the BS to fine-tune its synchronization to the transmission of the individual SS. As in the DL, the preambles may be 16- or 32-bit repetitions of 8- or 16-bit CAZAC sequences. The UCD message communicates the gap and preamble lengths. Figure 10–6 illustrates an example of the UL subframe.

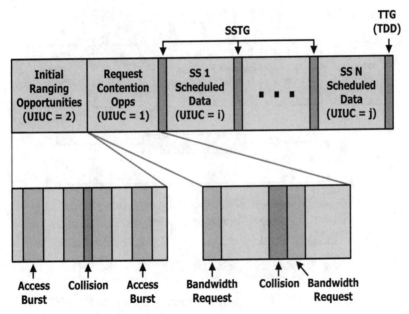

Figure 10–6:   UL subframe structure

Request bursts always use UIUC = 1, which always uses QPSK modulation and FEC type 1 or type 2 as defined for the DL. Request bursts are always at least as robust as the most robust unicast UL data bursts. Initial ranging bursts always use UIUC = 2, which specifies burst characteristics identical to those used in the DL in the frame control section. The UL MAP entries of WirelessMAN-SC are directed to the Basic CID (or multicast group or Broadcast CID) of a specific SS and are relative to the previous MAP entry. They specify the start of the burst, but the next MAP entry implies the end of the burst. The last burst of a UL subframe has an end symbol indicated by an end-of-MAP entry allocated to CID = 0 with UIUC = 10 (decimal).

The UL uses the same FEC and the same transmission convergence subframe as the DL. Also, like the DL, in addition to FEC and modulation, the UL data go through a randomization process and pulse shaping. The processing chains for transmit and receive are shown in Figure 10–7.

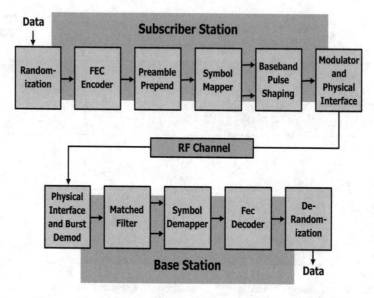

Figure 10–7:   Conceptual block diagram of UL PHY

As in the DL, the I and Q signals, prior to modulation, are filtered by square-root raised cosine filters with the excess bandwidth factor α set to 0.25.

This 0.25 roll-off factor defines the symbol rate, which, for a 25 MHz channel, would be 20 Mbd. This would give bit rates for QPSK, 16-QAM, and 64-QAM of 40 Mbit/s, 80 Mbit/s, and 120 Mbit/s, respectively. Other channel sizes scale linearly. For example, a 28 MHz channel has a 22.4 Mbd symbol rate.

CONTROL MECHANISMS

The DL demodulator must provide an output reference clock derived from the DL symbol clock. This reference is then used by the SS to provide timing-critical interfaces. One would expect that the BS reference is locked to an accurate (stratum 1) reference such as GPS. This highly accurate timing is used to provide frequency tuning to within $\pm 10 * 10^{-6}$.

The WirelessMAN-SC PHY requires accurate UL time-slot synchronization and UL transmit time advance to ensure that multiple users do not interfere with each other's transmissions. The ranging procedure outlined in Chapter 7 allows the fine-tuning of timing required for robust operation with efficient burst profiles.

For WirelessMAN-SC, the predominant dynamic impairment is rain fade. The typical extent of the fade varies geographically based on rain region and is affected by both carrier frequency and range. For a desired link availability and given set of transmission parameters, the rain region dictates the maximum cell radius. Additionally, uncorrelated rain fade between an interfering signal and the desired signal may also affect cell radius.

Power control is applied statically in the DL to effectively set cell size and dynamically in the UL to account for fading due to rain and foliage. This dynamic UL power control must handle fades of up to 20 dB/s with dynamic range of at least 40 dB. The specific algorithm is left for vendor differentiation. The BS's EIRP spectral density is capped by +28.5 dBmi/MHz or a lower limit if established by local regulations. The SS is limited to an EIRP spectral density of +39.5 dBmi/MHz or local regulation.

Received signal strength indicator (RSSI) and CINR signal quality measurements and associated statistics are used in burst profile selection. RSSI measurements do not necessarily require receiver demodulation lock and are, therefore, used for measurement of low signal levels. CINR requires receiver lock, but provides better information on the actual operating condition of the receiver, including interference and noise levels as well as signal strength.

WIRELESSMAN-SCA

This chapter concludes with a brief description of the WirelessMAN-SCa PHY specification, which is targeted at operation below 11 GHz. In many respects, it is very similar to the WirelessMAN-SC PHY. It was originally added to the standard in IEEE Std 802.16a-2003, with the tacit assumption that it would be used in conjunction with frequency-domain equalization. It has not found subsequent industry support and, at this time, appears to be an orphan.

The WirelessMAN-SCa FEC specification is similar to, but slightly different from, WirelessMAN-SC. The basic FEC is a concatenated FEC, with an outer RS code derived from a systematic RS ($N = 255$, $K = 239$) code using GF(256). The block may be shortened. An inner trellis-coded modulation (TCM) code derived from a $K = 7$, rate 1/2 convolutional code is used. Optionally, an interleaver may be placed between the inner and outer codes.

In addition to QPSK, 16-QAM, and 64-QAM, the WirelessMAN-SCa PHY also requires the implementation of spread binary phase shift keying (BPSK) and BPSK modulations. 256-QAM modulation is optional. Block turbo codes (BTC) and convolutional turbo codes (CTC) are also optional. When ARQ is implemented, the FEC can be disabled (through the "no FEC" option), and the ARQ can be used as a sole method for error correction.

Like WirelessMAN-SC, the WirelessMAN-SCa PHY is defined for both FDD and TDD operation. The WirelessMAN-SCa PHY includes explicit support for beam forming.

The system must handle fade rates of 30 dB/s with depths of 10 dB. Both RSSI and CINR are used for measurements of signal quality. Frequency accuracy must be within $\pm 15 * 10^{-6}$.

Chapter 11 PHY: WirelessMAN-OFDM

Multicarrier PHY for frequencies below 11 GHz

WirelessMAN-OFDM is one of the three PHY specifications defined in IEEE 802.16 for applications below 11 GHz. For brevity and convenience, the PHY of a WirelessMAN-OFDM system is abbreviated as the "OFDM PHY."

WAVEFORM CONSTRUCTION

Selection of the OFDM waveform

The choice of OFDM for use below 11 GHz was in general based on two considerations. The first of these was that, among the IEEE 802.16 members, a fair number had experience with OFDM systems due to the approval of the IEEE 802.11a amendment in 1999. In addition, a small number of participants were involved with companies developing proprietary OFDM-based systems for BWA. Selecting OFDM was for these members a natural choice. Aside from these external issues, however, OFDM was supported because of its favorable technical characteristics.

As noted in Chapter 3, robust BWA systems in spectrum below 11 GHz must deal with multipath without resorting to highly directional antennas, which would make deployment costly. As noted, one approach is to use redundant samples preceding transmitted symbols to ensure that any received samples affected by energy from the previous symbol can be discarded. This process is efficient only if the duration of the transmitted symbols is considerably longer than that of the expected multipath; otherwise, the redundancy would be constituting an unacceptably large relative overhead. In addition, the method used to modulate data onto these large symbols must be computationally simple. This is important because the number of bits per symbol will have to be very large to maintain a given data rate.

The data rate, in bits per second, can be expressed as [(bits/symbol) × (symbol/s)]. The symbol rate, (symbol/s), is the inverse of the symbol duration so that increasing the symbol duration requires modulating more bits per symbol to maintain the data rate (assuming a given channel bandwidth).

The preferred method of creating such symbols is the use of the IFFT, which effectively transforms a set of orthogonal narrowband transmissions into a single waveform. An additional advantage of the IFFT is that it is cyclical in the time domain so that when the redundant samples are copied from its tail, the transmitted symbol remains continuous in phase. OFDM provides for the subcarriers to be orthogonal and separable. When the subcarrier spacing is sufficiently small, the channel characteristics can be safely assumed to be constant across the subcarrier. This assumption drastically simplifies channel estimation because it can be performed independently for each subcarrier and, for each subcarrier, the channel can be represented by a single amplitude and phase.

Selection of FFT size

The FFT size is a critical parameter with implications in the time domain as well as the frequency domain. The designer must take into consideration a extraordinarily wide range of system factors, including the multipath propagation environment, the cost of precision electronics, and the nature of the network traffic. This job requires system-level thinking; it is not for narrow specialists.

The most fundamental frequency domain characteristic is the channel bandwidth. IEEE Std 802.16 is designed to accommodate a number of channel bandwidths, depending on the local regulatory environment and the operator requirements. For the purpose of this discussion, we will assume that the channel bandwidth is fixed.

In a given channel bandwidth, the subcarrier spacing is inversely proportional to the number of subcarriers and, therefore, to the FFT size. Meanwhile, the duration of the OFDM symbol is set to be the inverse of the subcarrier spacing. This setting is the necessary and sufficient condition to ensure that the subcarriers are orthogonal. As a result, the OFDM symbol duration is proportional to the FFT size.

The choice of FFT size is inextricably tied to assumptions about the physical environment in which the propagation takes place. This is because, to ensure that multipath propagation does not corrupt the received signal, a copy of the last T_g s of the symbol waveform with duration T_b is prefixed to the symbol before transmission (see Figure 11–1). This copy is termed the *cyclic prefix*. The receiver will sample only an interval T_b within this extended interval $T_b + T_g$. The prefix has no impact on the received symbol due to the cyclic nature of the FFT. Multipath components shorter than T_g s need not negatively affect the performance as long as the sampling interval is chosen correctly. How much smaller the multipath components need to be depends on the accuracy of sampling timing and the symbol windowing used because both negatively affect the protection against multipath. Multipath components in excess of the effective multipath protection duration, however, will significantly degrade the performance (see [B55]). It is, therefore, desirable to choose the cyclic prefix large enough to encompass the expected multipath of the application.

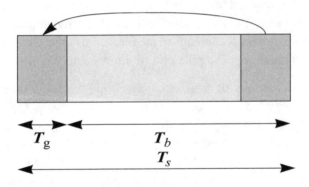

Figure 11–1: OFDM symbol structure

The addition of the cyclic prefix adds overhead due to the redundancy. It also increases the energy per symbol, resulting in an SNR drop of $10\log_{10}(1 - T_g/(T_b + T_g))$ dB. With the duration of the cyclic prefix dictated by the multipath, choosing a shorter OFDM symbol duration hence increases the proportion of time and energy lost on the cyclic prefix.

The designer must also consider the network traffic requirements. To understand why, begin with the fact that the FFT size determines the number of available subcarriers. Each subcarrier is modulated to carry data. For this discussion, assume QAM modulation, with each subcarrier delivering one QAM symbol during the duration of the OFDM symbol. The total number of bits carried in one OFDM symbol by all the subcarriers, therefore, scales with the FFT size (as does the symbol duration). This is where the network factors enter the problem. As the data capacity per symbol gets larger, the granularity may grow too coarse to efficiently carry variable-length data packets, particularly when the traffic consists of many small data packets that do not fill the symbol. For instance, if the UL is transmitting a few small packets, as typically found in VoIP and most other IP traffic, without a sufficient packet rate to take advantage of packing, then large symbols may need to be transmitted without a full load of data. In this case, a shorter symbol would be more efficient.

Finally, a longer OFDM symbol duration translates into narrower subcarriers in the frequency domain. This closer spacing makes the receiver performance more susceptible to phase noise, which corrupts the orthogonality of the subcarriers. With current technologies, the phase noise requirements have a direct and substantial impact on the cost of the system.

The above considerations all came into play during the discussions on the FFT size N_{FFT}, an issue complicated substantially by the targeted support of a large range of bandwidths (up to 28 MHz). In the early stages of development, the IEEE 802.16 Working Group considered many different FFT sizes, particularly 64 and 256 for OFDM and both 2048 and 4096 for OFDMA. Although it was realized that 64-FFT was suitable only to short links due to its low multipath protection capabilities, the opportunity to reuse IEEE 802.11a implementations was a strong consideration for many participants. Eventually, the 64-FFT option was dropped as an OFDM alternative, as performance issues outweighed issues regarding the reuse of legacy components; and OFDMA was restricted to 2048. With the IEEE 802.16e amendment, additional sizes of 128, 512, and 1024 were added for OFDMA. To provision in a more optimum fashion for deployments with short links and deployments with relatively little multipath, the IEEE 802.16 standard allows for several different cyclic prefix durations.

The OFDM waveform

The OFDM waveform as a function of time can be described as follows:

$$s(t) = \mathrm{Re}\left\{ e^{j2\pi f_c t} \sum_{\substack{k = -N_{used}/2 \\ k \neq 0}}^{N_{used}/2} c_k \cdot e^{j2\pi \cdot k \cdot \frac{n \cdot BW}{N_{FFT}} \cdot (t - T_g)} \right\} \qquad 0 < t < T_s$$

where

c_k is a data point out of the modulation constellation to be mapped onto the frequency offset index k

f_c is the RF

BW is the transmission bandwidth

n is the oversampling factor

N_{used} is the number of nonsuppressed subcarriers

In the early stages of the development of the OFDM and OFDMA PHYs, the working group attempted to maintain as much commonality between the two specifications as possible. This, along with the fact that IEEE Std 802.11a used 48 data subcarriers per OFDM symbol and the fact that 48 data subcarriers conveniently allowed mapping whole bytes to a single OFDM symbol for almost all modulations, led to a symbol structure containing a multiple of 48 data subcarriers. For the case of the 256-FFT, the working group settled on 192 data subcarriers. Eight additional "pilot subcarriers" are used mainly to track phase changes during a burst, making for a total of $N_{used} = 200$ subcarriers. The resulting frequency domain description is shown in Figure 11–2.

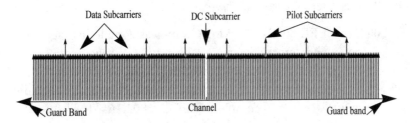

Figure 11–2: OFDM frequency description

At this point, it would have been possible to use the 200 subcarriers, along with a dc subcarrier in the center and guard subcarriers on either side (28 on one side and 27 on the other). However, the resulting spectral shape would end up unnecessarily conservative with regards to spectral masks established by relevant regulatory agencies; the result would be akin to a waste of spectrum. To take advantage of the opportunity to fully utilize the spectrum within the mask, the sampling factor n was introduced to create a sampling bandwidth somewhat larger than the actual bandwidth BW. This stretches the spectral shape of the signal to more fully fill the space with the spectral mask, shortening the symbol duration and hence leading to slightly higher throughput. The allowed values of n are as follows:

- 8/7 for BW multiples of 1.75 MHz
- 86/75 for BW multiples of 1.5 MHz
- 144/125 for BW multiples of 1.25 MHz
- 316/275 for BW multiples of 2.75 MHz
- 57/50 for BW multiples of 2 MHz
- And again 8/7 for any other bandwidth

These specific values ensure that there is an integer number of samples per frame for all likely bandwidths less than 28 MHz.

Subchannelization

The access method for the OFDM PHY is TDMA. This is, however, not an entirely complete description, as the PHY also provides an integrated optional OFDMA component. To avoid confusion between this component and the OFDMA PHY described in Chapter 12, the OFDMA component in the OFDM PHY is referred to as *subchannelization*.

In late 2002, the IEEE 802.16 Working Group was completing the IEEE 802.16a amendment, which added the new PHY specifications for frequencies below 11 GHz. The working group recognized the importance of specifying profiles as the basis of compliance testing for these alternatives, but the group was anxious to complete IEEE Std 802.16a. Therefore, it decided to develop an amendment, IEEE P802.16d, to add the missing profiles; the same project could be used to address any errors identified in

IEEE Std 802.16a. At about the same time, the working group, following formal study group activity, was initiating activities to expand IEEE Std 802.16 from a fixed-only to a fixed-and-mobile system, beginning with the opening of the IEEE P802.16e PAR in late 2002. Within the IEEE 802.16e effort, it was soon realized that the OFDM PHY as defined did not provision sufficient UL link budget to move from a fixed deployment with relatively large SS antenna gain to a mobile platform with typically very low SS antenna gain, while maintaining a reasonable cell size. Subchannelization was a known tool to extend the link budget. While it made sense for mobile terminals, the working group began to accept the idea that it would be applicable to fixed terminals also.

Another important factor in introducing subchannelization into the IEEE 802.16 OFDM PHY was the ETSI BRAN HiperMAN project. ETSI BRAN HiperMAN, insisting on the need to minimize alternatives in the standard, had at one point reached agreement on a standard that was fully compatible with WirelessMAN-OFDM. However, as IEEE Std 802.16a concluded, ETSI BRAN HiperMAN continue to progress with the introduction and development of subchannelization, even though ETSI BRAN HiperMAN was not chartered to consider mobile operation. The desire to maintain compatibility with ETSI BRAN HiperMAN was one reason that the IEEE 802.16 Working Group sought to introduce the feature. In 2003, the amendment project IEEE P802.16d was converted into a full revision project. This expanded the scope of the work and allowed for the introduction of features like subchannelization.

After the idea of adding subchannelization gained acceptance, the main contention point was how many subchannels to use. Some sought small subchannels to maximize the link-budget gain, while others sought larger subchannels due to concerns with the minimal achievable throughput per subchannel, losses due to the scrambling and encoding blocks getting very small, and the general increase in receiver complexity.

Eventually, the number of subchannels was chosen to be 16, each consisting of 12 data subcarriers. Subchannels may be combined in a predetermined fashion so that an SS can be allocated all of the data subcarriers or one-half, one-fourth, one-eighth, or one-sixteenth of them. Only eight pilot subcarriers are available for all of the subchannels. Therefore, for allocations of one-sixteenth of the data subcarriers, phase estimation must rely solely on data-driven approaches. The subchannel structure is shown in Figure 11–3.

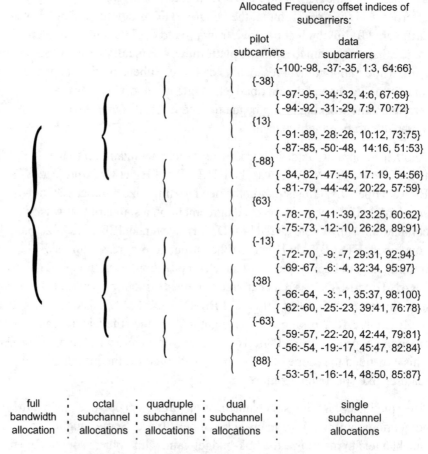

Figure 11–3:   Subchannel subcarrier allocations

The OFDM PHY can be applied to make use of the increased UL link budget provisioned by subchannelization, by limiting the maximum number of subchannels that are assigned in the UL to a given SS. This increased link budget is available from subchannelization because an SS can distribute its available power to fewer subcarriers, achieving a power gain per subcarrier (and hence link budget) of 3 dB every time the number of excited data subcarriers is halved. With a division into 16 subchannels, a total gain of 12 dB is hence possible (although implementation losses will somewhat reduce this gain). To exploit this opportunity, an SS with insufficient link budget to establish a link using full bandwidth can establish a link using less than the full bandwidth at the cost of peak throughput (but not at the cost of system capacity). To support this, special provisions allow efficient initial ranging and BW requesting for these cases. In order to make full use of this feature, however, sufficient link budget must be achieved on the DL, generally through to the use of a higher gain BS amplifier, to enable establishment of a connection. Especially in license-exempt bands, where allowed EIRP may be fairly restrictive, this may not be possible.

Aside from link range extension, subchannelization may additionally be used to exploit the lower overhead of the subchannelized preamble and to increase the modulation/coding efficiency. This may be easily understood from a simple example. A full bandwidth transmission of data filling x OFDM symbols requires $x + 1$ OFDM symbols (the extra OFDM symbol resulting from the preamble). The overhead ratio is hence $1/(x + 1)$. Transmitting that same data over eight subchannels (one-half the bandwidth) using the same data rate results in an overhead ratio of $1/(2x + 1)$ within those subchannels because the data now occupy eight subchannels of twice as many OFDM symbols, but the preamble only occupies eight subchannels of a single OFDM symbol. Because the remaining eight subchannels remain free for another SS, no loss of system capacity occurs (in fact, the reduced overhead slightly increases the system capacity). Because of the 3 dB gain achieved by using only half the subchannels, the SS may additionally be able to increase the data rate by increasing the modulation/coding efficiency, requiring less OFDM symbols (e.g., $1.6x$) to transmit the data. The total system resources used is then $(1 + 1.6x)/2 = 0.5 + 0.8x$, compared to the full-bandwidth $1 + x$.

The data subcarriers assigned to a single subchannel are distributed over the entire OFDM symbol, in four blocks of three subcarriers each. This spreading was introduced as a frequency diversity technique to reduce the suspectibility to partial band fading, whereas the grouping in blocks of three subcarriers results in slightly lower phase noise and frequency offset sensitivity. The allocation of blocks is such that combining subchannels still results in an evenly spread allocation of subcarriers. During a single OFDM symbol interval, it is possible to assign different numbers of subchannels to individual SSs as long as the frequency offset indices of the subchannels do not overlap.

One of the four major changes introduced in WirelessMAN-OFDM by the IEEE 802.16e amendment is the capability for provisioning subchannelization in the DL (the other three changes are the additions of private MAPs, private initial ranging opportunities, and open-loop power control). This DL subchannelization is provided as a separate zone, similar to the zone concept of the OFDMA PHY (see Chapter 12), with its own subchannelized preamble, FCH, and MAPs so that it is completely independent of the remainder of the frame. The DL subchannelization is restricted to the use of quadruple subchannel allocations (see Figure 11–3). The DL MAP in this construct provides allocations only for future frames, not for the current one, as it is assumed that mobile devices do not have the computational power necessary to decode this MAP quickly enough to efficiently avoid receiving data not intended for the station itself; this would unnecessarily drain the battery.

The addition of DL subchannelization is motivated by the desire for a larger coverage area for mobile devices, which have comparatively low antenna gain.

Preambles

The OFDM PHY specifies two 200-element sequences from which all preambles are derived. One sequence is used for all DL preambles and full-bandwidth UL preambles; the other is for UL allocations spanning less than the full channel. These two sequences are termed *mother sequences*. Computing an actual preamble from a mother sequence involves operations no more complex than multiplication with a constant, conjugation, and puncturing. This makes it practical to store only the two mother sequences.

All preambles are one OFDM symbol long, with exception of the DL preamble, initial ranging preambles, and mesh preamble, which consist of two OFDM symbols.

The first symbol of the two-symbol preambles, shown in Figure 11–4, consists of a cyclic prefix (CP) followed by four repetitions of a frequency domain derivative of the first mother code. Each repetition is said to be *length 64*, which means that it contains 64 samples and has a duration four times shorter than that of the 256-FFT OFDM symbol. This symbol primarily allows the receiver to perform large-gain adjustments of its amplifier, as well as coarse frequency and time correction. This is necessary on the DL to allow new SSs to properly track and adjust to the BS signal. On the UL, it is necessary for initial ranging, as the new SS's transmit gain and frequency/timing offsets have not yet been adjusted to within an acceptable range for the BS receiver. For other transmissions in the UL, the BS directs each SS to adjust its transmit power and frequency/timing offsets to within a desired range so that the preamble need not facilitate this functionality for these transmissions.

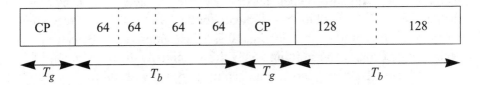

Figure 11–4: DL and network entry preamble structure

The second symbol of the long preambles as well as all of the one-symbol (short) preambles consists of a cyclic prefix followed by two repetitions of a length 128 frequency domain derivative of the mother codes, which is transmitted on the even subcarriers. Special cases exist for the DL STC-encoded region and for DL AAS transmissions. For the former, alternate subcarriers are transmitted from the two different antennas. For the latter, two identical "regular" OFDM symbols (each with cyclic prefix) are transmitted, where each beam (not each antenna) transmits on every fourth carrier. If a BS has more than four simultaneously active beams, the fifth beam transmits the same symbol as the first antenna, etc. This is possible because of the spatial separation.

For UL allocations smaller than the full bandwidth, subcarriers not part of the allocated subchannels are not excited when transmitting the preamble. If midambles are transmitted, they are identical to the short preamble or the second symbol of the long preamble that was used for the burst. Midambles are generally not necessary in fixed deployment, but the OFDM PHY provisions them to enable extension to mobile applications as defined in IEEE Std 802.16e. The purpose of the short preamble and the second symbol of the long preamble is to allow fine frequency/time estimation and channel estimation.

In the UL, SSs can be ordered through the UL MAP to cyclicly rotate the preamble by a given factor between 0 and 255. The idea behind this is to allow better simultaneous detection of preambles by AAS-enabled BSs. This is achieved by the concept that, if preambles are sufficiently rotated, correlation of the received signal with the nonrotated preamble and with the rotated preamble will result in two better estimates due to low cross-correlation of the preambles. Because this shift must be substantially larger than the main multipath components of any of the channels of involved SSs, the number of shift possibilities is limited. For the UL, the BS has a receiving beam in the direction of the individual SSs, which suppresses signals from other simultaneously transmitting SSs with substantially different angles of arrival (AoAs). However, this feature is only marginally useful for SSs that have relatively similar AoAs. Despite the fact that the cyclic shift method is a mandatory specification, it is of little use for non-AAS devices.

FRAME STRUCTURE

The OFDM PHY supports two noncompatible frame structures: a mandatory one for PMP architectures and an optional one for mesh-based architectures. Either frame structure allows for a number of frame durations between 2.5 ms and 20 ms.

Point-to-multipoint (PMP)

For PMP architectures, the duplexing method may be either TDD or FDD (with H-FDD SSs supported), except in license-exempt bands where only TDD is provisioned to ensure better coexistence with other IEEE 802 standards.

The TDD structure for PMP architectures is shown in Figure 11–5. The FDD structure is similar, except that TTG and RTG are omitted and the DL and UL subframes exist in parallel on different frequencies.

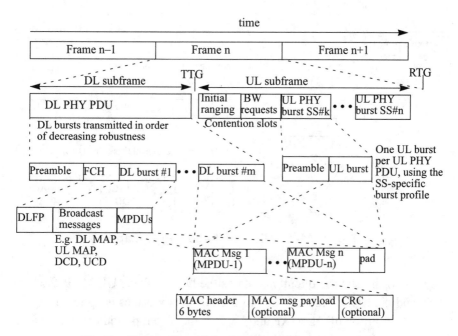

Figure 11–5: Example OFDM structure with TDD

The DL PHY PDU is initiated by the DL frame preamble, as previously described, followed by the FCH, which only contains the DLFP. The DLFP contains the lower 4 bits of the BS ID and the frame number (the full fields are present in the following DL MAP); this helps SSs with synchronization and identification of the right burst. The DLFP also contains a list of up to four DL bursts, indicating for each burst the DIUC (indicating, among other parameters, the modulation and coding methods), the existence of a preamble, and the length of each burst in OFDM symbols. For the first burst (Burst #0), the DIUC field is replaced by the Rate_ID, which indicates only the modulation and coding method. Lastly, the DLFP contains an HCS to identify errors in the FCH. It may also contain the start time of an STC-encoded region, if present, which lasts till the end of the DL frame. The FCH uses the most robust modulation method and coding (BPSK-1/2). The FCH hence allows the SS to individually demodulate each DL burst correctly.

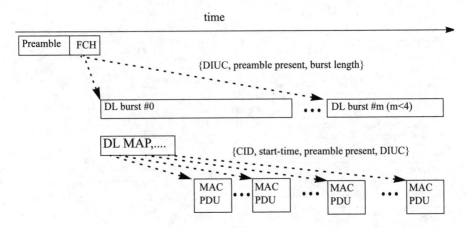

Figure 11–6: DL burst and MPDU mapping

The first DL burst subsequently contains the DL MAP, the UL MAP, the DCD, and the UCD. Transmission of each of the messages is optional (although insertion of DL MAP and DCD is mandated periodically to ensure that synchronization is maintained); but, when any of these four messages is present, the order is mandatory. The DL MAP may also contain markers to indicate that, from a certain starting point until the end of the DL subframes, STC, AAS, or DL subchannelization is enabled. These are the so-called *STC, AAS,* and *DL subchannelization zones*, respectively. If these zones are used to extend the cell range for devices, then all broadcast messages must be transmitted in these zones as well. Specifically, in the case of AAS, which has no broadcast PHY method, the FCH and unicast versions of the broadcast messages must be repeated on as many beams as necessary to reach all SSs. Similarly, in the UL, separate allocations must be inserted in these zones for the contention-based methods (ranging and BW requests).

The remainder of the first DL burst, as well as any subsequent DL burst within the frame, carries other MPDUs as indicated by the DL MAP.

The UL may contain contention slots for initial ranging and full and focused BW requests (explained later in this chapter). UL bursts may be received in parallel on nonoverlapping sets of subchannels (or on overlapping sets when exploiting the space division component through an antenna array). The main difference between the allocation mapping on the DL and UL is that on the DL, the duration of a burst is implicit from the start time of the next burst, whereas on the UL, the duration is explicitly stated. This was introduced to avoid the need to derive the duration from multiple parallel subchannelized subsequent allocations (i.e, the duration of an allocation could be derived from the earliest subsequent start time of any other allocation in any of the subchannels assigned to the allocation).

Mesh

For the mesh architecture, only TDD is specified. Each frame consists of two configurable fixed-duration subframes. The first period is the control subframe, which is used for network and schedule control, whereas the second is used for data transfer.

As illustrated in Figure 11–7, network and schedule control are fully separated, in that one out every few control subframes is dedicated to network control, whereas the remaining control subframes are dedicated to scheduling. Each control subframe consists of a number of transmission slots.

The first slot of a network control subframe is dedicated to contention-based network entry. The message used for network entry (MSH-NENT) is substantially shorter than the duration of the slot. This provides robustness with respect to timing errors due to unknown propagation durations, similar to the technique discussed later in this chapter for initial ranging (and is only used in PMP architectures). The remainder of the slot in the network control subframe is used for contentionless distribution of network configuration parameters (including neighbor lists) by the devices already part of the mesh network.

Three different messages are used in the scheduling control subframe, which manage the scheduling of bursts in the data subframe. The MSH-CSCF message is used to distribute the connectivity tree for centralized scheduling, whereas the MSH-CSCH contains the actual centralized transmission schedule. MSH-CSCF and MSH-CSCH messages are not bound by slot boundaries. MSH-DSCH is used to establish distributed schedules. MSH-DSCH may also occur during the data subframe to establish contention-based schedules.

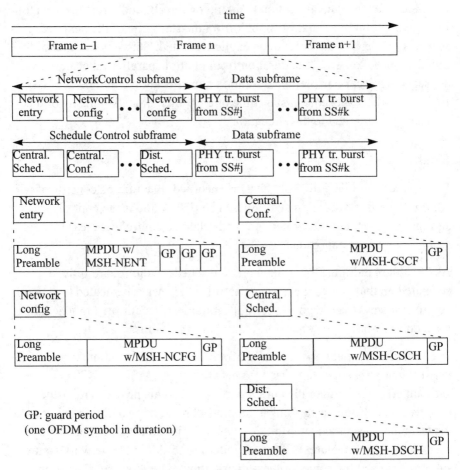

Figure 11–7:   Mesh frame structure

Each control message is terminated by a guard period. This is because, although all devices are synchronized, propagation durations will still vary between devices. The additional symbol avoids the end of a burst received over a long link suffering interference by the start of a subsequent burst received over a short link. Relying only on the cyclic prefix of the OFDM symbol for this would not have been sufficient.

Each of the (broadcast) control message formats, not only the network entry message, starts with a long preamble. This is because a device may, contention-free, conclude whether it is its time to transmit, but it cannot always determine who has the right to transmit otherwise. This is handled differently in the data subframe, where the transmission schedule is known. Although they use the long preamble by default, links can establish use of the short preamble in the data subframe.

The mesh architecture cannot make use of the subchannelization functionality. Although there is principally no obstruction to its use, the complexity of the scheduler would be prohibitive with current technology. The architecture does allow the use of patch antennas or antenna arrays, either with up to eight distinguishable directions. This allows for a substantially higher spacial reuse and hence substantial system capacity increases during the data subframe, at the cost of higher scheduling complexity. Extending link range through AAS as previously defined is, however, not possible due to the reliance on broadcasting for network control.

BW requesting is performed by use of the scheduling messages. The contention-based methods as described earlier are hence not used when using the mesh architecture.

CHANNEL ENCODING

The OFDM PHY channel encoding, which precedes the IFFT, consists of four steps; randomization, FEC, interleaving, and modulation.

Randomization

A specific sequence of data modulated on an OFDM symbol results in a given peak to average power ratio (PAPR) for that symbol. With a small probability of occurrence, this PAPR can be quite large (e.g., more than 7 dB). It is generally inefficient to back the transmitter power amplifier off far enough to avoid unrecoverable nonlinear distortion at the worst possible PAPR. Practical designs instead back off the amplifier only far enough to keep the probability of unrecoverable distortion acceptably low. However, retransmitting a data sequence with a high PAPR would result in a repeated receiver error. To avoid such persistent errors, the data are passed through a randomizer at the transmitter.

Randomization is performed by passing the block of data to be transmitted through a pseudo-random binary sequence (PRBS) with polynomial $1 + x^{14} + x^{15}$. For every data burst, the initial state of the PRBS memory is reset with a combination of the BS ID, the IUC for the allocation, and the frame number, except at the start of the frame, where it is reset with a fixed value, and on the first DL burst (i.e., the burst following the FCH), where it is not reset at all. Due to the fact that the FCH is fixed in duration, the state of the PRBS will always have the same value at the start of the first DL burst. Hence not resetting the PRBS at this point is equivalent to resetting it with a fixed value.

Forward error correction (FEC)

The OFDM PHY specified three methods of channel coding: Reed-Solomon concatenated with convolutional coding (RS-CC), block turbo coding (BTC), and convolutional turbo coding (CTC). RS-CC is mandatory, whereas BTC and CTC are optional. Although BTC and CTC generally provide on the order of 2 dB to 3 dB better coding gain, their substantial complexity made them less attractive as a mandatory requirement. The performance of BTC and CTC relative to each other varies slightly as a function of the coding rate and the size of the data to be encoded.

Reed-Solomon concatenated with convolutional coding (RS-CC)

The RS-CC encoding consist of a systemic RS(255,239,8) code over GF(2^8), which can be shortened and punctured to appropriate block sizes followed by a native rate 1/2 constraint-length 7 convolutional encoder (identical to the encoder defined in IEEE Std 802.11a), which can be punctured as well.

The randomized burst of data to be encoded is first appended by a byte of zeroes, as shown in Figure 11–8, in order to facilitate zero-tail termination of the CC encoder. Given the constraint length of 7, the usage of 6 tail bits would suffice (the last 2 bits will always remain zero after encoding). However, since data are generally passed to the PHY in whole bytes, the byte appendage is simpler. During development of the IEEE 802.16 standard, the use of tail-biting, which does not require the tail-byte overhead, in combination with per-OFDM symbol encoding, was favored. However, the disadvantage of having to process a substantial portion of the data twice was deemed to outweigh the incurred zero-tailing overhead. A separate RS block is computed for each OFDM symbol. The parity bits for each block are sent to the CC encoder before the block itself, ensuring that the added tail byte is passed last to the CC encoder. The CC encoder is employed over the entire burst. The RS encoder is bypassed entirely when less than the entire channel is allocated (i.e., when eight or fewer subchannels are allocated). This avoids performance deterioration due to the weakness of RS codes for small blocks.

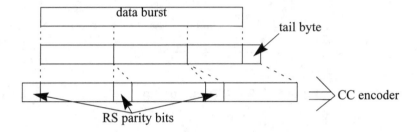

Figure 11–8: RS-CC encoding process

Block turbo codes (BTCs)

Unlike RS-CC, where the data are treated as a long string, BTC encodes the data in block form, where the block is filled row by row. Subsequently, a code polynomial is used to encode each row, with check bits added to the row. Subsequently, another code polynomial is used to encode each column, including the columns consisting of the row check bits. The polynomials are selected from a set of three extended Hamming codes and three parity check codes. The block (see Figure 11–9) may be shortened by not filling the first I_x rows, I_y columns, and B bits and zero-filling the Q bits until the next byte boundary. An encoding is hence fully described by its two constituent codes and the four parameters I_x, I_y, B, and Q. To specify blocks that fit into each of the five possible allocation sizes per OFDM symbol (see Figure 11–3), each of the four possible modulations and each of the two code rates per modulation—a total of thirty different encodings—are provided. The performance of BTC for an allocation of only a few subchannels was never thoroughly studied during the development of the subchannelization option. The performance may turn out to be problematic due to the small block sizes.

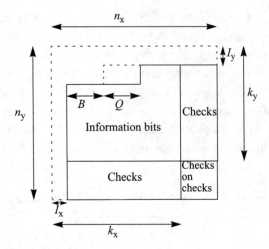

Figure 11–9: BTC encoding block

Convolutional turbo codes (CTCs)

The CTC encoder is most easily described from its depiction in Figure 11–10.

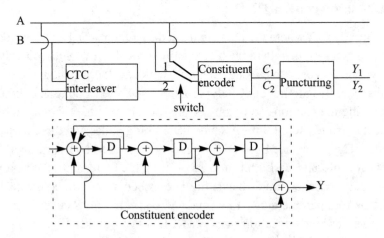

Figure 11–10: CTC encoder

Bits are alternately fed into A and B (MSB first), with the switch in position 1 as shown. The encoding process must be performed twice: one with the memory elements initialized to 0b000 with the switch in position 1 and one with the memory elements initialized with a value computed by passing the final value of the first run through a lookup table and the switch in position 2.

The final encoded stream is achieved by the sequence of alternated bits from A and B, followed by all punctured parity bits from the first encoding (switch position 1), followed by all punctured parity bits from the second encoding (switch position 2).

Interleaving

The interleaver is defined by a two-step permutation. The first ensures that adjacent coded bits are mapped onto nonadjacent subcarriers. The second permutation ensures that adjacent coded bits are mapped alternately onto less or more significant bits of the constellation, thus avoiding long runs of low-reliability bits. The interleaver is virtually identical to that used in IEEE Std 802.11a and IEEE Std 802.11g™, except that it uses 12 as computational base, rather than 16. The reason for this is that each subchannel has 12 data subcarriers. A base-12 permutation hence ensures that adjacent bits are mapped not only to different subcarriers but to different subchannels as well (if multiple subchannels are allocated).

Block sizes for the interleaver are defined so that the number of bits in a block fits into all allocated subchannels within an OFDM symbol. There is hence exactly one interleaving block per OFDM symbol for each transmitter.

Modulation

The OFDM PHY mandates BPSK as well as Gray-mapped 4-QAM (QPSK), 16-QAM, and 64-QAM constellations. The Gray-mapped 64-QAM constellation is optional for license-exempt bands, mostly to allow the use of IEEE 802.11 RF components that would typically not meet the 64-QAM performance requirements of IEEE Std 802.16.

Pilot tones are modulated with a BPSK signal. The values to be used are derived from passing fixed initialization sequences (one each for UL and DL) through a PRBS with polynomial $x^{11} + x^9 + 1$ clocked with the OFDM symbol rate. The values of the individual pilots within a single OFDM symbol are then defined as being either equal or the negative of the BPSK-modulated PRBS output.

CONTROL MECHANISMS

Of all control mechanisms that are available to the BS to control SS behavior, three have special provisions in the PHY: ranging, BW request, and power control.

Ranging

Ranging consists of two types: *initial* and *maintenance*. The first serves to initiate registration (or re-registration) of an SS with a BS. The second serves to periodically readjust PHY parameters of the SS and to verify that the SS is still on line.

Maintenance ranging is predominantly a MAC process, for which the PHY makes no special provisions in terms of scheduling or waveform creation. From the PHY's perspective, maintenance ranging is treated like any other data exchange.

Initial ranging is different in this respect in that it is, by necessity, a contention-based protocol with a nonzero probability of collision. The time period scheduled for initial ranging must be at least the total duration of the round-trip propagation delay and the transmission time of the initial ranging message, as a new SS is not yet able to correct its clock for propagation delays. The round-trip propagation delay is a function of the maximum cell size. An initial ranging attempt uses the long preamble. The initial ranging data fit in one OFDM data symbol (using the most robust modulation/coding) so that the total duration of an initial ranging allocation must be at least $3 + \lceil (n \cdot BW)/(256(1 + T_g/T_b)) \cdot (2r)/c \rceil$ OFDM symbols, where r is the cell radius in meters and c the speed of light. For typical deployments, this turns out to be a four-OFDM-symbol allocation. In fixed systems, contrary to mobile applications, registration is a relatively rare occurrence. It is, therefore, not necessary to facilitate initial ranging slots in every frame, and, as a result, the incurred overhead is negligible. The different initial ranging formats are shown in Figure 11–11.

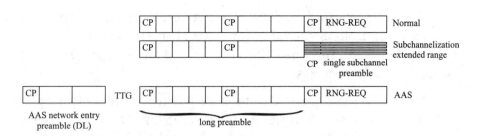

Figure 11-11: Initial ranging formats

As mentioned before, the PHY provides special provisions to allow SSs with insufficient link budget to use less than the full bandwidth. As a consequence of this design, a method is specified for these stations to start initial ranging with a slightly different method than the one described above. When an SS deduces, from estimating the required transmit power (for which the BS provides the necessary data in the UCD message), that it will not be able to close the link or when it has failed initial ranging attempts multiple times, it may replace the RNG-REQ message with the subchannelization preamble on one random subchannel so that the total initial ranging burst consists of the long preamble followed by the subchannelized preamble. The long preamble has sufficient inherent gain to be detected, whereas only an energy detect is needed to establish on which subchannel the subchannelization preamble was sent. Because the same total power is used, the spectral density of the single subchannel preamble will be roughly 12 dB higher, further improving the accuracy of the energy detect. Upon detection, the BS can issue a subchannelized UL allocation referencing the SS by the ranging slot and subchannel. For efficiency, SSs using the full bandwidth as well as those requiring a subchannelized initial ranging are using the same ranging opportunities. This avoids duplication of allocations.

An exception to this exists when the BS operates using AAS. Under the assumption that the cell range for AAS devices may be much larger than that of non-AAS devices, it cannot be assumed that AAS-enabled SSs can receive the broadcast DL MAP. They may, therefore, have no knowledge of the UL schedule, including initial ranging slots. To get around this problem, a BS must schedule an AAS initial ranging slot of eight OFDM symbols, the first symbol of which the BS uses to transmit the AAS network entry preamble, which has sufficient gain to be detected by AAS-enabled SSs. Following a TTG, an AAS-enabled SS may then transmit the long preamble and RNG-REQ message. The sum of SSRTG and round-trip delay may hence not exceed four OFDM symbols.

One of the few changes provided by the IEEE 802.16e amendment for the OFDM PHY is, as mentioned previously, the capability of the BS to allocate an SS-specific (and hence noncollision-based) initial ranging opportunity. This feature is useful when the BS receives notice (for example, from a network controller) that a MS will be attempting a handover to the BS in question. This shortens the handover time and avoids the possibility that many collision-based initial ranging opportunities need to be allocated because of handover frequency.

BW requests

There are multiple ways in which bandwidth can be requested: through polling, through piggybacking, and through two contention-based methods. From the PHY perspective, only the latter two are of interest. These two methods, of which only the first is mandatory, are *full contention* and *focused contention*.

Full contention may be used both with full bandwidth allocations and with subchannelized allocations. The only difference is that, using full bandwidth, the request is limited to two OFDM symbols to be filled with the short preamble and one data symbol containing the BW request, whereas using subchannelization, the width (in subchannels) and length (in OFDM symbols) is announced in the UCD. The BS may provide multiple request opportunities within a single allocation. The BS can identify the SS by the Basic CID in the BW request.

Focused contention is an optional feature. To request bandwidth using focused contention, an SS selects randomly a contention channel index out of the 48 available. This maps to four unique data subcarriers. These are arranged to form three sets of contention channels within each subchannel. On these four subcarriers, the SS transmits one of eight BPSK contention codes. On the following OFDM symbol, the SS again transmits the same contention code on the same subcarriers, with the difference that, for all subcarriers with contention element -1, the phase is inverted. The BS is now able to identify the SS by the contention channel index and contention code and can issue an UL allocation on that basis. The advantage of this scheme is that many SSs can simultaneously indicate a need for bandwidth with a low probability of collision (same contention channel) and an even lower probability of undetectable collisions (same contention channel and same contention code). A minor drawback of this scheme is that it introduces a single frame of delay, as the focused contention indication does not provide the BS with any knowledge about the size of the bandwidth needed. When a BS supports subchannelization, it may indicate through the UCD that a certain amount of the contention channel indices are reserved to indicate the SS needs a subchannelized allocation. In the case that it receives one of these codes, the BS does not know the identity of the requesting SS; therefore, it has no choice but to allocate the lowest number of subchannels required by any of its associated SSs.

Power control

Power control is needed and specified only in the UL, where it serves to ensure that the received signal power density at the BS falls within a narrow range. The SS transmit power level is adjusted under control of the BS. To allow this, each SS reports, during initialization, its current transmit power level and the maximum transmit power level for each of the modulations as part of its basic capabilities. This allows the BS to compute the feasibility of an SS changing its power level for full bandwidth allocations with the current modulation. It also allows the BS to compute the effects of a change in modulation and subchannelization. This can be used to improve bandwidth efficiency.

The BS can use both the maintenance ranging and fast power control MAC messages to force an SS to adjust its power levels. The SS must be capable of changing power at a rate of at least 30 dB/s.

An SS must maintain the same power spectral density when changing between transmissions with different numbers of subchannels. In other words, if the number of subchannels is double that of the previous allocation, the total transmit power, when feasible, must be doubled. Reciprocally, if the number of subchannels is halved compared to the previous allocation, the total transmit power, when feasible, must be halved.

Chapter 12 PHY: WirelessMAN-OFDMA

Multicarrier PHY for frequencies below 11 GHz

WirelessMAN-OFDMA is one of the three PHY specifications defined in IEEE Std 802.16 for applications below 11 GHz. This PHY has attracted much attraction in the past few years, particularly because of the expansion of the scope of the IEEE 802.16 standard, through the IEEE 802.16e amendment, to include mobile stations (MSs). IEEE Std 802.16e includes the expansion of all three of the lower frequency PHY specifications into the mobile arena (WirelessMAN-SC is excluded due the IEEE 802.16e upper limit of 6 GHz). However, the primary interest has centered around WirelessMAN-OFDMA, which is abbreviated as the "OFDMA PHY."

WirelessMAN-OFDMA uses orthogonal frequency-division multiple access (OFMDA), which is an extension of orthogonal frequency-division multiplexing (OFDM). The multiple access feature makes OFDMA considerably more complex. In OFDM, all of the transmitter's subcarriers are, at any given time, addressed to a single receiver; multiple access is provided solely by TDMA time slotting. OFDMA divides the subcarrier set into subsets, each of which can address a different receiver at any given time. This provides for a number of performance advantages, as discussed below, at the cost of considerable complexity. The complexity comes not only in the air interface, as defined in the IEEE 802.16 standard, but also in the control functionality outside the standard. For example, the design and optimization of an efficient scheduler will prove an interesting engineering challenge.

The description of WirelessMAN-OFDMA here is less comprehensive than those of the other PHYs. This is due to the large number of features and options present in the specification, as well as by the substantial changes and additions that have recently been added through IEEE Std 802.16e (which, as published, also includes an approved corrigendum to the base standard). This book is not intended to thoroughly cover IEEE Std 802.16e, but, on the other hand, neither do we wish to spend time explaining features that have changed significantly by virtue of that amendment. Therefore, while this chapter excludes a substantial number of features, mainly optional and mobility features added in IEEE Std 802.16e, care has been taken to ensure that the features described here are consistent with the changes in IEEE Std 802.16e.

INTRODUCTION

Selection of OFDMA waveform

As compared to the OFDM case discussed in the previous chapter, OFDMA involves different considerations regarding the complexity and implementation cost tradeoffs. Although WirelessMAN-OFDM has a limited capability for OFDMA through its subchannelization feature, its basic premise is TDMA with fairly large intercarrier spacing, which allows for an implementation hardly more challenging that the well-known Wi-Fi IEEE 802.11a or IEEE 802.11g OFDM products that have been readily available in the market since 2001. Those IEEE 802.11 amendments are based on a 64-FFT. Recall that WirelessMAN-OFDM uses a 256-FFT. WirelessMAN-OFDMA, as defined in IEEE Std 802.16-2004 [B20], uses a 2048-FFT. Therefore, for a given channel bandwidth, the OFDMA PHY presumes smaller intercarrier spacing and places more stringent demands on the hardware.

The issue of intercarrier spacing is made more complex by the decision, embedded in IEEE Std 802.16e, to move toward a "scalable" OFDMA providing for multiple FFT sizes: 2048, 1024, 512, and 128 (leaving out 256 presumably to maintain distinctiveness from WirelessMAN-OFDM). The idea of scalability is to match the FFT size to the channel bandwidth (i.e., larger FFT size with larger bandwidth), optimizing the subcarrier spacing. This idea of scalability was discussed earlier in the IEEE 802.16 Working Group, during the early Task Group 3 development of IEEE Std 802.16a-2003, but was rejected in order to simplify the interoperability problem. The IEEE 802.16 standard is rather ambiguous about channel bandwidth, as it is about carrier frequency, because of the lack of specific, harmonized worldwide spectrum in which to deploy.

In the remainder of this chapter, the 2048-FFT will be assumed, unless stated otherwise.

The OFDMA PHY using 2048-FFT is slightly more efficient than the OFDM PHY under the same conditions because the larger FFT size allows for a relatively smaller cyclic prefix (T_g/T_s).

The number of subchannels within the OFDMA PHY depends both on the FFT size and the time-frequency mapping (discussed below). As shown in Table 12–1, the number of data carriers per subchannel (per OFDM symbol) varies with the permutation method (discussed later in this chapter) and is 48 for full usage of subchannels (FUSC), optional full usage of subchannels (O-FUSC), and adaptive modulation and coding (AMC) and on average 24 and 16 for partial usage of subchannels (PUSC) and optional partial usage of subchannels (O-PUSC).

Table 12–1: Subchannel allocations

Method / FFT size	2048	1024	512	128
DL PUSC	60	30	15	3
UL PUSC	70	34	17	4
UL O-PUSC	96	48	24	6
DL FUSC	32	16	8	2
DL O-FUSC	32	16	8	2
AMC	32	16	8	2

As can be seen from Table 12–1, the power concentration that can be achieved, which is easily computed as 10 • log10(subchannels), varies widely with the time-frequency mapping and FFT size selected.

Time-frequency mapping

In the OFDMA PHY, the mapping of data to physical subcarriers occurs in two steps. In the first step, dictated by the scheduler, data are mapped to one or more data slots (in the time domain) on one or more logical subchannels. In the second step, for each data slot, each logical subchannel is then mapped to a number of physical subcarriers (in the frequency domain), with a mapping that may vary from one OFDM symbol to the next.

The first step, the logical mapping, makes use of certain basic terms in its description:

- *Slot:* The smallest unit of allocation in the time-frequency domain (except for mini-subchannels, which split a UL subchannel allocation further into two, three, or six pieces in the frequency domain). A slot always consists of a single subchannel, but, depending on the physical mapping, consists of one to three OFDM symbols.

- *Data region:* A contiguous allocation of slots in the time-frequency domain.

- *Subchannel group:* A single set of contiguous subchannels. The allocation of subchannels to subchannel groups is fixed.

- *Segment:* One or more subchannel groups that are controlled by a single instance of the BS MAC. As an example, a BS might segment the available subchannels among different sectors. There can be up to three segments where segment 0 must always contain subchannel group 0; segment 1, when present, must always contain subchannel group 2; and segment 2, when present, must always contain subchannel group 4.

Figure 12–1 displays these concepts in an example for 1024-FFT DL PUSC with three segments.

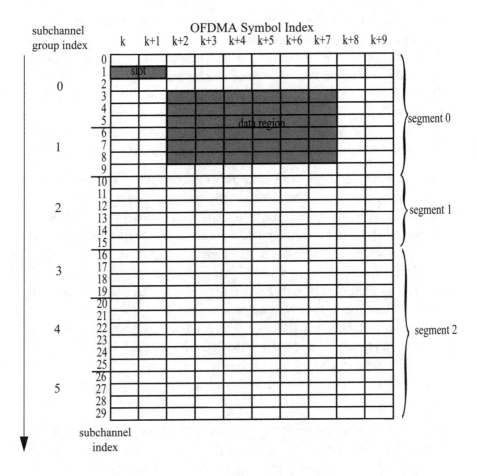

Figure 12–1:   Schematic indication of logical mapping elements

The second step, the mapping of logical subchannels to physical subcarriers, is termed *permutation*. The OFDMA PHY specifies seven permutation mechanisms: FUSC, PUSC, O-FUSC, O-PUSC, AMC, TUSC1, and TUSC2. A set of OFDM symbols over which the same permutation method is used (using more than one method in an OFDM symbol is not possible) is termed a *permutation zone*. A single frame may contain one or more permutation zones. These permutation alternatives are described below.

- **FUSC (DL only):** FUSC consists of a single segment containing all subchannel groups. Each subchannel is mapped onto physical subcarriers distributed throughout the entire physical channel as a function of IDCell (a number specific to the sector or cell). This mapping changes with each OFDM symbol. An example of this mapping is shown in the next section. Besides IDCell, which is used for the first DL zone in a frame, the BS may alter the permutations through the use of the PermutationBase parameter. A data slot consists of one subchannel over one OFDM symbol.

  The concept behind this method is that of interference mitigation through classical frequency hopping. Under light load, this allows the FEC to correct the few subcarriers on which a collision has occurred. As with classical frequency hopping, this scheme starts to lose its effectiveness if roughly one-third or more of the subchannels become occupied. The efficiency is hence roughly equal to that of system with frequency reuse of 3 (although some claim that it is comparable to a frequency reuse of 1), but the allocation of the instantaneous capacity (i.e., subchannels) to the individual sectors is more flexible. The disadvantage is that the BS needs to obtain a reasonable understanding (through network intelligence or channel measurements) of instantaneous load on other BSs, as throughput degradations are not a measure of only system load, but also of many other factors.

- **PUSC:** On the DL, each physical channel is split into clusters of fourteen subcarriers, which are then mapped to six major groups as a function of the IDCell. Subsequently, up to three segments are created out of these major groups. Within each group, pilots are allocated to each cluster in fixed locations that differ for odd and even symbols. The mapping of the remaining physical subcarriers within each major group to subchannels is similar to that of the FUSC case. A DL slot is defined as one subchannel over two OFDM symbols. On the UL, the subcarriers are split into groups of four consecutive physical subcarriers over a period of three OFDM symbols. Such a group is termed a *tile*. Six of those tiles are allocated to one subchannel, after which a UL slot is defined as one subchannel over that period of three OFDM symbols. The corners of each time-frequency tile are fixed for use as pilots. Except for subchannels allocated to UIUC 12 (CDMA BW request/CDMA ranging) and UIUC 13 (PAPR reduction allocation/safety zone), the mapping between subchannels and slots on the UL is rotated by a fixed number of slots every slot duration within each segment. A UL mini-subchannel is mapped to either one, two, or three tiles for the duration of the slot.

PUSC is in effect a compromise between the above, completely flexible subchannel allocation and the traditional frequency reuse schemes. It has as disadvantages both the load-balancing intelligence requirement and a somewhat large granularity in capacity allocation. This flexible method is hence valuable mostly when frequent, large shifts in capacity demand from one sector to the next are expected (which for WirelessMAN systems may not be typical) or when the deployment necessitates permanent unequal capacity per segment. PUSC is also claimed by some to be equivalent to a frequency reuse of 1, but segmenting the channel in three is similar in deployment to having a traditional reuse of 3 with one-third of the channel size. PUSC is, however, substantially advantageous over such a scheme due to the lack of need for guard bands between the segments.

- *O(ptional)-FUSC (DL only):* O-FUSC differs from FUSC solely in the manner by which pilots are allocated. Whereas FUSC's pilots are distributed in a pseudo-random way throughout the frame (after which the data subcarriers are pseudo-randomly allocated to subchannels), O-FUSC spaces the pilots always eight data subcarriers apart, rotating the locations three subcarriers for each OFDM symbol. After this, the data subcarriers are pseudo-randomly allocated to subchannels with the same algorithm used by FUSC. In synchronized networks, this setup has the effect of completely colliding the pilots when the same logical subchannel is used in nearby sectors.

- *O-PUSC (UL only):* O-PUSC differs from PUSC solely in the size of the tile and the allocation of pilots within it. PUSC in the UL has tiles sized four subcarriers by three OFDM symbols, where the four tile corners are used as pilots. O-PUSC instead uses tiles sized three subcarriers by three OFDM symbols, where only the center is allocated as a pilot.

- *AMC (AAS only):* The abbreviation AMC is not defined in the IEEE 802.16 standard, but it appears to have been derived from *adaptive modulation and coding.* It is also called *adjacent subcarrier permutation.* Using AMC, the entire channel is split in nonoverlapping groups of nine contiguous physical subcarriers. Subchannels are then mapped to a single such group over six contiguous OFDM symbols or to two such contiguous groups over three contiguous OFDM symbols. The middle subcarrier of each group is used as a pilot in each of the symbols. The allocation of physical subcarriers to these groups is the same on each OFDM symbol. Because AMC is used only with AAS, which inherently provides interference mitigation due to directionality, having such larger groups of carriers susceptible to group collision should not cause any degradation as compared to the use of smaller groups (where group collision would be substantially less recoverable by the FEC).

AMC allows for a somewhat simpler transceiver design than PUSC or FUSC. It is tailored specifically to AAS, where one can easily calculate a single weight per group, rather than having to compute individual weights for each carrier.

- **TUSC1 & TUSC2 (DL AAS only):** TUSC1 and TUSC2 (i.e., variations of tile usage of subchannels) are both optional and are similar to DL PUSC and O-PUSC, respectively. The sole difference is that pseudo-random allocation of subcarriers to subchannels follows a different equation.

The above permutation methods can in principle be combined at random into a single frame by indicating switches between permutation methods in the DL MAP (which, together with the FCH preceding, it is always transmitted with PUSC permutation).

Permutation examples

Because the above permutation mappings may be hard to visualize, a few example allocations are depicted. The allocation is derived for 128-FFT.

FUSC

The example given below assumes $IDCell = 0$, $PermutationBase_0 = [1\ 0]$, and $SymbolNumber = 0$.

1. Map physical carriers to logical data carriers by removing from the physical list all DC carriers, guard carriers, and pilots. Note that the pilots are specified relative to the first nonguard carrier. In Figure 12–2, the mapping of the first 40 physical carriers is shown.

2. Partition logical data carriers into subchannels using the *subcarrier(k,s)* formula. In Figure 12–2, the mapping is shown for the second subchannel ($s = 1$)

3. Map the logical data carriers k of the subchannels to physical carriers by combining step 1 and step 2. In Figure 12–2, the mapping of the logical indices k of subchannel 1 are shown for the first 40 physical carriers. It shows, for example, that physical subcarrier 12 is the 35th logical subcarrier of subchannel 1 because physical subcarrier 12 is logical subcarrier 0, which is the 35th element of *subcarrier(k,1)*.

4. Repeat step 1 through step 3 for the next OFDM symbol. The different allocation for *SymbolNumber* = 1 is shown as the last subplot in Figure 12–2, which is the result of the variable pilots being located in different places for odd and even symbols.

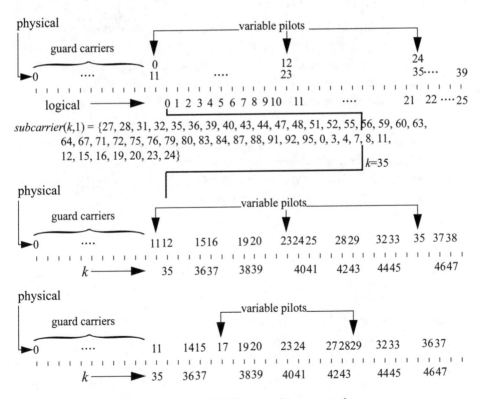

Figure 12–2: FUSC mapping example

PUSC, DL

The example given below assumes that this is the first DL zone and that there is only one segment. It also assumes that the *PermuationBase₃* = [0,1,2] since this parameter is not specified in the relevant table in the IEEE 802.16 standard.

1. Map the physical clusters consisting of 14 consecutive carriers to logical clusters, which are then assigned to the major groups and from there assigned to segments. The principle is shown in Figure 12–3. However, it is irrelevant to the subsequent mapping, as only one segment is used in this example containing all clusters.

2. Partition logical data carriers within the segment into subchannels, by use of the *subcarrier(k,s)* formula. In Figure 12–3, the mapping is shown for the second subchannel ($s = 1$).

3. Map the logical data carriers k of the subchannels to physical carriers by combining step 1 and step 2. In Figure 12–3, the mapping of the logical indices k of subchannel 1 is shown for the first 57 physical carriers. It shows, for example, that physical subcarrier 24 is the 11th logical subcarrier of subchannel 1 because physical subcarrier 24 is logical subcarrier 1, which is the 11th element of *subcarrier(k,1)*.

4. Repeat step 1 through step 3 for the next OFDM symbol. The different allocation for *SymbolNumber* = 1 is shown as the last subplot in Figure 12–3, which is the result of the variable pilots being located in different places for odd and even symbols

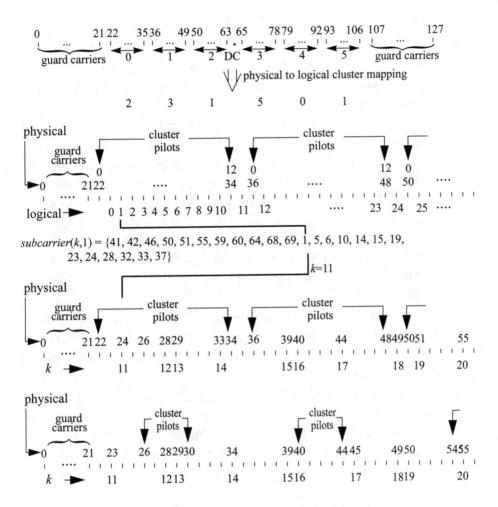

subcarrier(k,1) = {41, 42, 46, 50, 51, 55, 59, 60, 64, 68, 69, 1, 5, 6, 10, 14, 15, 19,
23, 24, 28, 32, 33, 37}

Figure 12–3: PUSC DL mapping example

PUSC, UL

The example given below assumes *UL_PermBase* = 0.

1. Map the physical tiles consisting of four consecutive carriers to per-subchannel logical tiles. The principle is shown in Figure 12–4, in which the carriers belonging to a single physical tile are stacked vertically. As shown, subchannel 1's six logical tile indexes 1 through 5 point to physical tiles 2, 4, 11, 13, 18, and 20, respectively.

2. Allocate the pilots within each tile and sequentially number the logical data carriers, starting with the lowest carrier in the lowest numbered logical tile in the earliest OFDM symbol. Number all carriers within that tile in the same OFDM symbol, and then do the same in the next to lowest logical tile. Repeat until all data carriers within that OFDM symbol for that subchannel have a logical number, and then repeat for the other two OFDM symbols in the slot. The resulting numbering is shown in the second subplot of Figure 12–4, where crossed-out locations are pilot locations.

3. The last step is the mapping of data symbols. Rather than mapping the first data modulation symbol to logical subcarrier 0, an offset of 13 times the subchannel number is applied. This data mapping rotation is shown in the third subplot of Figure 12–4.

4. Repeat step 1 through step 3 for the next slot. The different data mapping for *SymbolNumber* = {3,4,5} is shown as the last subplot in Figure 12–4. This is the sole result of the data subchannel rotation mechanism that must be applied on the UL, which in this case results in the previous subchannel 0 being renumbered subchannel 1.

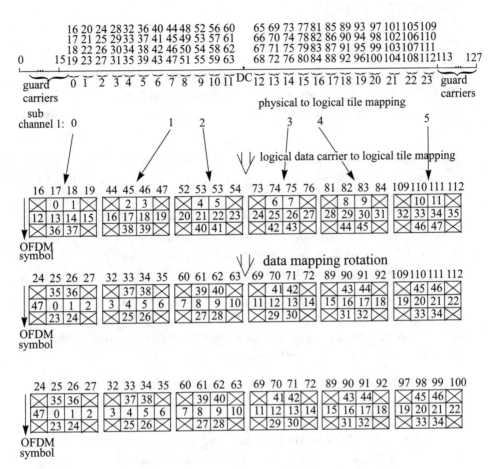

Figure 12–4: PUSC UL mapping example

FRAME STRUCTURE

Point-to-multipoint (PMP)

The only network configuration currently supported for the OFDMA PHY is PMP. It is, however, likely that the amendment project IEEE P802.16j, on mobile multihop relay (see Chapter 9), will target WirelessMAN-OFDMA only.

An example of the logical layout of a frame is shown in Figure 12–5, where the segment spans logical subchannels s through $s + L$. As in the OFDM PHY, the FCH, which has a fixed location and duration in the frame and contains only the DLFP, follows the DL preamble. The DLFP specifies the subchannel groups of which the segment is constructed, the length of the DL MAP, how much repetition coding is used on the DL MAP, and whether there is any change to the ranging allocation from the previous frame.

The MAPs in this example specify, among other things, that the frame spans two permutation zones on the DL, the first of which (PUSC) is mandatory. The MAPs also specify the location and content of all bursts (including burst profile; destination, if applicable; and so on).

As in the OFDM PHY, UL and DL bursts are separated by RTGs and TTGs to ensure that the transceiver at the BS has the time required to switch between receive and transmit. Furthermore (also as in the OFDM PHY), when establishing the MAPs, the BS must take into account the SSTTG and SSRTG, which is the minimum time in which an SS can switch between receive and transmit. Using Figure 12–5 as an example, the BS may be prohibited from providing an allocation to the station addressed in burst #6, as its SSTTG may exceed the BS's TTG. Similarly, the BS may be prohibited from addressing, in the next frame's DL, the station addressed in UL burst #1, as its SSRTG may exceed the BS's RTG, making it impossible for the SS to receive the necessary preamble of the following frame. Problems like these may result from the use of inexpensive SSs. The inability to address some SSs in a DL frame must be taken into account when scheduling broadcast messages.

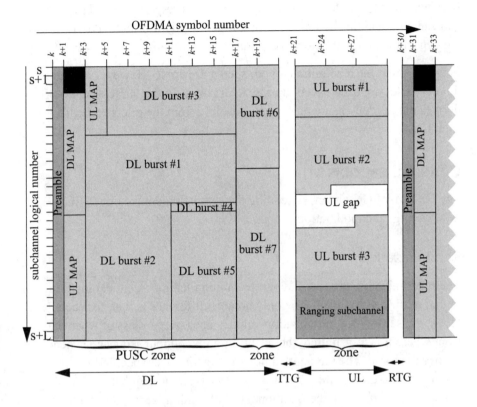

Figure 12–5:   Frame structure (TDD)

Preambles

In order to enable channel estimation at the receiver, two steps are taken:

- Preceding each burst with a preamble, as in the OFDM PHY
- Embedding pilots in the data stream, as in the DVB-RCT standard [B16]

In comparison to the OFDM PHY, therefore, the OFDMA PHY pilots are more abundant because they are not intended only for phase tracking.

To generate the preambles, the OFDMA PHY uses a series of 114 binary PN sequences of length 568. The sequence to be used is decided by the segment number and the *IDCell*. It is mapped to every third subcarrier except the DC carrier, starting with the subcarrier that equals the segment in number. In each segment, the SS can derive the complete channel response by interpolation and can augment this by updating the estimate with information derived from the constant and variable location pilots.

Each DL burst must be initiated with such a preamble. However, on the UL, no preambles are used except during AAS zones. In that direction, the BS derives the needed channel information solely from the pilots embedded in each tile.

CHANNEL ENCODING

The OFDMA PHY channel encoding consists of randomization, FEC, interleaving, and modulation.

Randomization

The randomization process is, apart from initialization, identical in definition to that of the OFDM PHY. The only notable difference is that, for each FEC block, the same fixed initialization value is used. Regardless of when and from where it is transmitted, the result for a given input sequence is hence always the same. The remnant value of the randomizer specified in this fashion is, therefore, limited. It is, however, worth noting that the PAPR concerns that stimulate the traditional use of randomizers is substantially mitigated in the OFDMA PHY due to the logical to physical subcarrier permutations and the different time-frequency allocation that may be used when retransmitting the data. Also note that a fixed randomizer seed is necessary with the use of HARQ to allow prerandomizer combining. When using chase combining, the transmitted bursts are hence necessarily identical. When incremental redundancy is used, the transmitted bursts are naturally different.

Forward error correction (FEC)

The OFDMA PHY specifies five methods of channel coding: convolutional coding (CC) with tail-biting, block turbo coding (BTC), convolutional turbo coding (CTC), low-density parity check coding (LDPCC), and CC with zero-tailing. Tail-biting CC is mandatory; the others are optional.

Convolutional coding (CC)

The CC encoder is identical to that used in the OFDM PHY and the IEEE 802.11a amendment. In the optional zero-tailing case, a single byte 0x00 is appended before encoding, as in the OFDM PHY. In the mandatory tail-biting case, the input data are used cyclically in that the first 6 data bits are fed into the encoder a second time to compute the last 6 output bits. This practice avoids the 1 byte of tail overhead employed by optional zero-tailing; however, the decoder needs to compute a substantial portion of the received sequence twice, mandating a faster decoder. This is an unusual example in that the optional specification is less complex than the mandatory implementation, but requires (slightly more) overhead. The motivation to specify this option is not entirely obvious.

In the IEEE 802.16a specification, the mandatory implementation required concatenation of RS-CC as in WirelessMAN-OFDM. However, it requires the use of tail-biting, rather than the zero-tailing used in the OFDM PHY. The small blocks into which the IEEE 802.16e amendment evolved the OFDMA specification made this concatenation less attractive due to the poor performance of small RS code words.

BTCs and CTCs

The BTC and CTC specifications in the OFDMA PHY are very similar to those in the OFDM PHY, with the minor differences stemming from the need to enable a finer granularity of blocks.

LDPCC

The low-density parity check (LDPC) is a coding method that has received substantial attention in the past few years, both academically and in industry. It is no surprise then that LDPC codes were also introduced into the OFDMA PHY through the IEEE 802.16e amendment. Some industry groups, however, appear to be leaning toward the CTC encoder as the primary optional codec.

The LDPC encoding process is the equivalent of the encoding of a data vector of size $n - m$ so that parity check matrix H multiplied by the resultant code word equals an all-zero vector. H is of size $m \times n$, where n is the length of the code and m indicates the number of parity bits. This results in a code rate of m/n. The matrix H is constructed out of submatrices of size $(n/24) \times (n/24)$, each of which is identified by a single rotation index. A value of -1 indicates an all-zero submatrix, whereas any larger value indicates an identity matrix right-rotated by the indicated number of times equal to the index. The IEEE 802.16 standard hence provides specifications of the H matrix by means of the reduced size $(m/24) \times 24$ rotation index matrix, also commonly termed the *shift index matrix*.

Specifically, the standard indicates the shift index matrix for the largest possible block size $n = 2304$. The shift index matrix for smaller block sizes $n^\dagger$ is obtained by replacing each non-negative rotation index $p(i,j)$ by $\lfloor p(i,j) \cdot n^\dagger/n \rfloor$ for all but one particular index matrix (for rate 2/3, code A) where $\mathrm{mod}(p(i,j), (n^\dagger/24))$ is used. The standard specifies five index matrices; one for rate 1/2 and two each for rate 2/3 and rate 3/4. The required dual implementation of index matrices for a particular rate has no technical advantages; it is purely a result of the political process.

The shift index matrix has the general structure shown in Figure 12–6.

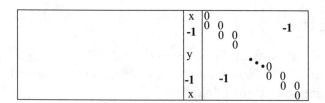

Figure 12–6:   General shift index matrix structure

Whereas the left side of the matrix may have almost any entries (subject only to column and row weight preferences), the remaining portions are strictly structured. The right side of the matrix consists of a so-called *dual diagonal staircase construction*, where the 0 indexes (which indicate identity submatrices) and the "-1" indexes (which indicate all-zero matrices, as previously explained) create an all-zero section in the H with two all-unity off-diagonals. The column in the shift index matrix preceding this section is defined so that the first and last indices (indicated by x) are identical, with only one non-"–1" entry (indicated by y) in the remainder of the column. This construction results in the last column in the H identified by this column in the shift matrix being of weight three.

The purpose of this strict general definition of the shift matrix (and hence of the H matrix) is that, for encoding schemes based on the general Richardson-Urbanke method (some specific implementations of which are explained as informative in the IEEE 802.16e amendment), a matrix that requires inversion results in the identity, making the inversion unnecessary.

Interleaving

The mandatory interleaver follows the IEEE 802.11a/g interleaver, with the identical base 16 (unlike WirelessMAN-OFDM, which uses base 12).

Because, as explained in the previous chapter, the interleaver's first step is intended to map adjacent coded bits to nonadjacent subcarriers, this does create an additional permutation on top of the logical to physical carrier mapping. In the late stages of the development of the IEEE 802.16e amendment, it was discovered (see [B12]) that the net effect of this combination of permutations could result in a performance loss as high as 5 dB. To resolve this problem, an optional interleaver was added. It has the same structure, but it uses as its base 16 times the number of slots per FEC block.

Modulation

As for the WirelessMAN-OFDM PHY, the modulation of the OFDMA PHY consists of square BPSK through 64-QAM constellations for all data. Exceptions exist for special channel feedback messages, which are encoded with various forms of circular constellations.

CONTROL MECHANISMS

Fast-feedback

The OFDMA defines two fast-feedback methods for the UL. These can be used to relate time-critical PHY parameters. Parameters for which this feedback is defined are SNR, specific MIMO coefficients for two antennas, and MIMO configuration parameters [space-time transmit diversity (STTD) vs spatial multiplexing, open- vs closed-loop spatial multiplexing, the permutation type, antenna selection, and precoding matrix selection]. Fast-feedback subchannels can be allocated by the use of a special MAC subheader or by the use of various DL MAP IEs.

The data to be transmitted within the fast-feedback channel are not encapsulated in an ordinary PDU message, but are instead in an orthogonal encoding of only a few bits of data using QPSK modulation. Each of the possible bit combinations is mapped to a sequence with one octal value per tile in the subchannel. This octal value represents a QPSK sequence for each of the eight data carriers per tile.

The specification allows for both full- and half-subchannel fast-feedback channels. The full subchannel can transfer 4 bits or 6 bits, depending on whether the fast-feedback is regular or enhanced. The half subchannel transfers 3 bits of data per slot. Although the IEEE 802.16 standard does not explicitly state so, the orthogonality of this setup allows for multiple SSs to transfer parameters during the same slot.

CDMA ranging and BW requests

To allow for initial ranging, maintenance ranging, handover ranging, and BW requests, the OFDMA PHY provides a collision-based method using PN sequences. The mandatory ranging allocation (called *ranging channel*) consists of six (and optionally eight) contiguous subchannels. Multiple channels may be allocated contiguously as well. The ranging and bandwidth channel are specified in the UL MAP.

Note that the specification of a minimum of six subchannels for initial ranging makes operation of this PHY with 128-FFT currently impossible, as, for 128-FFT, only four subchannels are defined in the uplink.

The initial ranging and handover ranging transmission may, on instruction from the BS, span either two or four OFDM symbols. When the two-symbol type is to be used, the transmission follows the depiction of Figure 12–7. This depicts the dual transmission of the same BPSK-modulated PN sequence, with the cyclic prefix (CP) and cyclic postfix replacement as indicated.

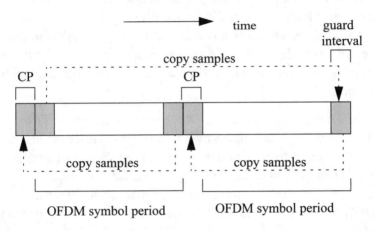

Figure 12–7: Single-slot initial ranging transmission for OFDMA

The transmission of the PN sequence in this fashion results in a phase-contiguous symbol and provides an addition gain on top of the spreading gain due to the repetition. To provide even more gain, the BS can instruct the use of the four-symbol alternative, in which the SS transmits two of the above symbols built from consecutive PN sequences out of the list of available sequences (in which the first PN sequence must be an even-numbered choice). Due to the spreading gain, this method can resolve multiple initial ranging transmissions simultaneously and is hence called *CDMA ranging*. In contrast to traditional CDMA, the PN sequences are modulated on the carriers in the frequency domain, rather that transmitted sequentially in the time domain.

For periodic (maintenance) ranging and BW requests, the same ranging channel is used, but the symbols are transmitted with the regular cyclic prefix instead of by the method shown in Figure 12–7. For these two methods, the transmission may be one or three OFDM symbols in duration (remember that a UL slot is three OFDM symbols in duration), but in the case of BW requests, three consecutive PN sequences must be used.

All PN sequences are created by a PRBS with generator polynomial $1 + x + x^4 + x^7 + x^{15}$ and an initial state that is partially fixed and partially consists of the IDcell so that different BSs can distinguish between the sequences. This PRBS results in 256 unique PN sequences, each 144 bits in length, a consecutive set of which is allocated to a BS through configuration. The BS can then divide its allocated sequences among the four functions (initial ranging, periodic ranging, BW requests, and handover ranging) by means of the UCD message.

Because the sequences for these methods are known, they can be used not only to identify the transmitting SS by the received sequence but also as a preamble, allowing extraction of (among other things) channel response for the used subchannels and timing and frequency offset information. Specifically, in response to an initial ranging sequence, the BS can immediately provide power, timing, and frequency offset corrections in its response.

PAPR reduction/safety zone

As discussed previously, a PAPR that exceeds the backoff of the amplifier results in nonlinear distortion at the transmitter that is unrecoverable at the receiver. Typically, this is tolerated by randomizing the data and backing the amplifier off until the error probabilities are tolerable. WirelessMAN-OFDMA also provides another method to minimize PAPR and allow for reduced backoff (and, therefore, higher average power or less expensive amplifiers). This method consists of allocating of subchannels, both in the DL and UL, in which the BS and SSs, respectively, can modulate arbitrary data to reduce the overall PAPR of their transmissions. These subchannels are filled with, in effect, random energy that can be freely ignored by the receiver. Naturally, this method comes with a cost: the transmitting party must first pass the data to be transmitted through the IFFT, compute the PAPR, and determine additional data that, when transmitted in the PAPR reduction subchannel, will ensure a low overall PAPR.

Note that the specification does not preclude the use of other PAPR reduction mechanisms, such as windowing and time-domain peak cancellation (see, for example, [B55]) as long as the transmitter relative constellation error does not exceed the required values.

The BS may also provide safety zones. These are gaps in the UL and DL schedule. To be used, they must be explicitly indicated because the MAPs specify only the duration of an allocation, with the remaining positioning information being implicit from the sequence of allocation. Gaps can be specified in coordination with other BSs to reduce interference on the indicated subchannels for the other BS or BSs.

HARQ support

To support HARQ feedback, the BS may allocate one or multiple pairs of UL acknowledgment channels, each of which consists of the even- or odd-numbered tiles that make up half a subchannel. As with the fast-feedback channel, data are QPSK-modulated directly on the channel, without packetization. The data consist of only a NACK or ACK bit, each of which is expanded to a specific orthogonal sequence to fill the three tiles. The location in the acknowledgment channels in which an HARQ–enabled SS returns its positive or negative acknowledgment is determined by the number of preceding SSs in the DL MAP that are to return such a signal.

HARQ is defined to support chase combining (implying simple retransmission of the data) for all coding methods as well as incremental redundancy methods for both the CC and CTC codec. For CC, the incremental redundancy is achieved through retransmission of the data using a different puncturing sequence.

Chapter 13 **Multiple antenna systems**

Support for advanced antennas

For operations below 11 GHz, the IEEE 802.16 standard provides support for advanced multi-element antenna systems, including adaptive antenna systems (AAS), space-time coding (STC), and multiple-input, multiple-output (MIMO) systems. These techniques are not supported above 11 GHz as they exploit spatial diversity not substantially present above 11 GHz. This chapter provides an overview of the IEEE 802.16 MAC and PHY support for these multiple antenna systems. This chapter subsequently discusses AAS, open-loop transmit diversity methods (i.e., all methods not requiring feedback), and closed-loop transmit diversity methods (i.e., all methods requiring feedback). AAS is also a closed-loop method, but it is described separately in this chapter, due to the fact that its use has, unlike all other methods, systemwide consequences.

The descriptions in this chapter refer to specific numbers of antenna elements, but should not be construed as limiting the number of physical antennas a system may have. A system may use more than the specified number of physical antennas as long as the resultant signal waveform meets the standard's specification. It is, for example, possible for a compliant system to use cyclic delay diversity (CDD) with multiple transmit antennas, resulting in a signal comparable to a single antenna's signal passed through a multipath channel. On the receive side, the number of antennas beyond the minimum (for example, four in the case of 4x4 MIMO) is completely transparent to the air interface and hence permissible as well.

ADAPTIVE ANTENNA SYSTEMS (AAS)

AAS technology, popularly known as a smart antennas (see [B2], [B32], and [B37]), accommodates the spatially diverse location of energy sources by discriminating signals based on their angles of arrival (AoAs) and angles of departure (AoDs) at an antenna array. In very general terms, the discrimination of signals, which is outside the scope of the IEEE 802.16 standard, is achieved by computing a complex weighted sum of the signals of all antenna elements, where the weights may be computed according to various metrics.

Through the AAS options, the IEEE 802.16 standard supports the use of smart antennas to perform beam forming. Beam forming can effectively create a narrower signal beam, resulting in increased gain and, therefore, higher range. This in turn increases capacity by increasing the range at which a particular PHY burst profile can be received. In addition, if a BS is capable of multiple simultaneous beams, capacity can be increased through spatial diversity. Theoretically, the spectral efficiency can be increased linearly with the number of antenna elements at the BS. This can lead to an in-cell frequency reuse proportional to the number of antennas.

AAS also allows for the suppression of noise sources, improving the SNR at the receiver; and discrimination on the AoD allows energy to be concentrated in the direction of the intended recipient, enabling larger cell ranges. In addition, nulls can be steered in particular directions, enhancing the interference resistance of the system. An example beam-forming network is shown in Figure 13–1. Obviously, beam forming toward mobile devices adds challenges.

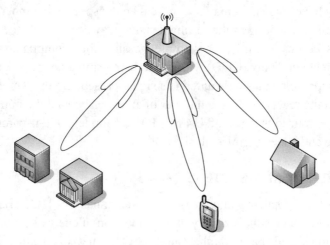

Figure 13–1:   Example beam-forming network

Drawbacks of these approaches include the increased system complexity and the inability to broadcast messages, reducing the spectral efficiency due to repetition of broadcast MAC messages to the various recipients.

AAS support in IEEE Std 802.16

While the AAS option requires some AAS-specific capabilities in an SS, the actual beam forming is defined only at the BS side of the link. Therefore, AAS systems in the context of the IEEE 802.16 standard are discussed purely as a BS feature, with both DL and UL AAS components. This limits the complexity of the AAS requirements in the SS to such functions as understanding the AAS zones, implementing AAS-specific preambles and permutations, and using AAS-specific messaging such as private DL MAPs and UL MAPs. At the BS, the algorithms and hardware designs necessary for AAS support are implementation-specific and left to vendor differentiation.

An AAS-enabled SS is an SS capable of establishing and maintaining connectivity with a BS that is actively using an antenna array. This predominantly requires synchronization and MAC capabilities. It does not pose any requirements on the number of SS antenna elements. An SS may also employ an antenna array, and could do so transparently from the standard's perspective, but SS arrays are unlikely due to complexity and size constraints. If AAS is used to extend the range of the cell, both the SS and BS need to be prepared to take additional steps to allow the SS to enter the network. The frame start preamble is a repetitive, well-known pattern. The SS may take advantage of processing gain associated with that preamble to synchronize with the DL in time and frequency, even if it is receiving the DL too weakly to decode the DL MAP and DCD messages.

The AAS option supported by IEEE Std 802.16 is defined so that a single BS can support both AAS-enabled and non-AAS SSs simultaneously. The key difference between pure PMP and AAS is that some AAS SSs may be outside the coverage area of some beams of an AAS BS. The basic framing needs to be modified to accommodate this; a single broadcast cannot be used to communicate information to all AAS SSs. For an IEEE 802.16 system, this results in a challenge, as IEEE Std 802.16 relies on broadcast data (UCD, DCD, DL MAP, UL MAP) to communicate the transmission schedules and to schedule contention slots. These messages must either be repeated to each individual SS or repeated to SS clusters in the same general direction, although the MAPs need contain only the schedule elements relevant for the targeted SS(s).

As AAS is optional, the BS must always support non-AAS SSs and, therefore, support broadcast transmissions. The coexistence of AAS and non-AAS SSs is achieved by partitioning the IEEE 802.16 frame in time into AAS and non-AAS zones, where the AAS zone is dedicated to AAS-capable SSs. In the AAS zone, AAS-capable SSs receive or transmit, knowing that the BS is directing a beam in their direction. This partitioning can be adjusted dynamically by the BS, as with dynamically asymmetric TDD. The IEEE 802.16 MAC and PHY also support additional functions necessary to support and manage AAS SSs, such as the ability to dynamically choose the best "beams" for each AAS SS and to support ranging of AAS-capable SSs. Since the AAS zone of the frame is explicitly indicated in the broadcast MAP by the BS, non-AAS SSs would not be confused by AAS-specific messages sent in the AAS zone.

We noted above that AAS in the IEEE 802.16 context has both a DL and UL component and provides range extension. However, the structure of the UL MAP allows for scheduling simultaneous or time-overlapping transmissions on the UL, as each scheduled transmission has both a start time and duration. Therefore, a BS may exploit this property to achieve spatial division multiple access (SDMA) transmissions in the UL. This approach can be used to substantially increase the UL capacity, with or without an increase in the range. Strictly speaking, these simultaneous transmissions can be achieved with non-AAS SSs, as it does not require the use of AAS zones or any other AAS-specific messages. On the DL, however, this is not possible, as each allocation implicitly ends with the start of the next allocation.

DL and UL framing

Figure 13–2 shows the TDD framing with AAS support, with both non-AAS and AAS zones. In the OFDM case, the AAS DL zone starts with an AAS-specific preamble; and, in the UL, all SSs require a separate preamble. Also, again in the OFDM case, the allocation start time in the UL MAP transmitted in an AAS zone is always relative to the start of that AAS zone.

It should be noted that the UL MAPs transmitted in the mandatory broadcast and in the AAS zone may overlap. Whenever such overlap occurs, the most recently received UL MAP with overlapping IE takes precedence.

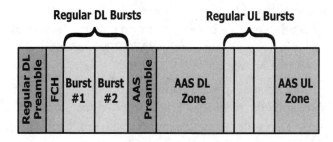

Figure 13–2: TDD framing for AAS Systems

The partitioning concept is very similar for FDD systems. Figure 13–3 shows the FDD framing of AAS systems.

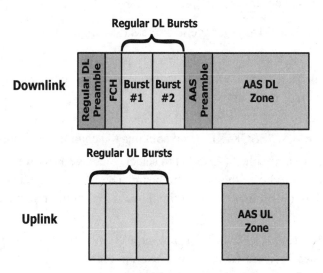

Figure 13–3: FDD framing for AAS systems

MAC service and control functions

The IEEE 802.16 MAC supports the following functions for AAS-capable SSs:

- The ability to transmit private management messages (such as UL MAP, DL MAP, UCD, and DCD) that may be sent directly to the AAS SS Basic CID, instead of to the Broadcast CID as in generic PMP. It is still possible for an AAS SS to receive broadcast information in certain deployments. During the initial ranging process, the AAS SS communicates to the BS its ability to receive broadcasts.

- Dynamic allocation of specific regions within the frame for initial ranging and BW requests for AAS SSs. The initial AAS ranging is achieved using a separate CID in order to avoid conflicts with the regular Initial Ranging CID used by non-AAS SSs.

- A set of AAS MAC management messages, such as those required for obtaining feedback from an AAS SS for channel estimation and for selecting the most optimal beam for an AAS SS.

AAS MAC management messages

The following AAS MAC management messages are defined:

- The AAS-FBCK-REQ (AAS feedback request) message is used to request channel measurement for a specific frequency. This may be used to adjust the direction of the adaptive array.

- The AAS-FBCK-RSP (AAS feedback response) message is sent in response to AAS-FBCK-REQ and contains the channel measurements.

- The AAS_Beam_Select (AAS beam select) message is used by an AAS-capable SS to indicate the preferred beam. The BS may use this information in deciding on the optimal beam for communication with this SS.

- The AAS_BEAM_REQ (AAS beam request) message is used to request channel measurement for a specific beam in order to adjust the adaptive array at the BS.

- The AAS_BEAM_RSP (AAS beam response) message is sent by an SS in response to an AAS_BEAM_REQ.

Channel state information

The adaptive array operation depends on the estimation of the channel state in both directions. The BS can estimate the UL channel state directly. In TDD systems, the BS may use this UL channel information to estimate the DL channel. However, for FDD systems this method is not accurate due to the different frequencies used for UL and DL. Therefore, for the FDD case, an AAS-FBCK-REQ message is used by the BS to instruct the SS to measure and send channel state estimation using an AAS-FBCK-RSP message. The BS may automatically allocate UL slots for transmitting AAS-FBCK-RSP messages after the measurement period.

AAS DL synchronization and initial ranging

Since a new SS trying to join the system may lie outside all beams, special measures are required to support DL synchronization of AAS SSs. Some SSs with high processing gain may be able to decode the broadcast DL frame boundaries for timing and frequency synchronization. The BS may also periodically perform active scanning by aiming at different directions or may use some PHY-specific diversity methods to support DL synchronization of such SSs.

An AAS-enabled SS that is outside broadcast range and wants to perform initial ranging has no information on the BS. It must, therefore, wait until it receives the AAS network entry preamble, which provides substantial processing gain, and respond immediately. The cell range in which SSs can join the BS is, therefore, limited by the directional gain of the AAS, which is related to the number of antenna elements, by the broadcast link budget, and by the processing gain of the AAS network entry preamble.

As with DL synchronization, some AAS stations may be able to decode MAP and channel descriptor information from the broadcast frame. If an SS is able to reliably decode this information, the initial ranging process can be performed as in the non-AAS case. Adjustments to the adaptive arrays can be made by the BS during this process. In order to support SSs that cannot decode the broadcast portion of the frame, a BS may use PHY-specific predefined AAS contention regions for the various general directions in order to allow AAS SSs to alert the BS of their presence. This approach has substantially higher overhead.

The SS always attempts initial ranging by trying to decode the broadcast DL MAPs and DCD. If it is unable to decode that broadcast, it tries the AAS slots. These AAS alert slots are similar to the regular contention slots for initial ranging. However, an AAS SS uses all contention slots, as opposed to just one slot used by a regular SS, so that the BS has sufficient time and processing gain to aim the beam at the SS's direction. If the BS is able to receive the alert message with sufficient processing gain, it may focus a beam on the SS and transmit the MAPs and channel descriptors. If the SS receives the DL parameters, it continues with the rest of the network entry process. If no DL MAP and DCD are received, the AAS SS performs an exponential backoff identical to the non-AAS case to select the next frame to send the alert.

AAS BW requests

An SS that wants to request bandwidth may lie outside a beam. When an SS performs ranging, its ability to receive broadcast allocations is communicated in the RNG-REQ message. The BS can instruct the SS to use broadcast allocations for BW requests, just like non-AAS SSs. If the SS is not capable of receiving broadcast or if the BS does not want the SS to use broadcast allocation, the BS is responsible for polling in the general direction of the SS to allow BW requests. The BS may modify the BW request method dynamically by instructing the SS in a RNG-RSP.

AAS support in OFDM PHY

Similar to DL subchannelization, AAS is supported in a zone logically separated from the remainder of the frame. Within the AAS zone, the BS can create what can be conceptualized as parallel instances of the PHY and lower MAC, each active on one or more beams in both SDM(A) and TDM(A) fashion. Each instance has its own preambles, FCH (termed *AAS-FCH*), and MAPs, with periodic use of a dedicated UCD and DCD. The IEEE 802.16e addition of private MAPs is particularly useful for this feature. Within each AAS-FCH, the BS transmits an AAS-DLFP that is different from the regular DLFP in that some of the functionality of the MAPs, such as provisioning UL allocations and contention-based opportunities, is embedded in it.

Two MAC management messages are specified to support the AAS PHY. The first allows the exchange of channel feedback information, which consists of CINR, RSSI, and the complex channel coefficient for a regularly spaced (on either side of the DC carrier) set of carriers with intervals of 4, 8, 16, or 32 carriers, depending on the request. The measurements are to be performed on a specific indicated burst. The second message is a similar exchange of channel information except that measurement information is requested and provided on one or more beams

For initial ranging, two separate messages are provided, one to be used with full channel AAS initial ranging opportunities and one to be used with subchannelized AAS initial ranging opportunities. Both messages violate the IEEE 802.16 MAC architecture in that they have no MAC header. It is unclear whether this omission is by design or by error; but, because they are the only responses allowed in the respective allocation, it is not a fatal violation.

The full channel initial ranging message allows the SS to provide the BS with measured amplitude values of up to four beams, providing the BS with in initial indication of which beam may best be used for the SS in question. The subchannelized initial ranging message, which may be used only if full channel initial ranging repeatedly failed, must be filled with the measured phase offset of three beams relative to a first beam (i.e., four beams total), providing the BS a measure of directionality for the SS and the average RSSI over those beams. In both messages, the SS sets a randomly selected network entry code, which the BS will use in its response to identify the SS.

AAS support in OFDMA PHY

AAS is supported by the OFDMA PHY for FFT sizes larger than 128 by means of the creation of an AAS diversity MAP zone (DMZ). The AAS DMZ can be allocated only on a specific set of subchannels of the DL: the highest two numbered subchannels in the segment ($N - 1$ and N) for FUSC and PUSC; and the fourth and $(N - 4)^{th}$ subchannel for AMC. The reason for this difference is that FUSC and PUSC provide frequency diversity through their permutations (logical subchannel to physical subcarrier mapping), whereas AMC does not.

The AAS DMZ contains pairs of the AAS DL preamble followed by the AAS-DLFP, each sent on a (directional) beam. The AAS-DLFP contains some configuration information (the beam used and type and presence of preambles) as well as a pointer to an initial ranging opportunity and a pointer to a DL allocation. The latter pointer may point to the broadcast DL MAP or to a private DL MAP or may constitute a private DL MAP in the form of a single allocation depending on the CID and DIUC in the pointer.

In addition to provisioning AAS as a zone within the general frame, the OFDMA PHY allows the use of so-called *superframes*. A superframe is a period of up to 1 s that must meet the following requirements:

- It must start with at least 20 regular frames, each containing UCD, DCD, UL MAP, and DL MAP, and ranging channel.
- It contains AAS-only periods no longer than 200 ms followed by at least one regular frame containing UCD, DCD, and DL MAP.

The motivation for this construct is that it allows the BS to spend up to 95.2% of its time in AAS-only mode, which is advantageous from a capacity and range perspective, without violating the timeout limitations on synchronization, which are defined as the ability to receive the DL MAP and the DCD. Because those timeouts have a maximum limit of 600 ms, the superframe construct provides at least three synchronization opportunities in that period and provides at least 20 regular frames to reacquire synchronization and network entry for non-AAS devices.

In terms of MAC support, the OFDMA PHY makes use of the same channel feedback message as discussed in regard to the OFDM PHY.

OPEN-LOOP TRANSMIT DIVERSITY

An "open-loop" transmit diversity method exploits the diversity provided by multiple antennas, but requires no knowledge at the transmitter of the channel. These methods require no MAC support.

The OFDM PHY specifies STC as its sole open-loop transmit diversity scheme. The OFDMA PHY, on the other hand, specifies three such methods. Although the IEEE 802.16 standard describes all these methods broadly as STC, a distinction is made in the proceeding between actual STC, *frequency-hopped diversity coding* (FHDC), and *spatial multiplexing* (SM), which is more popularly known as *open-loop MIMO*. The basic transforms for these methods are summarized in Figure 13–4.

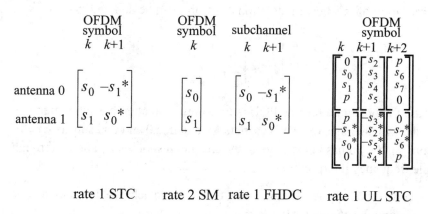

rate 1 STC   rate 2 SM   rate 1 FHDC   rate 1 UL STC

Figure 13–4:   Open-loop transmit diversity schemes

Each of the methods shown in Figure 13–4 requires two antenna elements at the transmitter or, if the BS is sectorized, two elements per sector.

STC uses the optimal space-time block code presented by Alamouti [B1], which achieves channel capacity (see [B44]). This STC is also known by the abbreviations *STTD* (space-time transmit diversity) or *STBC* (space-time block coding) and is, for example, also defined as part of the wideband CDMA (W-CDMA) standard [B17] (although in transposed form, which does not change its properties). This method requires that the channel be constant during the two-symbol interval over which the code is defined.

STC support in OFDM PHY

In the OFDM PHY, STC specifications are provided only for the DL.

The method is applied as follows: Data for two subsequent OFDM symbols s_0 and s_1 are collected after the modulation stage, where the pilot modulations are copied from the first symbol to the second. During the first OFDM symbol of the STC block, s_0 is transmitted from the first antenna and s_1 from the second antenna. During the second OFDM symbol of the STC block, the first antenna transmits $-s_1{}^*$ and the second antenna $s_0{}^*$.

The SS will hence receive $r_0 = h_0 \cdot s_0 + h_1 \cdot s_1$ during the first OFDM symbol, where h_0 and h_1 are the channel responses from antenna 0 and 1 to the SS antenna, respectively, and $r_1 = -h_0 \cdot s_1{}^* + h_1 \cdot s_0{}^*$.

The reception of this post modulation transform can be denoted as follows:

$$\begin{bmatrix} r_0 \\ r_1 \end{bmatrix} = \begin{bmatrix} s_0 & s_1 \\ -s_1{}^* & s_0{}^* \end{bmatrix} \begin{bmatrix} h_0 \\ h_1 \end{bmatrix}$$

Because the preamble for this PDU is transmitted on alternate subcarriers from the two antennas (even subcarriers for antenna 0 and odd subcarriers for antenna 1), the SS has knowledge of the channel responses $\hat{h}_0$ and $\hat{h}_1$ through simple frequency interpolation.

The receiver then applies the following transform to achieve the estimates $\{\hat{s}_0, \hat{s}_1\}$ using the channel estimates:

$$\begin{bmatrix} \hat{s}_0 \\ \hat{s}_1 \end{bmatrix} = \begin{bmatrix} \hat{h}_0{}^* & \hat{h}_1 \\ \hat{h}_1{}^* & -\hat{h}_0 \end{bmatrix} \begin{bmatrix} r_0 \\ r_1{}^* \end{bmatrix} = \begin{bmatrix} \hat{h}_0{}^* \cdot h_0 + \hat{h}_1 \cdot h_1{}^* & \hat{h}_0{}^* \cdot h_1 - \hat{h}_1 \cdot h_0{}^* \\ \hat{h}_1{}^* \cdot h_0 - \hat{h}_0 \cdot h_1{}^* & \hat{h}_1{}^* \cdot h_1 + \hat{h}_0 \cdot h_0{}^* \end{bmatrix} \begin{bmatrix} s_0 \\ s_1 \end{bmatrix}$$

The above equation shows the diversity combining on the diagonal terms and shows that the off-diagonal interference terms go to zero if the channel estimates are accurate. Note that this transform is also used for the modulated pilots.

The STC codec gain can be used to increase modulation/coding efficiency within the broadcast range. Also, to extend the cell range beyond the broadcast range, a separate STC zone is allocated, in which the BS sends an STC-encoded preamble and FCH as well as the MAPs and DCD/UCD, as needed. The inclusion of broadcast messages in the STC zone is necessary since a station outside broadcast range will be unable to decode the FCH and MAPs outside the STC zone.

STC, FHDC, and MIMO support in OFDMA PHY

The OFDMA PHY provides the same method of STC as the OFDM PHY for deployments in which the BS has two antennas (or four, where the second pair transmit the same signal as the first pair but with a respective phase rotation received through the fast-feedback mechanism so that directional gain is achieved). The method is defined only for FUSC and PUSC.

Using FUSC, the permutation during an STC zone changes only once per two OFDM symbols; an FUSC slot using STC can, therefore, be said to last two OFDM symbols. Within those two symbols, data are modulated from both antennas, as described for the OFDM PHY. For the pilots, different sequences apply to odd and even OFDM symbols, but the pilots are placed on the same subcarriers in both symbols. This method can be represented by the transfer matrix shown on the left in Figure 13–4. Because it encodes two data symbols over two OFDM symbols, this method is classified as a rate one encoding method.

Another method allowed by the OFDMA PHY is spatial multiplexing, in which a separate data symbol is transmitted from each antenna without repetitions in the subsequent OFDM symbol. Spatial multiplexing provides increased capacity, but provides no diversity gain and requires a more complex receiver including multiple receiver chains.

Another variant allowed is FHDC, which is conceptually the same as STC, but with encoding over two subchannels instead of over two OFDM symbols.

Unlike in the case of the OFDM PHY, STC and SM are also defined for the UL, specifically for PUSC. Recall that a tile for UL PUSC is defined as four contiguous carriers over a span of three OFDM symbols, with the corners reserved for pilots. Using STC and SM on the UL, this tile definition is slightly modified so that the pilots on the diagonal are not transmitted on antenna 0 and the pilots on the cross-diagonal are not transmitted on antenna 1. The STC encoding is subsequently applied per two subcarriers within the same OFDM symbol, as shown in Figure 13–4, where each vector indicates the encoding of the four subcarriers within the tile for a specific OFDM symbol per antenna.

PUSC also allows what is termed *virtual MIMO*, in which antenna 0 and antenna 1 belong not to the same SS but to completely different SSs. This allows SSs to transmit simultaneously with near certainty of the channels being uncorrelated and without requiring any additional complexity.

Lastly, a provision is established for STC and SM where the transmitter has four antennas, the matrices for which are shown in Figure 13–5.

$$
\begin{array}{l}
\\ \\
\text{antenna 0} \\ \text{antenna 1} \\ \text{antenna 2} \\ \text{antenna 3}
\end{array}
\quad
\begin{array}{c}
\text{OFDM symbol} \\
k \quad k{+}1 \quad k{+}2 \quad k{+}3 \\
\begin{bmatrix}
s_0 & -s_1^* & 0 & 0 \\
s_1 & s_0^* & 0 & 0 \\
0 & 0 & s_2 & -s_3^* \\
0 & 0 & s_3 & s_2^*
\end{bmatrix} \\
\text{rate 1 FHDC}
\end{array}
\quad
\begin{array}{c}
\text{OFDM symbol} \\
k \quad k{+}1 \quad k{+}2 \quad k{+}3 \\
\begin{bmatrix}
s_0 & -s_1^* & s_4 & -s_6^* \\
s_1 & s_0^* & s_5 & -s_7^* \\
s_2 & -s_3^* & s_6 & s_4^* \\
s_3 & s_2^* & s_7 & s_5^*
\end{bmatrix} \\
\text{rate 2 FHDC}
\end{array}
\quad
\begin{array}{c}
\text{OFDM symbol} \\
k \\
\begin{bmatrix}
s_1 \\ s_2 \\ s_3 \\ s_4
\end{bmatrix} \\
\text{rate 4 FHDC}
\end{array}
$$

Figure 13–5: Four-antenna STC and SM

CLOSED-LOOP TRANSMIT DIVERSITY

In addition to the open-loop schemes, the IEEE 802.16e amendment adds closed-loop transmit diversity methods for the OFDMA PHY. These methods rely on channel state information sent back by the SS on command from the BS. For this purpose, the BS uses the UL_Sounding_Command_IE, which provides a large array of possibilities for how the SS should parse the measurements it made.

Precoding

One of the closed-loop methods is so-called *precoding*, which is similar in application and motivation to STC but exploits knowledge of the channel. The coding in this case maps the orthogonal MIMO streams to the physical antennas.

There are principally two methods of conveying, from the SS to the BS, which precoding matrix to use. Using the first method, the SS simply returns quantized versions of the channel metrics. In the second method, both sides of the link have access to a set of so-called *codebooks*, each containing an indexed set of coding matrices. For each channel rank, which never exceeds the number of transmit antennas and dictates the number of spatial streams the channel can support, a different codebook is specified in the standard. The receiving station computes a precoding matrix to optimize throughput for the current channel rank, finds the nearest matrix in the codebook of the current channel rank, and sends the index of the selected matrix as well as the channel rank back to the transmitting station.

For each rank, the codebooks are designed so that the chordal distance between the code matrices is maximal. They are specified with 3-bit indexes (8 code matrices per codebook) and 6-bit indexes (128 code matrices per codebook). The choice allows the implementer a trade-off between complexity and performance. The receiver's decision about which index to select is implementation-dependent and may make use of various distance metrics between the computed optimal matrix and the available matrices in the codebook.

With the introduction of high-speed mobility as a consideration during the development of IEEE Std 802.16e, it was observed that, despite the use of the fast-feedback channel as discussed in Chapter 12, the protocol delays could in some cases exceed the coherence time of the channel. This would make the instantaneous channel metrics outdated by the time they were applied. To resolve this issue, a method was introduced to feed back not only the instantaneous channel metrics but also (long-term) first- and second-order statistics of the channel. The first-order statistic used is the LOS component of the Ricean fading channel approximation. The second-order statistics consist of the channel's spatial correlation information. These statistics may be used to improve the estimate of the channel condition at the time of transmission.

Antenna selection

Using antenna selection, the receiving SS can use the fast-feedback channel to identify the group of antennas it prefers the transmitting BS to use for full-rate transmissions. The spatial multiplexing is performed by the selection of a single vector (similar to the rate 4 FDHC shown in Figure 13–5) where one or more elements are nulled (the vector being of a length equal to the number of transmit antennas, wherein the number of non-nulled elements is equal to the number of spatial streams desired). Antennas for which the element reads null are not used during the transmission.This method is defined only for systems with three or four transmit antennas.

Antenna grouping

The antenna grouping method uses the methods described in the open-loop section of this chapter. These methods are less than full-rate. Instead of disabling antennas as with antenna selection, the antenna grouping method groups antennas into STC pairs. This may be most trivially visualized as row permutations of the space-time coding matrix. For example, for four transmit antennas, these are row permutations of the rate 1 and rate 2 FHDC shown in Figure 13–5. The receiving SS uses the fast-feedback channel to indicate the index of the preferred permutation to use. This method is also defined only for systems with three or four transmit antennas.

Chapter 14 Performance analysis

MAC and PHY performance and throughput

INTRODUCTION

This chapter reviews several published simulations that assess the performance of WirelessMAN systems. Such simulations are complex and depend on a vast number of parameters, assumptions, and algorithms that are typically not fully detailed in the publications. However, due to the great interest among both industry and academia in the WirelessMAN technology, the research base is growing as the community has more time to study and evaluate the IEEE 802.16 standard. As more research becomes available, confidence in the results will begin to grow.

Due to the complexity of the simulations and the time required to run them, most studies chose to treat certain aspects of the system in a general way, while modeling other aspects in great detail. While simulation results may indicate potential performance, they cannot do so definitively.

WIRELESSMAN-OFDM, FIXED OPERATION

Capacity analysis

A 2005 paper by Ghosh, et al., [B18] reports the performance of the WirelessMAN-OFDM system with different frequency reuse patterns and cell sectorization. Figure 14–1, reproduced from that paper, illustrates the aggregate MAC-level throughput with a 3 km cell radius and a 50 W BS transmitter. The traffic model is not detailed, but the marginal differences between the PHY and MAC throughput presented suggest a full-buffer model. The calculations use three and six sectors per cell, with reuse factors 1 and 3. The reuse factor in this case is defined from cell to cell; for example, with six sectors per cell and a reuse factor of 3, a set of six channels, chosen from among three available sets of six channels, is used at each cell. According to

---

the paper, the results show the "average throughput over an entire cell per 5 MHz carrier." However, it appears that, in calculating the average, the throughout has been divided by the number of channels per cell, not the total number in all cells. This is suggested by the paper's conclusion that "if at all possible, 1/1 reuse (frequencies reused every cell) should be employed since the gain from going to 1/3 reuse (frequencies reused every third cell) does not come close to compensating for the associated tripling of consumed bandwidth." To calculate the throughput per channel used in the system, it appears that the bars labeled "1/3 reuse" in Figure 14–1 should be divided by a factor 3.

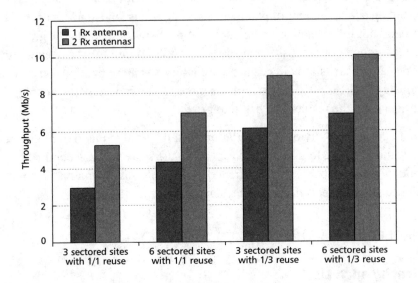

Figure 14–1: WirelessMAN-OFDM throughput per 5 MHz channel

The figure also shows the effect of adding a second receiver in the SS, with half a wavelength (7 cm) separation.

The performance of the system can also be improved by the use of STC. The paper shows that, in the three-sector case (with 1/1 reuse) and one receive antenna, adding STC with a second transmit antenna provides performance similar to the six-sector single-in, single-out (SISO) case, but with only half the channels.

Figure 14–2 shows the outage relative to 1.5 Mbit/s; that is, the fraction of cell area with maximum data rate less than this value. This figure indicates that narrower sectorization cuts down drastically on bad channel conditions and interference. Also the use of receive diversity improves the outage substantially.

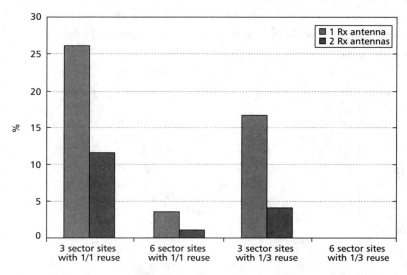

Figure 14–2:   WirelessMAN-OFDM outage relative to 1.5 Mbit/s

MAC performance

MAC performance is assessed under more realistic traffic conditions (although less realistic environmental conditions) in a 2006 paper by Cicconetti, et al., [B10]. It considers an interference-free FDD WirelessMAN-OFDM system with 7 MHz bandwidth and 10 ms frame duration. The UL allocation start time is chosen to be 10 ms, meaning that all UL allocations point at a minimum to the next frame. This introduces a minimum delay of 10 ms, but it provisions terminals with very low processing power. The analysis is performed on a single cell (i.e., no interference) and assumes no transmission errors apart from contention-based collisions. The authors assumed that the SSs employ the following modulation methods and coding rates, which are evenly partitioned among SSs: QPSK 3/4, 16-QAM 3/4, and 64-QAM 3/4.

Two scenarios are studied in the paper. In the residential scenario, the authors consider offering broadband Internet access. The Internet traffic is modeled as a Web traffic source (see [B39]). Packet sizes are drawn from a Pareto distribution with cutoff (shape factor = 1.1, mode = 4.5 kB, cutoff threshold = 2 MB), while packet interarrival times are distributed exponentially (average = 5 s), which yields an average load of 25 kB/s. The resulting throughput delay curves are shown in Figure 14–3. In addition to the residential scenario, the paper covers a small- and medium-enterprise scenario. For details, please consult the paper.

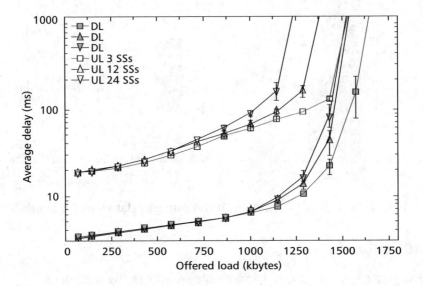

Figure 14–3:   Average throughput delay vs offered load

The following sections deal with previously unpublished results generated by the authors of the previously discussed paper using the same simulator. The assumption for these results are, however, slightly different. In this set of results, a frame duration of 5 ms was used. The allocation start time was set to be equal to the length of the frame. The channel bandwidth is 7 + 7 MHz.

The system is loaded with a number of applications spanning all four defined service types (UGS, rtPS, nrtPS, BE) and resulting in eight connections in each direction:

- DL

    MPEG video: 128 kbit/s

    IP traffic: 200 kbit/s [10% gaming, 9% mail, 9% FTP, 4.5% Telnet, and 67.5% Hypertext Transfer Protocol (HTTP)]

    Videoconference: 80 kbit/s

    VoIP: 16 kbit/s

- UL

    MPEG video: 128 kbit/s

    IP traffic: 50 kbit/s (same mix as DL)

    Videoconference: 80 kbit/s

    VoIP: 16 kbit/s

The UL allocation start time is chosen to be 5 ms, meaning that all UL allocations point at a minimum to the next frame. This introduces a minimum delay of 5 ms, but it provisions terminals with very low processing power. The analysis is performed on a single cell (i.e., no interference) and assumes no transmission errors apart from contention-based collisions (except in the ARQ analysis, where a fixed 10^{-6} BER is assumed on each link).

Full-duplex with single burst profile

The SSs in this example are all full-duplex, and all use the same 64-QAM-5/6 burst profile, which implies a 28.23 Mbit/s raw PHY data rate. The DLFP and MAPs are transmitted at their required rates.

The most interesting delay curve is that for the IP traffic. Because this traffic is BE, it provides a good indicator of behavior as a function of load. This is because an increase in load increases the demand for the other services and hence reduces the capacity left over for BE traffic. In Figure 14–4 and Figure 14–5, an example of this performance is shown for the DL and UL, respectively.

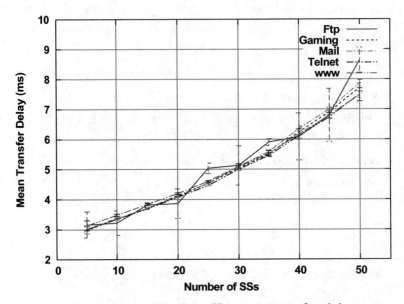

Figure 14–4:   DL, IP traffic mean transfer delay

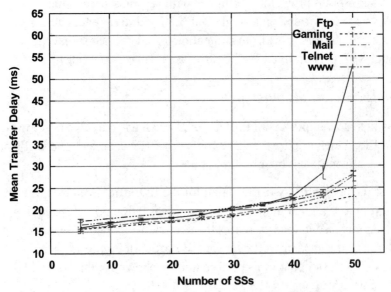

Figure 14–5:   UL, IP traffic mean transfer delay

Besides the data transfer, the UL must carry contention-based request transmit opportunities. The absolute number of these, as well as the the the utilization (*S*) and load (*G*) of request transmit opportunity (UIUC = 2), is shown in Figure 14–6 and Figure 14–7, respectively. Notice that only the mandatory full contention method is used.

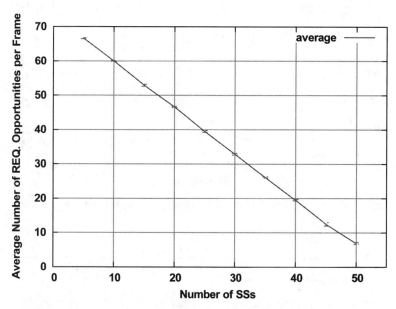

Figure 14–6:   UL, Average transmit opportunities per frame

Lastly, the MAC overhead within a frame is of interest. This overhead, which consists of the FCH and MAPs, is shown in Figure 14–8. It can be seen that the overhead increases to 4.7% (7 OFDM symbols per frame) when supporting 400 connections (50 SSs) in both directions.

The loads of the DL and UL are roughly linear with the number of SSs. For 50 SSs, the load on the DL is roughly 75%, whereas the load on the UL approaches 90%, driving the UL faster into saturation than the DL, despite carrying only two thirds the net throughput.

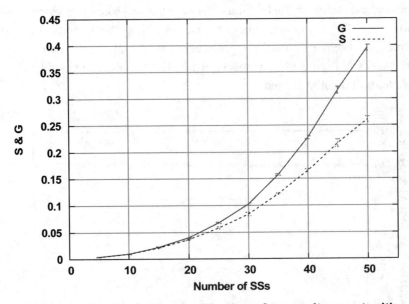

Figure 14–7:   UL load and utilization of transmit opportunities

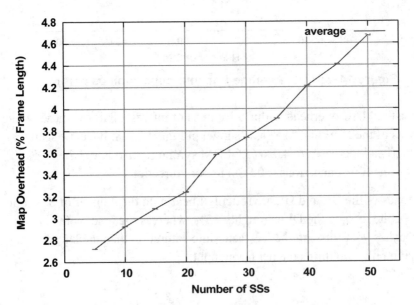

Figure 14–8:   FCH + MAP overhead

Mixed full-duplex/half-duplex with burst profiles case

In this scenario, the DIUC and UIUC span the QPSK-1/2 to 64-QAM-5/6 range. Performance is evaluated for SSs having the same DIUC and UIUC, respectively. Half the SSs are full-duplex, and half are half-duplex. As the average PHY rate is substantially lower (on average around 10.8 Mbit/s) than in the previous case, simulations are limited to 30 SSs. Because the overhead is not depended on the used IUC and whether an SS is full-duplex or half-duplex, the overhead shown in Figure 14–8 remains valid for these cases.

Figure 14–9 and Figure 14–10 show the difference between full-duplex and half-duplex SSs with regard to the transfer delay for the VoIP application. It shows that the transfer delay for half-duplex SSs is around 20% higher than that of full-duplex devices, which is predictable as the uplink cannot be used by these devices while actively listening to the DL and during SSRTG and SSTTG.

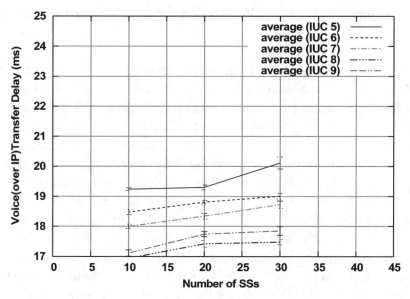

Figure 14–9:   UL, full-duplex SSs, VoIP mean transfer delay

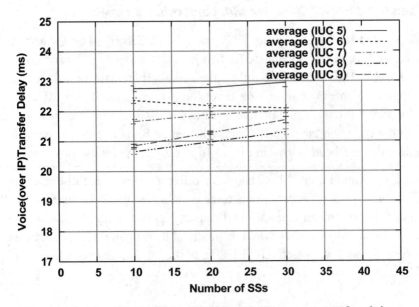

Figure 14–10:   UL, half-duplex SSs, VoIP mean transfer delay

Figure 14–11 and Figure 14–12 show the transfer delay difference between DL and UL for the same best effort HTTP service. Figure 14–11 shows that, on the DL, the traffic is on average allocated either to the immediately following frame or to the next frame after that. On the UL, shown in Figure 14–12, an SS may first have to use a contention slot to request bandwidth or, if it has other data scheduled for transmission, can use piggy-backing. The allocation for the burst will then appear in the DL MAP no sooner that the next frame with an allocation for the frame following that. Because of this, the average transfer delay for any service will typically exceed 10 ms. Lastly, in Figure 14–13, the same scenario is shown with ARQ enabled, assuming a 10^{-6} BER on each link. It can be seen that the transfer delay roughly doubles, which is to be expected from retransmissions and the increase in system load. For 30 SSs, the UL can be seen to go rapidly into saturation.

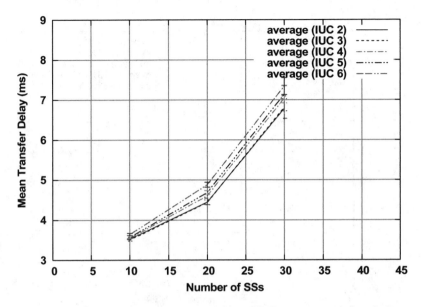

Figure 14–11:   DL, half-duplex SSs, HTTP mean transfer delay

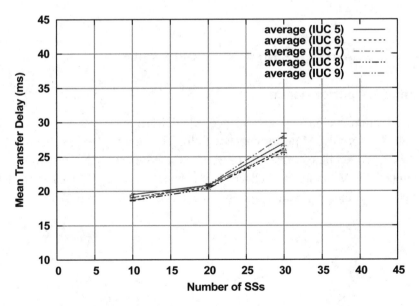

Figure 14–12:   UL, half-duplex SSs, HTTP mean transfer delay

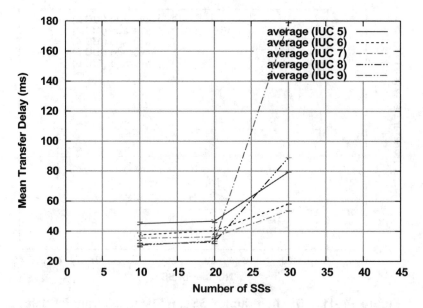

Figure 14–13: UL, half-duplex SSs, HTTP mean transfer delay w/ARQ

WIRELESSMAN-OFDM, MOBILE OPERATION

Basic PHY performance

In this section, we present unpublished calculations regarding the
WirelessMAN-OFDM performance with limited mobility. Figure 14–14
shows the coded block error rate (CBLER) for WirelessMAN-OFDM
using the ITU Vehicular A (VehA) 3 km/h channel model [B30]. These results
include all of the synchronization loss, which comes to 1 dB to 2 dB.

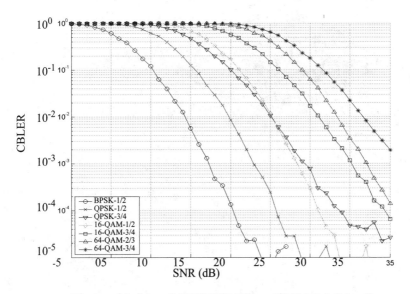

Figure 14–14: OFDM CBLERs—ITU VehA 3 km/h

Figure 14–15 shows the CBLER as a function of OFDM symbol index and SNR using the ITU VehA 30 km/h channel model with QPSK modulation and 1/2 rate coding. The symbol index is the count of data OFDM symbols following the preamble within the same burst, and the CBLER is computed individually over each OFDM symbol (as the outer RS code is a per-OFDM-symbol block code). Due to the rapidly varying channel condition, the channel estimate computed from the preamble will be increasingly less accurately matched to the instantaneous channel condition for the OFDM symbol as the OFDM symbol index in the burst increases. For BPSK 1/2, the degradation will be slower, while for higher order modulations, it will be more severe. This figure shows that, even at this modest speed, the degradation is severe due to the changing channel conditions. For this reason, the nonsubchannelized bursts in mobile operation must be relatively short. To address the problem, midambles, for subchannelized operation, have been added through IEEE Std 802.16e. The degradation can also partially be mitigated by updating the channel estimates by means of data-driven decision feedback, which is not included in Figure 14–15.

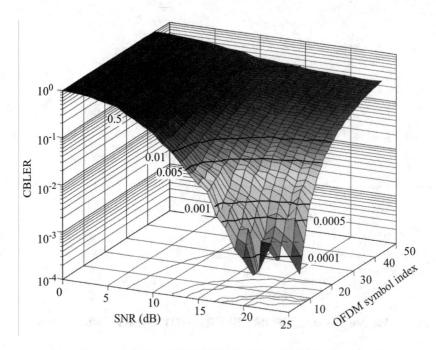

Figure 14–15:   OFDM CBLERs—ITU VehA 30 km/h

Capacity analysis

In a 2005 paper, Ball, et al., [B5] report a study of a mobile network using WirelessMAN-OFDM in a 3.5 MHz channel at 3.5 GHz with cell radius of 300 m. The 2 W BS uses three sectors with 17.5 dBi antennas. The analysis considers a full-buffer FTP model with up to 300 randomly distributed nomadic clients per sector. The analysis is performed with reuse factors of both 1 and 3.

Figure 14–16 depicts the channel utilization of each sector utilization as a function of the offered load, showing a limit of around 3 Mbit/s for a reuse factor of 1 and at 4.25 Mbit/s for a reuse factor of 3. This corroborates the findings of Ghosh, et al., [B18] that the spectral efficiency is higher with a reuse factor of 1. Figure 14–17 shows, as a function of the offered load (which is varied identically in each cell), the fraction of SSs served with each modulation/coding rate in this simulation. For reuse factor 1, the modulation/coding rate efficiency is substantially lower due to the increased interference (lower C/I); roughly 80% of the SSs are using modulation/coding no more efficient than QPSK-3/4.

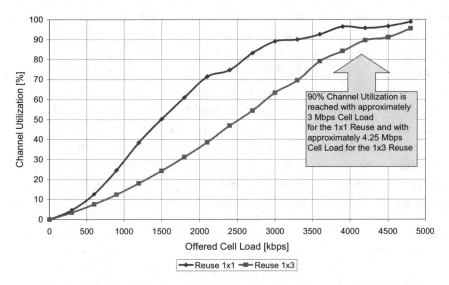

Figure 14–16:   Channel utilization as function of offered load

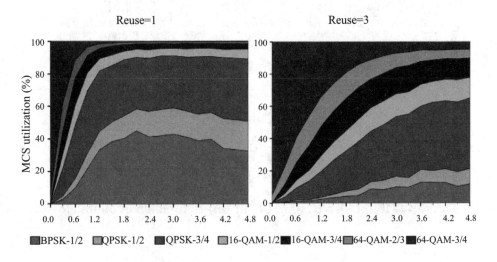

Figure 14–17:   Modulation/coding utilization as function of offered load

The Ball paper further analyzes (see Figure 14–18) the occurrence of call blocking (i.e., rejection by admission control due to insufficient C/I). In the reuse factor 1 scenario, a cell load of 1 Mbit/s results in 10% of call blocking, whereas the reuse factor 3 scenario experiences no substantial call blocking up to the channel utilization limit shown in Figure 14–16. The spectral efficiency is hence 1 Mbit/s / 3.5 MHz = 0.29 (bit/s)/Hz per sector for the reuse factor 1 case and 4.25 Mbit/s / (3 * 3.5 MHz) = 0.40 (bit/s)/Hz per sector for reuse factor 3. The reuse factor 3 scenario hence provides higher spectral efficiency when both channel utilization and call blocking limitations are observed. It should be noted that the analysis considers none of the optional WirelessMAN-OFDM features to improve C/I performance, such as STC, diversity techniques, advanced coding, and subchannelization. Applying these options would likely provide a substantial performance improvement, particularly for the reuse factor 1 scenario.

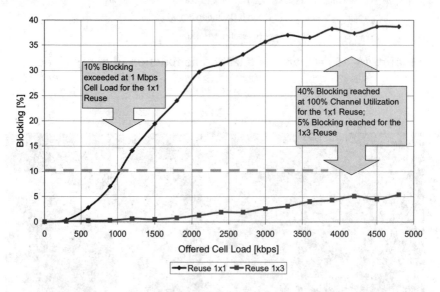

Figure 14–18: Call blocking (C/I < 1 dB) as function of offered load

WirelessMAN-OFDMA, MOBILE OPERATION

Overhead and capacity

In a 2005 paper, Lee, et al., [B35] report on studies of a system considered for "WiBro" services. The system studied is "somewhat similar" to a draft of IEEE Std 802.16e at some stage prior to approval. According to the paper, the differences include "the structure and performance of preamble" and "subcarrier permutation in FCH/DL MAP." The results may shed some light on WirelessMAN-OFDMA performance. Recall that WirelessMAN-OFDMA is defined in IEEE Std 802.16-2004 [B20] using a 2048-FFT. The system under study uses a 1024-FFT, which is an option for WirelessMAN-OFDMA in IEEE Std 802.16e.

The system studied uses a channel of 8.447 MHz and a 10 MHz sampling bandwidth. The system is TDD, with a 2/1 DL/UL ratio. The overhead, including all PHY overhead as well as control subchannels for channel quality information (channel feedback), acknowledgment, and initial ranging, was found to be between 22% and 31%.

This paper also analyzes performance as a function of cell radius. It assumes a fixed modulation/coding and maximizes the number of subchannels to satisfy the CINR required to achieve a 1% PER. Figure 14–19, redrawn from the published data, illustrates the results. Combining the results by selecting the maximum capacity at a given range, the solid line in Figure 14–19 shows the DL sector capacity if the only modulation/coding options available for the link adaptation algorithm to choose were the three indicated in this figure. Introducing the remaining modulation/coding combinations would have the effect of largely removing the jaggedness of the curves.

Figure 14–20 is the parallel to Figure 14–19 for the UL.

Figure 14–19 and Figure 14–20 suggest that, in order to guarantee a single user 512 kbit/s DL and 256 kbit/s UL, the cell radius must be limited to 2.3 km.

Using the same approach with an outdoor-to-indoor channel model, Figure 14–21 and Figure 14–22 show the 512 kbit/s DL / 256 kbit/s UL cell radius to be limited to approximately 0.7 km.

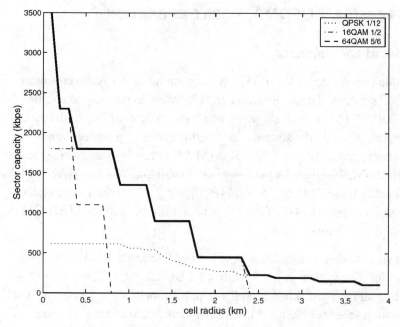

Figure 14–19: DL sector capacity per cell radius—ITU VehA (60 km/h)

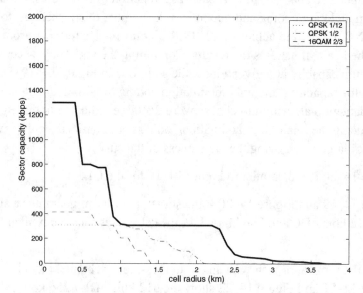

Figure 14–20: UL sector capacity per cell radius—ITU VehA (60 km/h)

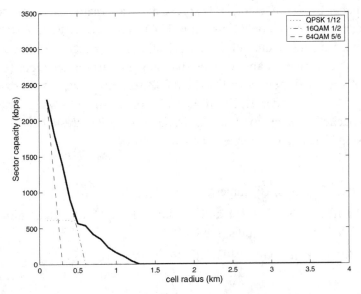

Figure 14–21: DL sector capacity per cell radius (ITU outdoor-indoor)

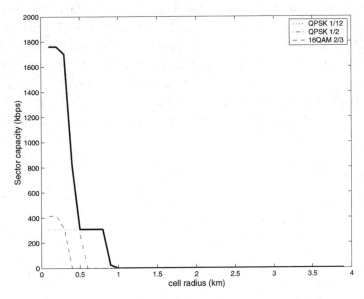

Figure 14–22: UL sector capacity per cell radius (ITU outdoor-indoor)

A 2005 paper by Kwon, et al., [B34] presents an alternative study of the system capacity. The modeled system is TDD using a 1024-FFT, 9 MHz bandwidth, 5 ms frame duration, and a 2/1 DL/UL frame split. The network is loaded with a number of applications spanning all four service types (UGS, rtPS, nrtPS, and BE) defined in IEEE Std 802.16-2004 as well as the extended rtPS (ertPS) introduced in IEEE Std 802.16e for real-time flows generating variable packet sizes on a periodic basis, such as VoIP with silence suppression. The traffic load per user is as follows:

- DL

    Video: 32 kbit/s, 10 frames/s

    IP traffic: ~90 kbit/s FTP and ~15 kbit/s HTTP traffic (according to 1xEV-DV traffic models; see [B8])

    VoIP: 9.6 kbit/s with 0.403 activity factor

- UL

    VoIP: 9.6 kbit/s with 0.403 activity factor

    IP traffic: As needed to support UL FTP, HTTP, and video traffic

The analysis is performed in a hexagonal cellular environment considering interference from first and second tier cells with identical traffic distributions. Figure 14–23 shows the delay in transmitting VoIP data using rtPS, UGS, and ertPS. On the DL, the number of VoIP calls is limited to 92 when the delay limit is 60 ms. This figure shows that the introduction of ertPS in IEEE Std 802.16e is justified for voice with silence suppression since the call limit is similar to the DL limit using ertPS, but is significantly lower using rtPS and UGS.

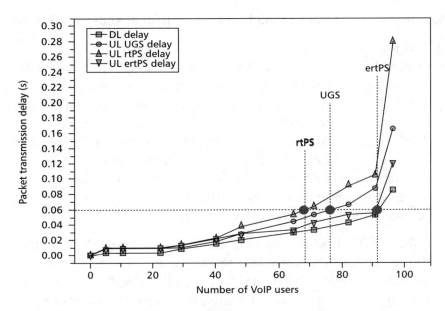

Figure 14–23: Delay for VoIP using rtPS, UGS, and ertPS

Figure 14–24 shows the average cell throughput both for UL and DL using various scheduling mechanisms, permutations, and feedback.

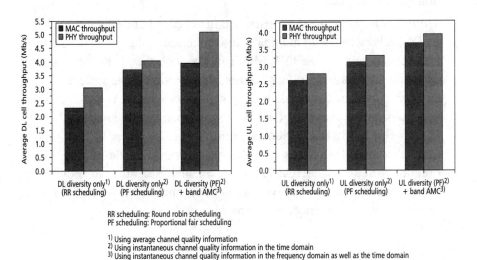

Figure 14–24: Average cell throughput for various schemes

WirelessMAN-OFDMA vs high-speed DL packet access (HSDPA)

A 2005 paper by Shin, et al., [B48] compares the capacity of a WiBro-compatible version of WirelessMAN-OFDMA (based on a preapproved draft of IEEE Std 802.16e) with that of HSDPA. The study concludes that the OFDMA approach has "much better performance." In particular, it finds that the spectral efficiency of WirelessMAN-OFDMA is close to double that of HSDPA for the ITU VehA channel model and also for the ITU Pedestrian A (PedA) channel at 10 km/h [B30]. For the ITU PedA channel model at 3 km/h, the spectral efficiency of WirelessMAN-OFDMA is found to be about one third greater than that of HSDPA. The authors ascribe the WirelessMAN-OFDMA advantage to two main factors:

- *TDD vs FDD:* Both systems work within a system bandwidth of 10 MHz. However, the OFDMA system under study uses TDD (with a fixed 2/1 DL/UL ratio) with signal bandwidth of 8.3 MHz. HSDPA uses FDD carried in two 5 MHz channels, but its effective signal bandwidth is just 3.84 MHz times two. The wider bandwidth of OFDMA is naturally not all positive, as it can result in a shrinkage of the coverage area due to the 3.4 dB higher thermal noise. In a noise-limited scenario, this coverage shrink is estimated at 19%, but the real situation may be interference-limited in both cases.

- *Multipath performance:* As discussed in Chapter 3 and Chapter 11, OFDM is very robust to multipath due to the use of the cyclic prefix and the cyclical nature of the FFT. Multipath channels that do not exceed the cyclic prefix duration will not significantly deteriorate the system performance (provided the sampling is adequate). CDMA, on the other hand, typically makes use of a rake receiver to extract the multipath energy. This results in deteriorated performance. The paper by Shin reports as much as 40% HSDPA capacity loss going from a single-path (multipath-free) channel to a three-path channel. Naturally, the move away from conventional rake receivers to more advanced receiver designs provides a means to reduce that capacity loss.

Basic WirelessMAN-OFDMA PHY performance

This section reports on unpublished studies of the basic WirelessMAN-OFDMA PHY performance. It assumes 10 MHz channelization at 2.5 GHz with a 1024-FFT; a 1/8 cyclic prefix; and practical levels of phase noise, mixer I/Q imbalance, and frequency offset. The reported E_sN_0 is equivalent to the SNR per subcarrier. For the repetition codes, postdetection combining of soft decisions is employed. All simulations shown use the mandatory CC.

Figure 14–25 and Figure 14–26 present the CBLER for the OFDMA modulation using the ITU VehA 3 km/h channel model. The first figure represents ideal synchronization and channel estimation, whereas the second represents a practical system of moderate complexity. The poor performance of QPSK-3/4 is a result of the mandatory interleaver problem discussed in Chapter 12.

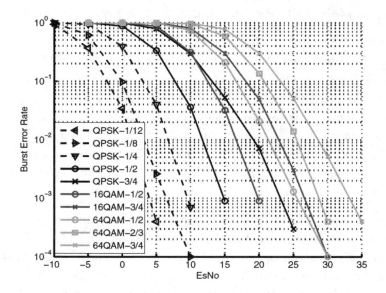

Figure 14–25: CBLER, UL PUSC, ITU VehA 3 km/h, CC, ideal

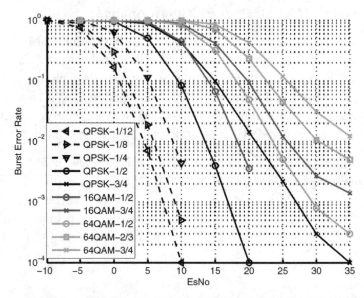

Figure 14–26: CBLER, UL PUSC, ITU VehA 3 km/h, CC, estimated

Figure 14–27 and Figure 14–28 provide similar data using the ITU VehA 120 km/h channel model. The ideal performance suffers only moderately as a result of the higher speeds, even for the higher modulation orders. However, the practical challenges of synchronization and channel estimation cause the performance at the higher modulation orders to degrade substantially, with error floors developing. For the lower order modulations, including the omitted repetition rates, the PHY is fairly robust to the rapidly changing channel.

Similar analysis for VehA 30 km/h channel shows no significant degradation compared to the VehA 3 km/h model, showing that the OFDMA PHY is substantially robust to practical speeds. On the other hand, the performance degrades substantially using the VehB 10 km/h model due the insufficient cyclic prefix duration compared to the impulse response in that model.

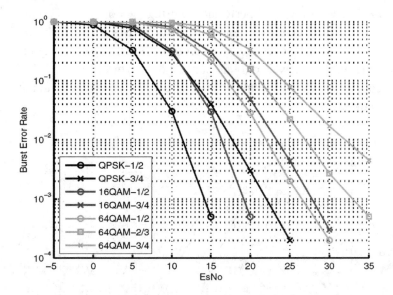

Figure 14–27:   CBLER, UL PUSC, ITU VehA 120 Km/h, CC, ideal

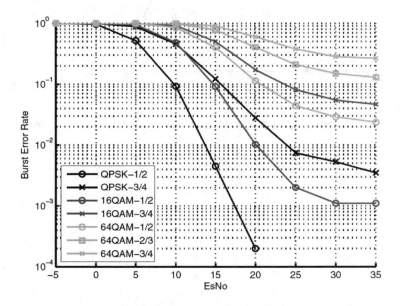

Figure 14–28:   CBLER, UL PUSC, ITU VehA 120 km/h, CC, estimated

Figure 14–29 shows the ideal performance using the optional CTC codec, which is the optional codec of choice within the WiMAX profiles. The improvement over the CC codec varies from 1 dB to 1.5 dB although it need be noted that in this analysis only the longest code word is used. Use of shorter code words can result in a loss of up two 2 dB (compared to the longest code word).

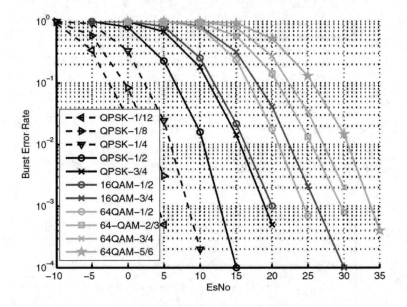

Figure 14–29:   CBLER, UL PUSC, ITU VehA 3 km/h, CTC, ideal

Figure 14–30 shows the ideal performance of the DL modulations, including that of the DLFP. It is best to protect the DLFP since it constitutes a single point of failure within the frame.

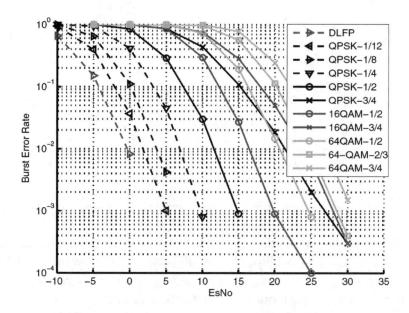

Figure 14–30: CBLER, DL PUSC, ITU VehA 3 km/h, CC, ideal

In Figure 14–31, the performance of the DL repetition coding as a function of channel impulse response (CIR) is shown. This result is important to determine which repetition code can be used at the cell edge. As may be observed from the figure, the use of QPSK-1/4 (QPSK-1/2, repetition rate 2) or QSPK-1/8 (QPSK-1/2, repetition rate 4) has sufficient CIR performance to make usage at the cell edge feasible. The large gap between the solid (ideal synchronization) and dashed (practical synchronization) curves is the result of the general difficulty of achieving good synchronization under low signal to noise-plus-interference ratio, regardless of the method used.

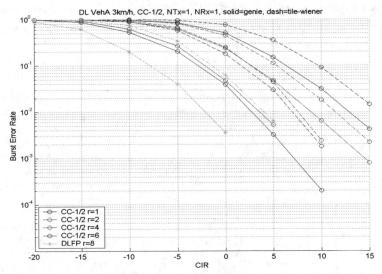

Figure 14–31: CBLER, DL PUSC, ITU VehA 3 km/h, CC, ideal

Finally, Table 14–1 shows the cell ranges that may be achieved with the DL modulations with a 20 W transmitter, 10 dB backoff, 15 dB antenna gain, and a 10 dB noise figure. For this case, the mobile is assumed to have two antennas on which it performs MRC. Achievable cell ranges are shown both with and without a 10 dB shadowing margin using the Cost 231 Hata channel model [B11].

Table 14–1: Cell-ranges DL PUSC

Modulation/ code rate	Max. PHY throughput (Mbit/s)	SNR @ 0.01 CBLER (dB)	Cell range (m)	Cell range with fade margin (m)
QPSK-1/12	1.19	−1.1	1586	825
QPSK-1/8	1.79	0.7	1410	733
QPSK-1/4	3.57	2.0	1259	674
QPSK-1/2	7.14	6.0	997	512
QPSK-3/4	10.71	9.2	809	421
16-QAM-1/2	14.28	10.3	753	392
16-QAM-3/4	21.42	14.9	557	290
64-QAM-1/2	21.42	15.5	536	279
64-QAM-2/3	28.56	19.0	426	222
64-QAM-3/4	32.13	21.2	369	192

MIMO capacity

This section discusses the capacity gain achieved by the use of MIMO. The capacity is analyzed at link level for a full-buffer FTP stream in an OFDMA network with 19 cells, each having three sectors, using 10 MHz bandwidth at 2.5 GHz. The channel models are a mix of stationary/pedestrian, pedestrian, mobile, and highly mobile models in analogue to the 1xEV-DV evaluation methodology (see [B8]). Some capacity results, in megabits per second, are shown in Table 14–2, where the notation *mxn* indicates *m* transmitter chains and *n* on the receiver. Note that the results are indicative only of the relative capacity gain, as a full-buffer FTP analysis provides an overly optimistic result for achievable capacity. The use of 2x1 STC on the DL, for which no simulation data were available for this specific scenario, would be expected to result in values between those of 1x1 SISO and 2x2 MIMO, but relatively closer to the 2x2 MIMO results.

Table 14–2: Relative performance benefit of MIMO and virtual MIMO[a]

DL/UL ratio		DL			UL		
	Reuse	1x1	2x2 MIMO	2x4 MIMO	1x2 MRC	1x2 V-MIMO	2x2 MIMO
1/2	1	3.80	7.97	11.76	3.20	7.04	6.72
	3	1.52	3.98	7.06			
1/1	1	5.75	12.08	17.83	2.40	5.28	5.04
	3	2.30	6.04	10.70			
2/1	1	7.67	14.00	20.67	1.60	2.40	3.36
	3	3.07	7.00	12.40			

[a] Data courtesy of Nortel Networks.

In Table 14–3, the spectral efficiency and capacity are shown for various antenna configurations in the same system, with DL/UL ratio of 2/1 and using the ITU PedA 3 km/h channel model.

Table 14–3: Performance using MIMO[a]

	1x1	1x2	2x2	2x4	4x2	4x4
(bit/s)/Hz per sector	1.2	1.8	2.8	4.4	3.7	5.1
Aggregate sector throughput (Mbit/s)	7.6	11.6	17.8	27.7	23.6	32.6

[a] Data courtesy of Nortel Networks.

Chapter 15 Conformance and interoperability

Conformance standards and testing

In this chapter, we discuss the IEEE 802.16 conformance standards and the WiMAX Forum® interoperability and certification procedures. The conformance standards are essentially comprehensive test plans to evaluate whether a particular vendor's implementation meets the requirements defined in IEEE Std 802.16. The WiMAX Forum is an industry alliance, analogous to the Wi-Fi Alliance for IEEE Std 802.11, that defines interoperability tests and certifies IEEE 802.16 BSs and SSs that pass these tests.

THE WiMAX FORUM

By the end of 2000, even before the original IEEE Std 802.16-2001 was approved, it became clear that the standard would include a plethora of options. For instance, the high-frequency WirelessMAN-SC mode has literally more than 4000 possible combinations of FEC and modulation alternatives, while in practice four or five would suffice. While most options have at least one corresponding proponent who can argue for its benefits under certain circumstances, implementing all of the options would clearly be an onerous and expensive task. Conversely, if equipment vendors independently chose which options to implement, full interoperability would be precluded, and IEEE Std 802.16 would be a standard in name only. Systems would be, for all intents and purposes, proprietary through the choice of options. In March of 2001, the outline of Clause 12 was added to the IEEE 802.16 draft standard.

It also became clear that full interoperability would require proper conformance testing, lest two vendors pick the same profile but interpret it differently. In IEEE 802, standards rarely include test specifications; in most cases, they stop with functional specifications. In the 2000 time frame, intense debate surrounded IEEE Std 802.16 versus ETSI BRAN HiperACCESS as the standard of choice for PMP broadband wireless access above 10 GHz. ETSI BRAN HiperACCESS was a simpler but less efficient standard than IEEE 802.16 WirelessMAN-SC. ETSI BRAN HiperACCESS advocates criticized IEEE Std 802.16, citing the plethora of options and the impact on interoperability. They also cited the ETSI mandate for a full set of test documentation based on the ISO/IEC 9646 standards series (identical to the ITU x.290 series) specifying test documentation. In response, several companies involved in IEEE Std 802.16 were determined to eliminate this as a source of discredit to IEEE Std 802.16 so the two standards could be compared on a head-to-head technical basis.

These companies decided to take the concept a step further and initiate a certification body and process modeled as a hybrid of Wi-Fi and GSM type approval. In April 2001, representatives of Nokia, Ensemble Communications, Harris Corporation, and Crosspan (a division of Raytheon) met in Antibe, France, alongside an ETSI BRAN HiperACCESS meeting in neighboring Sophia Antipolis; and the WiMAX Forum was born. Unlike the WiMAX Forum of today, the original group was concerned with single-carrier systems above 11 GHz. This was due to the facts that this aspect of IEEE Std 802.16 was much further developed than the lower frequency specifications and that it was ETSI BRAN HiperACCESS to which they were reacting.

By the way, "WiMAX" is derived from the expression *worldwide interoperability for microwave access*. Jay Klein, who chaired the WirelessMAN-SC PHY Task Group, has been credited with creating the name.

The initial task of the early WiMAX Forum was to develop system profiles for WirelessMAN-SC operation. These were submitted during the development of IEEE Std 802.16c, which specified detailed system profiles in Clause 12. Following draft development and balloting, these profiles, with minor changes, were approved; and IEEE Std 802.16c was approved in December 2002. The WirelessMAN-SC profiles are shown in Table 15–1.

Table 15–1:   WirelessMAN-SC system profiles

Identifier	Description
profM1	Basic ATM MAC profile
profM2	Basic packet MAC profile
profP1	25 MHz channel PHY profile
profP1f	25 MHz channel PHY profile—FDD
profP1t	25 MHz channel PHY profile—TDD
profP2	28 MHz channel PHY profile
profP2f	28 MHz channel PHY profile—FDD
profP2t	28 MHz channel PHY profile—TDD

Once this task was accomplished, the next step was to deal with the issue of test specifications. With the aid of other companies such as Nera, Intracom, Hewlett-Packard, and Hughes Network Systems, the founders of the WiMAX Forum started to create the test specification necessary to be on par with the ETSI test specifications. Four documents were identified:

1. Protocol implementation conformance statement (PICS) proforma: A document that serves two purposes. First, it lists the features that are mandatory, optional, conditionally mandatory, forbidden, etc., for each of the system profiles. Second, it serves as a form that vendors fill out to specify which features they implement and for what sets or ranges of allowed values.

2. Test suite structure and test purposes (TSS&TP): A test plan for covering the features outlined in the PICS proforma.

3. Abstract test suite (ATS): Equipment-independent test procedures for implementing the tests identified in the TSS&TP.

4. Radio conformance test (RCT) specification: Tests for the radio portion of the system that could not be specified as protocol tests.

CONFORMANCE TEST STANDARDS FOR WIRELESSMAN-SC

The WiMAX Forum played a key role in the development of PICS proforma, TSS&TP, and RCT documents in IEEE Std 802.16. In each case, a new IEEE 802.16 project was proposed and initiated for each test specification, each to lead to an independent published standard. Once the project was underway, an open call for contributions was issued; in each case, a single proposal, developed within the WiMAX Forum, was received.

These proposals were unchallenged primarily because of the involvement of the same participants in both the WiMAX Forum and the IEEE 802.16 Working Group. Nonetheless, the documents proceeded through the official IEEE 802 process of working group review, working group letter ballot, and sponsor ballot, with comments addressed along the way. The result was three official IEEE standards: IEEE Std 802.16/Conformance01-2003 [B21], IEEE Std 802.16/Conformance02-2003 [B22], and IEEE Std 802.16/ Conformance03-2004 [B23], each in accordance with ISO/IEC 9646.

By the time the RCT document was initiated, relations between companies participating only in ETSI BRAN HiperACCESS and those interested in both ETSI BRAN HiperACCESS and IEEE Std 802.16 had improved, and the ETSI BRAN HiperACCESS RCT was used as the foundation of the WirelessMAN-SC RCT document. Unfortunately, before the ATS document could be initiated, the bottom fell out of the market for high-frequency broadband wireless access equipment. Consequently, the WirelessMAN-SC ATS standards project was never initiated.

THE WiMAX FORUM'S MOVE TO LOWER FREQUENCIES

In April 2002, as the specifications for systems below 11 GHz were solidifying, entities interested in operation below 11 GHz began to join the WiMAX Forum. The first was the OFDM Forum in April, followed by Wi-LAN and Fujitsu in November 2002. Then, in January of 2003, the WiMAX Forum jumped to twelve "charter member" companies and, fueled by the efforts of Intel Corp., started expanding rapidly. The WiMAX Forum, with well over 300 member companies, is now certifying fixed-access WirelessMAN-OFDM systems through independent laboratory conformance testing and overseeing interoperability "plug-fests." The WiMAX Forum is devoting great energy to the development of system profiles and test specifications for mobile systems based on the enhancements in IEEE Std 802.16e.

LOWER FREQUENCY PROFILES AND TEST SPECIFICATIONS

The WirelessMAN-OFDM, WirelessMAN-OFDMA, and WirelessMAN-SC PHY specifications were the first foray of IEEE Std 802.16 into the lower frequencies (less than 11 GHz). These were first defined in IEEE Std 802.16a. In late 2002, as IEEE Std 802.16a was nearing completion, the IEEE 802.16 Working Group opened the amendment project IEEE P802.16d, which was intended to develop profiles in support of the low-frequency modes. Eventually, IEEE P802.16d was dropped as a project to amend IEEE Std 802.16, replaced by a broader revision project that eventually led to IEEE Std 802.16-2004 [B20]. That revised standard does include the low-frequency profiles.

IEEE Std 802.16-2004 specifies profiles for all three PHY mode alternatives. However, for this fixed-access case, WirelessMAN-OFDM was of the greatest interest for two reasons:

1. The parallel ETSI BRAN HiperMAN standard initially supported a mode compatible with WirelessMAN-OFDM only.

2. The WiMAX Forum contributed only to the development of the WirelessMAN-OFDM and developed fixed-access interoperability tests only for that mode.

The low-frequency profiles are considerably more complex set than those for WirelessMAN-SC. In the case of WirelessMAN-OFDM, a system profile consists of four components: a MAC profile, a PHY profile, a duplexing selection, and a power class. The IEEE 802.16 standard defines two MAC profiles (PMP and mesh), six PHY profiles (1.75, 3.5, 7, 3, 5.5, and 10 MHz channel bandwidth, with the last one specified in the license-exempt case only), two duplexing alternatives (TDD and FDD), and five power classes.

In the case of WirelessMAN-OFDMA, only one MAC profile (PMP-based) is specified, along with nine PHY profiles (1.25, 3.5, 7, 8.75, 14, 17.5, 28,

10, and 20 MHz channel bandwidth, with the last two specified in the license-exempt case only), two duplexing alternatives (TDD and FDD), and four power classes.

The ATM CS is not included in either the WirelessMAN-OFDM or the WirelessMAN-OFDMA profiles. In both cases, RF profiles are mentioned, but not specified except in the license-exempt case.

The IEEE P802.16/Conformance04 project, to specify PICS proforma for the low-frequency modes, was initiated in March 2004. Draft 6, which entered IEEE-SA sponsor ballot in January 2006, includes no specification for WirelessMAN-SCa, which had lost support by the time the project got underway. The draft does include a detailed PICS specification for WirelessMAN-OFDMA. It should be noted that this PICS proforma concerns only fixed systems, not any mobile systems. In the case of WirelessMAN-OFDM, the draft simply makes normative reference to ETSI 102 385-1 V1.2.1, entitled "Broadband Radio Access Networks (BRAN); HiperMAN; Conformance testing for the Data Link Control Layer (DLC); Part 1: Protocol Implementation Conformance Statement (PICS) proforma." Since ETSI BRAN HiperMAN has long supported a PHY compatible with WirelessMAN-OFDM and since ETSI is recognized for its expertise in conformance test specifications, it is natural to defer to the ETSI standard on this issue. It is worth noting, however, that ETSI defers to IEEE standards as well. In fact, the ETSI test specifications have, in many cases, pointed not to ETSI air interface standards but to IEEE Std 802.16 for their references.

WiMAX Forum conformance testing for fixed access

The only publicized industry activities related to the testing of conformance to IEEE Std 802.16 have been developed in the WiMAX Forum. In January 2006, the WiMAX Forum announced that four products (two BSs, one BS "solution," and one SS "solution") had been certified for 3.5 GHz operation following testing at its laboratory in Malaga, Spain. It also announced plans to test an additional 26 products. Protocol conformance, radio conformance, and basic interoperability were certified. Some additional capabilities, including QoS and security, were not tested. Certification testing of these features is expected later in 2006. Testing of systems operating in the 5.8 GHz bands is also in development.

WiMAX FORUM AND MOBILE BROADBAND WIRELESS ACCESS

With the progress of the amendment project IEEE P802.16e, the WiMAX Forum developed an interest in this work also. The WiMAX Forum eventually came to the conclusion that it would strongly support the evolution of the IEEE 802.16 standard from fixed to mobile applications and would commit itself to the development of test specifications for mobile systems defined in IEEE Std 802.16e. Most of the WiMAX Forum effort on mobility has been directed at the WirelessMAN-OFDMA mode. Work to create the test specifications is well underway. This work is being undertaken jointly between ETSI and the WiMAX Forum based on agreement between the two. An activity related to the mobile enhancement of WirelessMAN-OFDM is also underway, but this has seen considerably less support among the member companies of the WiMAX Forum.

Chapter 16 Related standards
Other wireless standards with similar applications

This chapter opens by comparing the IEEE 802.16 (WirelessMAN) standard to the IEEE 802.11 (WLAN) standard. It next overviews progress in two IEEE 802 working groups, IEEE 802.20 and IEEE 802.22, that are engaged in developing standards, but have not at this time completed any. The chapter reviews some history and the status of related projects in ETSI's BRAN activity, briefly adding reports on the relation between the IEEE 802.16 Working Group and two other regional standardization activities: the Korean Telecommunication Technology Association (TTA) and the China Communications Standards Association (CCSA). We close the chapter with a review of the important relationships between IEEE Std 802.16 and the ITU, particularly those leading to the international recommendation of IEEE Std 802.16 through the ITU.

IEEE STD 802.11

The IEEE 802.11 standard defines MAC, PHY, and management protocols for WLANs. A LAN typically has a range on the order of 100 m and serves a space the size of a home, an office building, or a small campus. The distinction between LANs and MANs is especially important in the wireless case because the choices for MAC and PHY protocols and parameters essentially dictate the type of network that can be built. WLANs and wireless MANs have widely different requirements, and a MAC or PHY designed for one type of network may be unsuitable for the other.

Since its initial ratification in 1997, IEEE Std 802.11, and its later amendments, have become the foundation for a large and growing market. WLANs based on IEEE Std 802.11 are an essential network technology. With their standardization and subsequent widespread adoption in homes and enterprises, IEEE 802.11 devices—known as *Wi-Fi* to the general public, due to the certification mark of the Wi-Fi Alliance—have become commodities in recent years. The following subsections summarize the IEEE 802.11 MAC and PHY and discuss the similarities and differences between IEEE Std 802.11 and IEEE Std 802.16.

IEEE 802.11 MAC

Medium access control layer (MAC)

The IEEE 802.11 MAC is based on the carrier sense multiple access with collision avoidance (CSMA/CA) protocol, in which IEEE 802.11 devices primarily follow a "listen before talk" rule, with random backoff to avoid collisions. The fundamental premise of CSMA is that all devices sharing the channel can hear each other. Most IEEE 802.11 devices have omnidirectional antennas and, in the local area of primary application, can hear each other. Some mechanisms [e.g., the request to send and clear to send (RTS/CTS) mechanism] do exist, albeit at a higher overhead, to partially address the "hidden node" case in which some nodes are within range of the AP (analogous to an IEEE 802.16 BS), but not within range of each other.

The IEEE 802.11 CSMA/CA protocol is similar to the CSMA/CD protocol of Ethernet. In the wireless case, carrier sensing is possible, except in the hidden node situation. However, collision detection (the "CD" of Ethernet) is not. An Ethernet node continues to sense the medium during transmission and is thereby able to detect collisions. IEEE 802.11 devices are half-duplex and, therefore, cannot listen while transmitting. In fact, given the fact that IEEE Std 802.11 uses the same frequency for UL and DL, it would be a significant challenge to design an IEEE 802.11 device that could accurately sense the medium for collisions during transmission.

In lieu of collision detection, IEEE 802.11's collision avoidance process uses a physical carrier sense and a virtual carrier sense mechanism are used to avoid potential collisions. The physical carrier sense mechanism listens for the presence of any other transmissions on the channel before attempting a transmission of its own. The virtual carrier sense mechanism utilizes the explicit indication from another transmitter that essentially reserves the channel for a specific amount of time.

Since CSMA/CA is a fully distributed scheme with no central coordinator, collisions are still possible. The absence of collision detection makes it essential to use an acknowledgment scheme to ensure successful reception. IEEE Std 802.11 uses immediate MAC acknowledgments to confirm that an IEEE 802.11 frame has reached its destination. This is essentially a stop-and-wait ARQ scheme with synchronous acknowledgment, where the transmitter waits for the acknowledgment before attempting the next transmission or repeating the previous one. In the IEEE 802.11 MAC, the immediate acknowledgment is mandatory for all unicast frames[7]. Note that, in IEEE Std 802.11, the word *frame* is used in both the MAC sense (as a single MAC PDU, carrying data, management, or control content) and in the PHY sense (as a unit of PHY transmission) because each PHY frame carries one MAC frame. In contrast, IEEE Std 802.16 uses *frame* only as a PHY transmission unit; an IEEE 802.16 frame may carry one or more MAC PDUs or partial MAC PDUs. Although IEEE Std 802.11 supports fragmentation, it is optional and seldom implemented by IEEE 802.11 devices. The optional fragmentation is not dynamic and uses a fixed, configurable fragment size.

The duplexing method used in IEEE Std 802.11 is TDD, which is half-duplex by definition. However, in IEEE Std 802.11, no centralized coordinator allocates time slots; instead, transmission decisions are taken independently at each transmitter. The IEEE 802.11 AP, which is analogous to the IEEE 802.16 BS, has no special rules for accessing the media. The AP has to contend for the media just like any other IEEE 802.11 client communicating with it.

MAC overhead

The IEEE 802.11 MAC requires a minimum of 24 bytes of MAC header information per frame. Note that an IEEE 802.11 MAC frame can carry only one higher layer SDU; i.e., multiple higher layer SDUs such as IP packets cannot be packed into a single MAC frame. For example, each 40 byte TCP acknowledgment would require at least 42 bytes of MAC overhead (24 bytes of MAC header + 4 bytes of CRC + 14 bytes of acknowledgment) in addition to the per-frame PHY overhead. The PHY rates must be considered in this

---

[7]   IEEE Std 802.11e™-2005 includes an option to omit immediate acknowledgments in favor of delayed acknowledgments.

MAC discussion because the IEEE 802.11 acknowledgment is typically sent at one of the more robust PHY rates. A large frame sent at a lower rate can occupy the channel for a significant amount of time.

The transmission rate to be used is also chosen independently by the transmitters. The standard does not specify an algorithm to choose the transmission rate, and vendors implement their own rate-selection algorithm. However, an IEEE 802.11 AP can restrict the number of transmission rates that can be used in the network.

The IEEE 802.11 MAC supports two alternative channel access modes, the distributed coordination function (DCF) and the optional point coordination function (PCF). The DCF is the fully distributed scheme discussed in the previous paragraphs. PCF offers some centralized polling, although it is not scalable and the polling does not take the frame size or transmission rate into account. For these reasons, PCF is not widely supported and is rarely used.

MAC summary

The CSMA/CA protocol with DCF works reasonably well in a WLAN environment. However, it is not without problems in some situations. The following is a partial list of concerns in CSMA/CA WLAN:

- *Nondeterminism:* CSMA/CA is nondeterministic, and the stop-and-wait ARQ can make successful reception significantly unpredictable if frames need to be resent often. The situation can become drastically worse as the network load increases. Deterministic behavior is required for many QoS-sensitive applications.

- *Hidden nodes:* Two CSMA devices that cannot hear each other may simultaneously initiate transmission. A collision will result. Retransmission will take time and waste future transmission opportunities. This problem is somewhat addressed by the RTS/CTS-based reservation protocol, but its effectiveness declines with a large number of hidden nodes.

- *Near-far problem:* The near-far problem refers to a scenario in which devices closer to an IEEE 802.11 AP can overpower devices farther away, thus unfairly affecting the performance of those devices.

- **Slowest node:** Since transmission decisions are taken independently, one or more nodes choosing to transmit at the lowest rate will, by using a less efficient communication scheme, occupy a disproportionate fraction of airtime. This adversely affects other devices that can communicate at a higher rate.

- **Capture effect:** The capture effect is a fundamental problem in any CSMA protocol, including wired networks such as Ethernet. Since the CSMA devices make independent decisions on transmissions and backoff, the device that successfully completes a transmission resets its contention window and starts afresh for the next contention. This provides an unfair advantage to the device with a successful transmission when multiple device are contending for access.

- **Overhead:** In general, the IEEE 802.11 MAC has high overhead, especially for smaller frames such as voice or TCP acknowledgments.

IEEE 802.11 PHY

IEEE Std 802.11 has standardized several PHY modes, including the following:

- Frequency-hopping spread spectrum (FHSS), up to 2 Mbit/s
- Direct sequence spread spectrum, up to 11 Mbit/s, in IEEE Std 802.11b™
- OFDM, up to 54 Mbit/s, in IEEE Std 802.11a™ and IEEE Std 802.11g™

The quoted data rates are raw PHY rates. For instance, the maximum IEEE 802.11a PHY rate is 54 Mbit/s, but the highest actual UDP throughput with maximum packet size is only about 30 Mbit/s due to the MAC and PHY overheads.

The OFDM PHY of IEEE Std 802.11a and IEEE Std 802.11g is based on a 64-FFT. The channel size is approximately 20 MHz. The maximum delay spread is about 800 ns, which is sufficient for a LAN with reasonable multipath.

IEEE 802.11 extensions

The IEEE 802.11 Working Group has recently developed and is currently developing many amendment projects, including the following:

- *IEEE Std 802.11e:* IEEE Std 802.11e-2005 defines MAC extensions to CSMA/CA in order to improve QoS. Some MAC efficiency improvements are also included. The QoS extensions improve both DCF and PCF by defining a new hybrid coordination function (HCF), which includes enhanced distributed channel access (EDCA) and the HCF coordinated channel access (HCCA) functions. Both EDCA and HCCA schemes are backward compatible with IEEE 802.11 DCF and frame formats. HCCA offers some centralized scheduled functions. Some of the efficiency improvements include the option of not having immediate acknowledgments and allowing block (i.e., delayed) acknowledgments. IEEE Std 802.11e works for any of the underlying PHY modes.

- *IEEE P802.11n:* The primary objective of the amendment project IEEE P802.11n is to improve the throughput of IEEE Std 802.11 beyond 100 Mbit/s, using techniques such as SDMA and MIMO. Both PHY and MAC enhancements are under consideration. This project is expected to make use of most of the IEEE 802.11e features and to define additional efficiency mechanisms. The project is expected to achieve most of the throughput improvements by doubling the channel size. Completion of the standard is expected in 2007.

- *IEEE P802.11p:* The objective of the amendment project IEEE P802.11p is to support communication between vehicles, and from vehicles to the roadside, while operating at speeds up to 200 km/h for communication ranges up to 1000 m. This is specifically targeted for the 5 GHz bands and in all forms of surface transportation. Both MAC and PHY enhancements are being considered, including a smaller channel size of 10 MHz.

- *IEEE P802.11s:* The amendment project IEEE P802.11s is working on defining a mesh MAC extension to IEEE Std 802.11. The objective is to use the existing IEEE 802.11 addressing format (especially the four-address format) to fully define a multihop network with autoconfiguration capabilities.

Using IEEE Std 802.11 as a MAN

Although IEEE Std 802.11 and its extensions were designed to operate in a LAN environment, many IEEE 802.11 systems, sometimes with modifications, are being operated in long-range outdoor MAN environments. To extend the range, most of these systems make use of high-gain directional antennas to support clients communicating with a (standard or modified) IEEE 802.11 AP. This situation massively exacerbates the hidden node problem, forcing every terminal to be hidden from every other. MAC customization becomes particularly important. Many proprietary and nonstandard extensions to IEEE Std 802.11 have been applied, including the following:

- *IEEE 802.16-like MAC over IEEE 802.11 PHY:* Some IEEE 802.11 chipsets allow an implementor to use the IEEE 802.11 PHY, but bypass the IEEE 802.11 MAC. Some vendors have used such an approach to implement a TDMA scheme. However, the PHY overhead and some of the MAC overhead cannot be eliminated using this approach. Similarly, the limitations of the IEEE 802.11 PHYs, such as the limited ability to address the delay spread typical of long-range outdoor environments, cannot be eliminated either.

- *Modified IEEE 802.11 MAC:* Some of the proprietary modifications include modification of basic IEEE 802.11 timing or channel access parameters, the turning off of acknowledgments, and modification of the acknowledgment timeout. The IEEE 802.11 standard requires that the immediate acknowledgment be received within a specific time. This time is sufficient to deal with smaller propagation delays in a LAN. However, long-range communication such as a MAC requires longer propagation delays.

- *FDD:* Some systems use an FDD-like approach, with communications in opposite directions using different IEEE 802.11 channels.

Quantitative comparison of IEEE Std 802.11 and IEEE Std 802.16

In general, the IEEE 802.11 and IEEE 802.16 technologies are complementary, as one is targeted for LANs and the other for MANs. However, we need to summarize the differences between the two standards in order to stress the differences in the target markets, which is reflected in the design choices made. Table 16–1 shows some of these differences.

Table 16–1:   Comparison of IEEE Std 802.11 and IEEE Std 802.16

Feature	IEEE Std 802.11	IEEE Std 802.16
Medium access	CSMA/CA Some TDMA-like functions in IEEE Std 802.11e	TDM/TDMA OFDMA (optional)
Channel size	~22 MHz (IEEE Std 802.11b) 20 MHz (IEEE Std 802.11a and IEEE Std 802.11g) 10 MHz (IEEE Std 802.11j™)	1.25–28 MHz
Maximum PHY data rate	54 Mbit/s	72 Mbit/s (10 MHz channel)
Maximum (bit/s)/Hz	~2.7	~5
Delay spread tolerance	Up to 800 ns	Up to 10 µs
Minimum MAC overhead	28 bytes (header + CRC)	6 bytes
QoS	Limited QoS support (up to eight priorities per system) in IEEE Std 802.11e	Per-connection QoS

Although IEEE Std 802.11e supports a form of QoS, with eight priorities (or classes), the addition of QoS to IEEE Std 802.11 is an afterthought. On the other hand, the IEEE 802.16 MAC is a QoS-centric MAC that has been designed with QoS as an integral part of the MAC functions. Moreover, the IEEE 802.16 MAC is much more efficient than IEEE Std 802.11 for outdoor applications, with packing, ARQ, dynamic fragmentation, and smaller header overhead. The problems associated with the IEEE 802.11 CSMA/CA protocol, such as hidden nodes, capture effect, and the near-far problem, are eliminated by the use of a centrally scheduled IEEE 802.16 MAC with TDM/TDMA (and optionally OFDMA).

Ultimately, IEEE Std 802.11 and IEEE Std 802.16 are two parts of a team that can form the basis of a flexible network with an extensive range of applicability.

IEEE 802.20 WORKING GROUP

The IEEE 802.20 Working Group, chartered to develop a standard for mobile broadband access supporting vehicular speeds, first met formally in March 2003. The goal is to support operation in licensed bands below 3.5 GHz allocated to the mobile service. The main performance targets given in the IEEE P802.20 PAR are shown in Table 16–2.

Table 16–2:   IEEE 802.20 performance targets

Characteristic	Target value
Mobility	Vehicular mobility classes up to 250 km/h (as defined in ITU-R M.1034-1)
Sustained spectral efficiency	>1 (bit/s)/Hz per cell
Peak user data rate	> 1 Mbit/s in paired 1.25 MHz channel
Peak aggregate data rate per cell in the DL	>4 Mbit/s in 1.25 MHz channel
Peak aggregate data rate per cell in the UL	> 800 kbit/s 1.25 MHz channel
Airlink MAC round-trip time	< 10 ms
Duplex	FDD, TDD
Channel bandwidth	1.25 MHz, 5 MHz

At the moment, very little can be said about the technology itself, and whether it meets the targets, as the group has addressed mostly the technology selection process and system requirements rather than developing draft specifications. The working group first considered proposals for draft text in

November 2005, but the process is in its early stages. It is too soon to speculate on the content of the future standard.

IEEE 802.22 WORKING GROUP

The IEEE P802.22 PAR on wireless regional area networks (WRANs) was approved in September 2004. The PAR covers "Cognitive Wireless RAN Medium Access Control (MAC) and Physical Layer (PHY) specifications of fixed point-to-multipoint wireless regional area networks operating in the VHF/UHF TV broadcast bands between 54 MHz and 862 MHz." This range of operation is based on U.S. frequency allocations and implies 6 MHz channelization. For other scenarios, the broader frequency range of 41 MHz to 910 MHz may be considered a target, with 6 MHz, 7 MHz, or 8 MHz channelization. However, the PAR specially addresses U.S. rules promulgated by the FCC.

The aim of the project is to develop a fixed broadband access standard for use in licensed but unused broadcast television spectra. Such a network is envisioned to have a long range, yet prevent interference with incumbent licensed services within the same spectrum. A primary objective is to provide rural, low-population markets with performance levels similar to those of broadband access technologies servicing urban and suburban areas. The WRAN must support QoS, enabling simultaneous support for data, voice, and audio/video applications, and must be available 99.9% of the time.

The target markets to be addressed by such a system are residential, small office/home office (SOHO), small business, multitenant dwellings, and campuses. IEEE P802.22 is intended to provide these customers with capabilities similar to those provided by digital subscriber line (ADSL) and cable modems, but more economically in sparsely populated rural settings. The system is expected to have a typical range of 33 km, with a maximum range of 100 km. The minimum peak throughput at cell edge is required to be 1.5 Mbit/s per subscriber in the DL and 384 kbit/s per user in the UL. To achieve this, the PAR proposes a spectral efficiency goal of 5 (bit/s)/Hz.

Use of the VHF/UHF TV broadcast frequency range is possible in near-LOS and NLOS situations. In the rural settings targeted, operation may include partial or complete blockage by foliage, contributing to multipath and attenuation of the signal. We may expect that macrodiversity, repeaters, and other techniques will be combined with the natural propagation characteristics of the frequency band to solve the issues of range, multipath, and blockage.

The problems faced by systems operating in this spectrum are compounded by the fact that they are operating in an license-exempt mode in spectrum that technically is licensed to other uses. These other uses are not limited to broadcast TV services that would be relatively easy to map and avoid. They also include FCC Part 74 devices, such as licensed wireless microphones, wireless intercoms, etc. Regulations require that license-exempt users detect these licensed users and avoid interfering with them by moving to a different frequency. This may, however, be very challenging because some of the licensed users are transient and low power in nature.

To better ensure the proper avoidance of incumbents, designers envision that subscriber devices (and repeaters) will be in a master-slave relationship with the BS, with the BS providing centralized power and spectrum management and centralized scheduling control. They also imagine distributed sensing of the RF environment by the subscriber devices for the purpose of detecting incumbents, with the sensing also controlled by the BS.

In terms of services supported, the clear intent of the IEEE P802.22 functional requirements document (see [B46]) is to support variable-length IP-based services with some level of QoS.

There is currently some interest in using a version of IEEE Std 802.16, modified to provide the intended detection and avoidance of incumbents, in order to meet the needs of the IEEE P802.22 PAR.

ETSI BRAN

Within the European Telecommunications Standards Institute (ETSI), the Broadband Radio Access Networks (BRAN) Technical Committee has been active since 1997. Its three primary activities include ETSI BRAN HiperLAN/2 (a WLAN standard that is, in effect, defunct) and two broadband wireless access standards: ETSI BRAN HiperACCESS and ETSI BRAN HiperMAN.

The IEEE 802.16 Working Group has had an active liaison relationship with ETSI BRAN since 1999; for example, the two groups have exchanged over 60 documents and reports. The initial collaboration strove for harmonization between the WirelessMAN-SC and ETSI BRAN HiperACCESS projects, in which ETSI was the first mover. Although this effort made some progress, it ultimately failed. Later, the attention turned to the lower frequency work, in which IEEE Std 802.16 went first. In this case, the harmonization results were remarkably successful. ETSI BRAN HiperMAN, as discussed below, is compatible with the WirelessMAN-OFDM and WirelessMAN-OFDMA modes in IEEE Std 802.16.

ETSI BRAN HiperACCESS

ETSI BRAN HiperACCESS defines PMP broadband wireless for use in the frequency bands between 11 GHz and 66 GHz. In this respect, it attempts to fill the same niche as IEEE 802.16 WirelessMAN-SC.

The ETSI BRAN HiperACCESS standard was developed by many of the same participants involved in creating WirelessMAN-SC. Some parties, however, participated in only one. The result was less harmonization than would ultimately have been desirable. Both standards assume LOS operation and specify a single-carrier PHY. Many other similarities exist also. However, the two also have significant differences and are not compatible.

ETSI BRAN HiperACCESS shares the following features with IEEE 802.16 WirelessMAN-SC:

- Single-carrier modulation, with QPSK, 16-QAM, and 64-QAM
- Similar FEC
- Same basic request/grant scheme
- Same CoSs and QoS concepts
- Same basic ranging, power control, and PHY maintenance
- Similar UL and DL MAP structure
- Similar frame structure, including support for TDD, FDD, and H-FDD
- Similar RCT specification

ETSI BRAN HiperACCESS differs from IEEE 802.16 WirelessMAN-SC in the following ways:

- Fixed-length packet ATM cell transport only
- No packing
- No fragmentation
- Far fewer options of FEC combinations (a few options are defined, tailored for the size of an ATM cell)
- No incremental BW requests
- No piggybacked BW requests

The result was a standard that is simpler to implement than WirelessMAN-SC, but considerably less efficient (even for ATM cells) and not forward looking to IP as the preferred network protocol.

ETSI BRAN HiperMAN

When ETSI BRAN initiated the HiperMAN activity to consider fixed broadband wireless access below 11 GHz, the amendment project IEEE P802.16a was well on its way to adding the new PHY specifications WirelessMAN-OFDM, WirelessMAN-OFDMA, and WirelessMAN-SCa, all operating under the IEEE 802.16 MAC, first defined in IEEE Std 802.16-2001. Thanks to active cooperation between the IEEE 802.16 Working Group and the ETSI BRAN HiperMAN activity, a unified solution was obtained. One key player in this success was Mariana Goldhamer, who served as the IEEE 802.16 Working Group's liaison official to ETSI BRAN while also chairing the ETSI BRAN HiperMAN activity and later serving as vice chair of the ETSI BRAN Technical Committee. Another key player was Nico van Waes, the technical editor of the IEEE 802.16a and ETSI BRAN HiperMAN drafts as well as the area coordinator (ETSI terminology for chair) of ETSI BRAN HiperMAN during that period.

Rather than adapting the MAC from ETSI BRAN HiperACCESS, the ETSI BRAN HiperMAN group came to the conclusion that it would instead adopt the IEEE 802.16 MAC. However, it sought a single PHY solution, rather than three of them. It came to choose WirelessMAN-OFDM, eventually supporting a standard that was a subset of WirelessMAN-OFDM. The ETSI project was not passive in the collaboration; it contributed ideas and comments to IEEE Std 802.16 during the refinement of that standard. Notably, ETSI moved first to specify UL subchannelization; this was not added into WirelessMAN-OFDM until IEEE Std 802.16-2004 [B20].

More recently, in 2005, the scope of the ETSI BRAN HiperMAN project was extended to include nomadic wireless access. At this stage, ETSI BRAN HiperMAN moved to extend its support of IEEE Std 802.16 and incorporated not only WirelessMAN-OFDM but also WirelessMAN-OFDMA.

As discussed in Chapter 15, ETSI is very active in the development of conformance test specifications for IEEE Std 802.16, particularly through its agreement with the WiMAX Forum. ETSI conformance test documents have referenced IEEE Std 802.16 as their source standards. IEEE Std 802.16 has already moved to accept the notion of making normative reference to ETSI conformance test documentation.

OTHER REGIONAL STANDARDS ACTIVITIES

Beyond its active relationship with ETSI, the IEEE 802.16 Working Group has developed and maintained relationships with some other regional organizations that are interested in broadband wireless access standardization.

Korean Telecommunication Technology Association (TTA) and WiBro

The Korean TTA approached the IEEE 802.16 Working Group in January 2004 with information concerning its effort to develop a parallel standard in the PG05 project, later renamed PG302. Unlike the ETSI BRAN interest, which was on fixed access, the TTA focus was on mobile access. Also, in contrast to ETSI BRAN HiperMAN's interest in WirelessMAN-OFDM, it was clear that the TTA focus was on WirelessMAN-OFDMA. Early in the process, it appeared that the TTA was headed toward a standard similar to, but different from, the WirelessMAN-OFDMA that was eventually approved in IEEE Std 802.16e. However, a number of letters and reports were exchanged, and meetings were held to coordinate activity. Eventually, the two projects began to see significant alignment.

In July 2004, the Korean Ministry of Information and Communication announced that service in the 2.3 GHz spectrum, which had been set aside for portable Internet service, "must comply with IEEE 802.16-2004 and IEEE 802.16e/Draft3 or later versions." The Korean regulations also stipulate some additional specifications, including use of the WirelessMAN-OFDMA mode with a 1024-FFT. The Korean government named this service *WiBro*, for wireless broadband. It is important to understand that WiBro is a service, not a technology. The regulations required that the technology be WirelessMAN-OFDMA.

In retrospect, the IEEE 802.16 Working Group's decision to support multiple modes turns out to have been well justified. The working group, in trying to develop a single standard for worldwide use, had a major challenge because the world includes many diverse interests. In this case, the Europeans leaned toward an OFDM solution and the Koreans toward OFDMA. The multimode decision included in IEEE Std 802.16a formed a basis to satisfy both.

China Communications Standards Association (CCSA)

The CCSA has been pursuing broadband wireless access standards through Working Group 3 of its Technical Committee 5. CCSA has been communicating actively with the IEEE 802.16 Working Group since IEEE Std 802.16's Session #31, which was held in Shenzhen, China, in May 2004. Since then, the organizations have remained in contact through letters, reports, and liaison exchanges. In November 2005, CCSA reported a number of standards projects in fixed access related to IEEE Std 802.16, including test specifications. Work regarding mobile broadband wireless access was at an earlier stage.

INTERNATIONAL TELECOMMUNICATIONS UNION (ITU)

Historically, some of the IEEE 802 standards have been entered into the more formal arena of "international" standardization through IEEE 802's engagement with Joint Technical Committee 1 (JTC1), the committee on information technology standards administered by two private Geneva-based organizations, the International Organization for Standardization (ISO) and the International Electrotechnical Commission (IEC). Although the original collaborations covered only wired networks, the partnership was extended to wireless networks also by the adoption of IEEE Std 802.11 as ISO/IEC 8802-1 in 1999.

In spite of the successful relationship between IEEE 802 and ISO/IEC JTC1, the IEEE 802.16 Working Group has never shown an interest in following that precedent. The working group has always recognized the value of engaging with the formal international process in Geneva. However, instead of ISO/IEC, IEEE Std 802.16 looks to partner with the ITU, particularly the ITU Radiocommunications Sector (ITU-R). This is partly because IEEE Std 802.16 is the only IEEE 802 standard expected to operate significantly in licensed spectra, the administration of which is significantly dependent on the ITU. It is also because the ITU has always been a focal point of the telecommunications industry. IEEE Std 802.16 is a convergent technology, drawing significantly from the data communications community, but targeting applications that have historically been supported by the telecommunications establishment. Many of the potential customers for IEEE 802.16 systems have a natural home in ITU. The IEEE 802.16 community naturally seeks to engage that community and encourage its participation.

From the beginning, IEEE Std 802.16 has looked to engage with ITU. The original IEEE 802.16 PAR, submitted in March 1999, in answering the question "Is this standard planned for adoption by another international organization?" responded "Yes," referencing ITU. Since then, IEEE 802.16 PARs have done so consistently.

Central to building the IEEE 802.16 Working Group's relationship with ITU-R was Dr. José Costa, who has served since Session #1 in 1999 as the working group's ITU-R liaison official. Dr. Costa originally helped to connect the working group with ITU-R's Joint Rapporteurs Group 8A/9B, on Wireless Access Systems, where Dr. Costa served as one of two principle rapporteurs.

The relationship strengthened significantly in October 2003, when ITU-R named IEEE as a Sector Member, under the same membership classification as ISO. Shortly before, the ITU-R's Strategy and Policy Unit had published its seminal document *The Birth of Broadband* [B28], which made numerous references to the role of IEEE Std 802.16, particularly regarding the expansion of Internet access in rural and developing areas. ITU, perhaps as a consequence, invited the IEEE 802.16 Working Group chair to speak at its major conference, ITU Telecom World, in Geneva during October 2003. During that meeting, the working group chair, with the assistance of ITU-R Counsellor Fabio Leite, met with staff and various study group officials, discussing opportunities to develop the relationship and, in particular, on drafting new ITU-R recommendations based on the IEEE 802.16 standard and its amendments, which were seen to be valuable because of ITU-R's "global approach and visibility." Following those discussions, the working group began to pursue contacts with several ITU study groups, including some in ITU's Telecommunications Bureau (related to network access and cable modem technology).

In November 2003, the IEEE 802.16 Working Group followed up these discussions with a letter to ITU-R indicating its support for the idea of developing ITU-R recommendations referring to IEEE Std 802.16. This work was slightly delayed by the termination of Joint Rapporteurs Group 8A/9B. However, with the assistance of Dr. Costa (who had assumed the chairmanship of Working Party 8A), the work progressed. IEEE, using its ITU-R membership, contributed a series of documents to Working Party 9B, coordinated with ETSI, on the development of a recommendation on "radio interface standards for broadband wireless access (BWA) systems in the fixed service operating below 66 GHz," pointing to IEEE Std 802.16 and the compatible ETSI BRAN HiperMAN standard. After many discussions and coordination actions, Study Group 9 approved this recommendation in December 2005. The ITU-R recommendation awaits a formal written consultation of the member administration in mid-2006.

In March 2005, the IEEE contributed a parallel proposal to ITU-R Working Party 8A, proposing a preliminary draft new recommendation on "radio interface standards for broadband wireless access systems in the mobile service operating below 6 GHz." The proposal would point to IEEE Std 802.16, including the IEEE 802.16e mobility amendment. The proposal has not progressed rapidly in Working Party 8A, mainly because supporters of a number of alternative technologies have become vocal. It is far from certain where this will lead.

Appendix A IEEE 802.16 headers, subheaders, and management messages

The list of IEEE 802.16 MAC headers, subheaders, and MAC management messages is provided here for quick reference. Table A–1 shows the list of MAC headers, subheaders, and special payloads.

Table A–1: MAC headers, subheaders and special payloads

Description	Size in bytes
Generic MAC header	6
BW request header	6
Fragmentation subheader (FSH)	1
FSH (ARQ)	2
FSH (Extended)	2
Grant management subheader (GMSH)	2
Packing subheader (PSH)	2
PSH (ARQ)	3
PSH (Extended)	3
Mesh subheader (MSH)	2
Fast-feedback allocation subheader (FFSH)	1

Table A–2 lists the management messages and the type of connections that may be used to send them.

Table A–2: Management messages

Type	Name	Description	Connection
0	UCD	UL channel descriptor	Fragmentable Broadcast
1	DCD	DL channel descriptor	Fragmentable Broadcast
2	DL-MAP	Downlink access definition	Broadcast
3	UL-MAP	Uplink access definition	Broadcast
4	RNG-REQ	Ranging request	Initial Ranging or Basic
5	RNG-RSP	Ranging response	Initial Ranging or Basic
6	REG-REQ	Registration request	Primary Management
7	REG-RSP	Registration response	Primary Management
8		Reserved	
9	PKM-REQ	Privacy key management request	Primary Management
10	PKM-RSP	Privacy key management response	Primary Management
11	DSA-REQ	Dynamic service addition request	Primary Management
12	DSA-RSP	Dynamic service addition response	Primary Management
13	DSA-ACK	Dynamic service addition acknowledgment	Primary Management
14	DSC-REQ	Dynamic service change request	Primary Management

Table A–2: Management messages *(Continued)*

Type	Name	Description	Connection
15	DSC-RSP	Dynamic service change response	Primary Management
16	DSC-ACK	Dynamic service change acknowledgment	Primary Management
17	DSD-REQ	Dynamic service deletion request	Primary Management
18	DSD-RSP	Dynamic service deletion response	Primary Management
19		Reserved	
20		Reserved	
21	MCA-REQ	Multicast assignment request	Primary Management
22	MCA-RSP	Multicast assignment response	Primary Management
23	DBPC-REQ	DL burst profile change request	Basic
24	DBPC-RSP	DL burst profile change response	Basic
25	RES-CMD	Reset command	Basic
26	SBC-REQ	SS basic capability request	Basic
27	SBC-RSP	SS basic capability response	Basic
28	CLK-CMP	SS network clock comparison	Broadcast
29	DREG-CMD	De/Re-register Command	Basic
30	DSX-RVD	DSx received message	Primary Management

Table A–2:   Management messages *(Continued)*

Type	Name	Description	Connection
31	TFTP-CPLT	Config file TFTP complete message	Primary Management
32	TFTP-RSP	Config file TFTP complete response	Primary Management
33	ARQ-Feedback	Stand-alone ARQ Feedback	Basic
34	ARQ-Discard	ARQ Discard	Basic
35	ARQ-Reset	ARQ Reset	Basic
36	REP-REQ	Channel measurement report request	Basic
37	REP-RSP	Channel measurement report response	Basic
38	FPC	Fast power control	Broadcast
39	MSH-NCFG	Mesh network configuration	Broadcast
40	MSH-NENT	Mesh network entry	Basic
41	MSH-DSCH	Mesh distributed schedule	Broadcast
42	MSH-CSCH	Mesh centralized schedule	Broadcast
43	MSH-CSCF	Mesh centralized schedule configuration	Broadcast
44	AAS-FBCK-REQ	AAS feedback request	Basic
45	AAS-FBCK-RSP	AAS feedback response	Basic
46	AAS_Beam_Select	AAS beam select	Basic
47	AAS_BEAM_REQ	AAS beam request	Basic
48	AAS_BEAM_RSP	AAS beam response	Basic
49	DREG-REQ SS	SS deregistration message	Basic
50–255		Reserved	

Bibliography

[B1] Alamouti, S. M., "A simple transmit diversity technique for wireless communications," *IEEE Journal on Selected Areas of Communication,* vol. 16, pp. 1451–1458, Oct. 1998.

[B2] Alexiou, A., and Haardt, M., "Smart Antenna Technologies for Future Wireless Systems: Trends and Challenges," *IEEE Communications Magazine,* vol. 42, no. 9, pp. 90–97, Sept. 2004.

[B3] Ayanoglu, E., Eng, K. Y., and Karol, M. J., "Wireless ATM: Limits, Challenges and Proposals," *IEEE Personal Communications,* Aug. 1996.

[B4] Anderson, H. R., *Fixed Broadband Wireless System Design,* John Wiley & Sons, 2003.

[B5] Ball, C. F., Humburg, E., Ivanov, K., and Treml, F., "Performance analysis of IEEE 802.16 based cellular MAN with OFDM-256 in mobile scenarios," *Proceedings of IEEE Vehicular Technology Conference 2005-Spring,* vol. 3, pp. 2061–2066, May 2005.

[B6] Bertsekas, D., and Gallager, R., *Data Networks,* Prentice Hall, 1992.

[B7] Bing, B., *Broadband Wireless Access,* Kluwer Academic Publishers, 2000.

[B8] C50-20010820-026, "1xEV-DV Evaluation Methodology Addendum [V6]," July 2001, (http://ftp.3gpp2.org/TSGC/Working/2001/TSG-C_0108/ TSG-C-0801-Portland/WG5/C50-20010820-026%201xEV-DV-Evaluation-Methodology-Addendum-V6.doc).

[B9] Chandran, S. R., "A Selective Repeat ARQ Scheme for Point-to-Multipoint Communications and its Throughput Analysis," *ACM SIGCOMM Conference on Communications Architecture and Protocols,* 1986.

[B10] Cicconetti, C. , Lenzini, L. , Mingozzi, E., and Eklund, C., "Quality of Service Support in IEEE 802.16 Networks", *IEEE Network*, March/April 2006.

[B11] COST Action 231, "Digital mobile radio towards future generation systems, final report," tech. rep., European Communities, EUR 18957, 1999.

[B12] Doron, A., et al., "Problem in Interleaver and Permutation Combination in OFDMA," *IEEE C802.16maint-05/068r23*, July 2005.

[B13] Eklund, C., Marks, R. B., Stanwood, K. L., and Wang, S., "IEEE Standard 802.16: A Technical Overview of the WirelessMAN Air Interface for Broadband Wireless Access," *IEEE Communications Magazine*, pp. 98–107, June 2002.

[B14] ETSI TS 101 761-1 v1.2.1, Broadband Radio Access Networks (BRAN): HIPERLAN Type 2—Data Link Control (DLC) Layer—Part 1: Basic Data Transport Functions, 2000.

[B15] ETSI TS 101 350 v6.1.0, Digital Cellular telecommunication System (Phase 2+): General Packet Radio Service (GPRS)—Overall description of the GPRS Radio Interface Stage 2, 1998.

[B16] ETSI EN 301 958 v1.1.1, Digital Video Broadcasting (DVB); Interaction channel for Digital Terrestrial Television (RCT) incorporating Multiple Access OFDM, 2002.

[B17] ETSI TS 121 101, Universal Mobile Telecommunications System (UMTS); Technical Specifications and Technical Reports for a UTRAN-based 3GPP system.

[B18] Ghosh, A., Wolter, D. R., Andrews, J. G., and Chen, R., "Broadband wireless access with WiMax/802.16: Current performance benchmarks and future potential," *IEEE Communications Magazine*, vol. 43, no. 2, pp. 129–136, Feb 2005.

[B19] Housely, R., Polk, W., Ford, W., and Solo, D., "Internet X.509 Public Key Infrastructure Certificate and Certificate Revocation List (CRL) Profile," *RFC 3280*, Apr. 2002.

[B20] IEEE Std 802.16-2004, IEEE Standard for Local and Metropolitan Area Networks—Part 16: Air Interface for Fixed Broadband Wireless Access Systems.

[B21] IEEE Std 802.16/Conformance01-2003, IEEE Standard for Conformance to IEEE Std 802.16—Part 1: Protocol Implementation Conformance Statement (PICS) Proforma for 10–66 GHz WirelessMAN-SC Air Interface.

[B22] IEEE Std 802.16/Conformance02-2003, IEEE Standard for Conformance to IEEE Std 802.16—Part 2: Test Suite Structure and Test Purposes for 10–66 GHz WirelessMAN-SC Air Interface.

[B23] IEEE Std 802.16/Conformance03-2004, IEEE Standard for Conformance to IEEE Std 802.16—Part 3: Radio Conformance Tests (RCT) for 10–66 GHz WirelessMAN-SC Air Interface.

[B24] IEEE Std 802.16e-2005, IEEE Standard for Local and Metropolitan Area Networks—Part 16: Air Interface for Fixed and Mobile Broadband Wireless Access Systems—Amendment 2—Physical and Medium Access Control Layers for Combined Fixed and Mobile Operation in Licensed Bands.

[B25] IEEE Std 802.16f-2005, IEEE Standard for Local and Metropolitan Area Networks—Part 16: Air Interface for Fixed Broadband Wireless Access Systems—Amendment 1—Management Information Base.

[B26] IEEE Std 802.16.2-2004, IEEE Recommended Practice for Local and metropolitan area networks—Coexistence of Fixed Broadband Wireless Access Systems.

[B27] Ishimaru, A., *Electromagnetic Waves, Propagation, Radiation and Scattering*, Prentice Hall, 1991.

[B28] ITU Internet Report, "The Birth of Broadband," ITU Sales Service <http://www.itu.int/osg/spu>.

[B29] ITU-T Recommendation X.200, Information technology—Open Systems Interconnection—Basic Reference Model: The basic model, 1994.

[B30] ITU-R Recommendation M.1225, "Guidelines for valuation of radio transmission technologies for IMT-2000," 1997.

[B31] Jeon, W. S., and Jeong, D., "Improved Selective Repeat ARQ Scheme for Mobile Multimedia Communications," *IEEE Communications Letters*, vol. 4, no. 2, 2000.

[B32] Johnson, D., and Dudgeon, D., *Array Signal Processing*, Prentice Hall, 1993.

[B33] Keshav, S., *An Engineering Approach to Computer Networking*, Addison-Wesley, 1997.

[B34] Kwon, T., Lee, H., Choi, S., Kim, J., Cho, D.-H., Cho, S., Yun, S., Park, W.-H., and Kim, K., "Design and implementation of a simulator based on a cross-layer protocol between MAC and PHY layers in a WiBro Compatible IEEE 802.16e OFDMA system", *IEEE Communications Magazine*, pp. 136–146, Dec. 2005.

[B35] Lee, R., Kim, J., Yu, J., and Lee, J., "Capacity analysis considering channel resource overhead for wireless mobile Internet system," *Proceedings of IEEE Vehicular Technology Conference 2005-Spring*, May 2005.

[B36] Li, H., Lindskog, J., Malmgren, G., Miklos, G., Nilsson, F., and Rydnell, G., "Automatic Repeat Request (ARQ) Mechanism in HIPERLAN/2," *The IEEE Semiannual Vehicular technology Conference - VTC2000*, Spring 2000.

[B37] Liberti, J. C., and Rappaport, T. S., *Smart Antenna for Wireless Communications*, Prentice Hall, 1999.

[B38] Marks, R. B., "IEEE Standard 802.16 for Global Broadband Wireless Access," ITU Telecom World 2003, Geneva, Oct. 12–18, 2003.

[B39] Motorola, "Evaluation Methods for High Speed Downlink Packet Access (HSDPA)," TSG-R1 document, TSGR#14(00)0909, 2000.

[B40] NIST 800-38C, Recommendation for Block Cipher Modes of Operation: the CCM Mode for Authentication and Confidentiality, May 2004.

[B41] Pang, Q., Bigloo, A., Leung, V. C., and Scholefield, C., "Performance Evaluation of Retransmission Mechanisms in GPRS Networks," *IEEE WCNC'00*, Sept. 2000.

[B42] Partridge, C., *Gigabit Networking*, Addison-Wesley, 1993.

[B43] Rosen, E., Viswanathan, A., and Callon, R., "Multiprotocol Label Switching Architecture," *IETF RFC 3031*, Jan. 2001.

[B44] Sandhu, S., and Paulraj, A., "Space-time block codes: a capacity perspective," *IEEE Communication Letters,* vol 4, pp. 384–386, 2000.

[B45] Schneier, B., *Applied Cryptography: Protocol, Algorithms and Source code in C*, 2d ed., John Wiley & Sons, 1994.

[B46] SCTE DSS 00-05, Radio Frequency Interface Specification, v.1.1, *DOCSIS*, July 2000.

[B47] Seybold, J. S., *Introduction to RF Propagation*, John Wiley & Sons, *Inc.,* 2005.

[B48] Shin, S., Kang, C., Kim, J.-C., and Oh, S.-H., "The Performance Comparison between WiBro and HSDPA," *2nd International Symposium on Wireless Communication System*s, pp. 346–350, Sept. 2005.

[B49] Sklar, B., "Rayleigh Fading Channels in Mobile Digital Communication Systems: Part I: Characterization," *IEEE Communications Magazine*, Sept. 1997.

[B50] Stallings, W., *Data & Computer Communications*, Prentice Hall, 2000.

[B51] Stallings, W., *Wireless Communications and Networks*, Prentice Hall, 2002.

[B52] Stanwood, K., and Piggin, P., "WiMAX and Mesh Networking in the Home," *WiMAX and Mesh Networks Forum,* IEE, Savoy Place, London, June 2005.

[B53] Stevenson, C. R., Cardeiro, C., Sofer, E., Chouinard, G., "Functional Requirements for the IEEE 802.22 WRAN Standard," *IEEE 802.22 Working Group Document*, IEEE 802.22-05/0007r45, Mar. 2006.

[B54] Tanenbaum, A. S., *Computer Networks*, Prentice Hall, 1996.

[B55] Van Nee, R., and Prasad, R., *OFDM for Wireless Multimedia Communications*, Artech House, 2000.

Index

NOTE—When phrases have been abbreviated in the text of this book, the abbreviations have generally been used when creating this index. See the Acronyms and Abbreviations list at the beginning of the book for the complete phrase for each abbreviation.

A

AAS, 45–46, 95, 311–320
AAS subchannelization zone, 274
ABR, 183
ACID, 216, 218
acknowledgments, 64–65
active networks, 237
adaptive CAC, 187–189
adaptive modulation and coding, 293
adaptive PHY, interactions with QoS and CAC, 181–189
adjacent subcarrier permutation, 293
admission control and provisioning (QoS function), 59
AES-CCM, 96, 221–224
aggregate request, 191
AISN, 216
AK management, 224
allocation start time, 161
AMC, 293
angle diversity of antennas, 45
antenna
    adaptive systems of, 45
    basic concepts of, 40–48
    design, 47
    directional and sectorized, 42
    diversity, 42–45
    grouping method, 326
    impact on protocol design, 47
    MIMO, 46
    parameters, 41

selection, 326
ARQ
    basic concepts of, 64–66, 90–92, 97
    blocks, 127, 137–139
    BSN, 128
    cumulative acknowledgments, 141
    feedback, 64, 92, 125–126, 134, 143, 210, 213
    fragmentation and, 129–131
    GBN retransmission policy, 65, 91
    immediate acknowledgment, 64
    interaction with scheduler, 215
    packing and, 134
    parameters and timers, 203–205
    piggybacked payload, 125, 134
    protocol messages, 206
    receiver, 210–213
    reset and resynchronization, 213–215
    selective acknowledgments, 141
    SR retransmission policy, 65, 68, 86, 91, 204
    stop-and-wait retransmission policy, 65, 90, 93, 144, 218, 367–368
    transmitter, 207–210
    WirelessMAN-SCa, 260
ARQ Feedback IE, 139
ATDD, 51, 62, 84, 87
ATM CS, 99–101
ATS, 359
authorization, 152

B

bandwidth
    allocation, 247
    antenna, 41
    channel, 95
    determining available, 181
    on demand, 182–186
BE service, 97, 178, 183, 185–186, 196–198
beam steering adaptive antenna, 46
beam width of antenna, 41

---

bit ordering, 98
block-based retransmission, 91–92, 137
BPSK, 260
broadcast polling, 199
BS, 71–73
BSN comparison, 207
BTC, 256, 260, 279, 302–303
burst profile, 75, 84, 167–170, 248,
    331–338
BW request
    AAS, 318
    CDMA, 96, 292
    contention-based CDMA, 200
    focused contention, 96, 200, 284
    full contention, 284
    header, 96, 114
    piggyback, 97, 111, 122, 192, 198
    WirelessMAN-OFDM, 284
    WirelessMAN-OFDMA, 306–308
BW request/grant mechanism, 190–193
byte-based retransmission, 91

C
CAC, 174, 181–189
CBC, 89, 220
CBR, 183
CCSA, 379
CDMA, 55, 96, 167, 200, 292, 306–308,
    348
channel bandwidth, 95
channel descriptors and MAPs, 156–163
channel encoding
    WirelessMAN-OFDM, 277–282
    WirelessMAN-OFDMA, 302–305
channel state information for AAS, 317
channels, frequency, 30, 50–52
chip rate, 55
CID
    AAS Initial Ranging, 111, 113
    Basic, 108, 110, 113, 145–147, 149,
        316
    basic concepts of, 70, 107–113, 173
    Broadcast, 112–113, 146, 199
    CSs and, 99, 102
    Fragmentable Broadcast, 112–113

header field, 115–117
Initial Ranging, 108–109, 113
mesh, 234
Multicast Polling, 111, 146
Padding, 111, 113
Primary Management, 110, 113, 149
Secondary Management, 110, 113
Transport, 110, 113
classification, 97, 102–104, 173
classifiers, 73
clock comparison, 201–203
clock skew, 192
closed-loop transmit diversity, 324–326
coarse synchronization, 237
coarse-grain QoS, 61
coexistence, 27, 32, 314
Compact MAP IE, 144
concatenation, 135
constant average power, 256
constant peak power, 256
continuous PHY, 85
control mechanisms
    WirelessMAN-OFDM, 282–286
    WirelessMAN-OFDMA, 306–309
    WirelessMAN-SC, 259
control plane, 69
CPE, 71
CPS, 69
CRC, 96, 136
CS, 69–70, 97
CSMA, 47, 52, 366
CSMA/CA, 86
CTC, 260, 280, 302–303
cyclic prefix, 38, 94, 263, 270, 307, 348

D
DAMA, 247
data/control plane, 68
DCD, 75, 78, 144, 163
delay, 58
DES, 89, 96, 220
DFS, 31, 96
differentiated services, 69
directed mesh, 88, 244
discarded state, 209

DIUC, 75, 156, 163, 199, 252, 273, 335
diversity of antennas, 42–45
DL and UL framing for AAS, 314
DL burst, 78
DL channel encoding, 252–256
DL MAP, 75, 77–78, 248
DL subchannelization zone, 274
DL subframe, 77–78
DL synchronization for AAS, 317
Doppler frequency, 40
DSA, 154, 171
DSC, 172
DSCP, 69
dual diagonal staircase construction, 304
duplexing, 50–52, 74–83, 95
dynamic fragmentation, 62, 129
dynamic service, 154, 171

E

EGC, 44
encryption, 96, 219–224
equalization, 39
equalizer, 38–39, 49, 93–94
error correction methods. See ARQ, FEC
ertPS, 346
Ethernet CS, 70
ETSI's BRAN activity, 267, 358, 375–378

F

fading, 39–40
fast-feedback control mechanism, 306, 323, 326
fast-receive diversity, 44
FDD
    basic concepts of, 50–52, 74, 158–162, 194
    framing, 78–80, 250
    half-duplex, 158
    WirelessMAN-SC, 247, 250
FDE, 93
FDM, 52
FDMA, 53
FEC
    basic concepts of, 48
    WirelessMAN-OFDM, 278–281

WirelessMAN-OFDMA, 302
WirelessMAN-SC, 252–258
WirelessMAN-SCa, 260
feedback, ARQ, 64, 92, 125–126, 134, 210, 213
FFSH, 118, 124, 127, 383
FFT, 95, 262–264
FHDC, 323
fine-grain QoS, 61
fixed access and WiMAX Forum, 362
fixed TDD, 84
fixed wireless networks, 33
fixed-length SDU and packing, 131
focused contention BW request, 96, 200, 284
fragmentation, 56, 62–64, 87, 97, 129–132
fragment-based retransmission, 91
frame duration, 95
framed PHY, 85
framing, 74–83
frequency bands, 30–33
frequency diversity of antennas, 43
Fresnel zone, 36
FSDD, 50
FSH, 118, 120, 126–127, 383
full contention BW request, 284
full-duplex. See duplexing
FUSC, 292, 294

G

gain, antenna, 41
GBN retransmission policy, 65, 91
GFR, 183
GMSH, 97, 118, 122, 127, 383
granularity, 60, 76, 248, 264, 293, 303
guard interval, 95

H

half-duplex. See duplexing
HARQ
    acknowledgments, 147, 216
    Compact MAP IE, 144
    overview, 92–93, 97, 143
    packet construction, 145
    parameters, 216

performance and QoS implications, 217
RCID, 145—147
subpacket transmission, 216
WirelessMAN-OFDMA, 309
HARQ Control IE, 144, 216
head-of-line blocking problem, 63
H-FDD, 50—51, 79—80, 95, 250—251
hidden nodes, 368
HiperACCESS, ETSI BRAN, 358, 360, 376
HiperMAN, ETSI BRAN, 267, 361, 377,
381
HSDPA, 348

I

IEEE 802.20 Working Group, 373
IEEE P802.22 Working Group, 374
IEEE Std 802.11, 365—373
incremental request, 192
initial ranging, 149—150, 283, 317
interleaving, 85, 281, 305
IP connectivity, 153
IP CS, 70
ISI, 39, 49
ISO/IEC JTC1 and IEEE Std 802.16, 379
ITU and IEEE Std 802.16, 379—381
IUC, 84, 278

J

jitter, 58, 179

K

key, security, 152, 224
Korean TTA, 378

L

latency, 58
LDPCC, 303—305
licensed spectrum, 9, 30
license-exempt spectrum, 9, 30
light licensing, 31
logical mesh, 232—244
LOS, 36, 38
loss, 58

M

MAC
address basics, 107
connections, 383—386
CPS, 70
CS, 70
efficiency, 85—87
headers, subheaders, and special
payloads, 113—124, 383
management, 136
management messages, 316, 383—386
service and control functions for AAS,
316
management plane, 68, 71
mandatory IEEE 802.16 components, 94—97
MAPs and channel descriptors, 156—163
master-slave relationship, 71
maximum latency, 179
maximum sustained traffic rate, 176
maximum traffic burst, 177
MBS, 184
medium access overhead, 86
mesh
antennas, 42
centralized scheduling, 241—244
connections and addressing, 234
directed, 88, 244
distributed scheduling, 239
logical, 232—244
network configuration, 234—236
network entry, 236—239
overview, 35, 87, 97, 229—232
WirelessMAN-OFDM, 275—277
midamble, 271
MIMO, 46, 323, 355
minimum reserved traffic rate, 177
minimum tolerable traffic rate, 177
mini-slots, 76
mobile broadband wireless access, 363
mobile networks, 33
modulation, 39, 96, 281, 305
mother sequences, 270
MPDU, 70, 86, 107, 116, 125—126
MPDU header, 86—87, 115
MPLS, 101

MRC, 44, 46
MSDU, 70
MSH, 118, 123, 127, 383
multicast connections, 189
multicast polling group, 200
multihop networks, 35
multipath
    basic concepts of, 38
    delay spread, 39
    fading, 38
    mitigation, 94
    propagation, 38
multiple access, 52
multiplexing, 52

N

near-far problem, 368
NLOS, 36, 38
nomadic service, 34
not-sent state, 208
nrtPS, 97, 178, 183, 185–186, 196–198
nrtVBR, 183

O

OFDM. See WirelessMAN-OFDM
OFDMA. See WirelessMAN-OFDMA
O-FUSC, 293
open-loop transmit diversity, 321–324
optimum combining antenna, 46
optional IEEE 802.16 components, 94–97
O-PUS, 293
outstanding state, 209

P

packing, 56, 63, 87, 97, 101, 121, 131–134
PAPR, 49, 277, 308
path loss, 39
PCR, 184
PCS, 101–105
PDU, 56
per-class QoS, 60
per-flow QoS, 60
periodic ranging, 163–167
permutations for WirelessMAN-OFDMA,
    291–299

PHS, 70, 97, 104–105
PHSI field, 101
PHY
    maintenance, 155–170
    overview, 93–94
PICS, 27
PICS proforma, 27, 359, 362
piggyback BW request, 97, 111, 122, 192,
    198
PKM, 89, 152, 219
PM bit, 123, 193
PMP network, 35, 42, 97, 272, 300
polarization diversity of antennas, 44
polarization, antenna, 41
policing, shaping and (QoS function), 60
polling
    broadcast, 199
    multicast, 200
    unicast, 197–199
portable networks, 34
power control, 285
preamble
    WirelessMAN-OFDM, 270–272
    WirelessMAN-OFDMA, 301
precoding, 325
provisioning, admission control and (QoS
    function), 59
PS, 76
PSH, 91, 118, 121, 126, 383
PtP and directed mesh, 244
PtP network, 34, 42
PUSC, 292, 296–299

Q

QAM, 76
QoS
    basic concepts of, 57–62, 88–89
    DOCSIS model, 172
    fragmentation and, 63
    interactions with CAC and adaptive
        PHY, 181–189
    model, 172
    parameters, 173–181, 184
    per-class/flow, 60
    service flows and, 171–201

wireless, 61–62
QoS parameter set type, 174–176

R

radiation pattern, antenna, 41
randomization, 277, 302
ranging
    CDMA, 306–308
    channel, 306
    initial, 149–150, 283, 317
    periodic, 163–167
    WirelessMAN-OFDM, 282–284
RCID, 145–147
RCT, 27, 359
reassembly, 97
receiver-initiated reset, 214
registration, 153
request/transmission policy, 179
retransmission, 65, 86, 91
RS-CC, 278
RTG, 77, 156
rtPS, 97, 178, 183, 185–186, 196–198,
    346
rtVBR, 183

S

SA, 152, 219–224
SAP, 56
SC. See WirelessMAN-SC
SCa. See WirelessMAN-SCa
scheduling
    centralized, 241–244
    distributed, 239
    generally, 193–197
SCR, 184
SDU, 56, 131–132, 180
security, 62
    AES-CCM, 221–224
    encryption, 219–224
    key management, 224–227
    sublayer, 69–70, 89–90, 97
See also DIUC, UIUC
service class name, 173
service flow
    basic concepts of, 109

dynamic, 96
encodings, 173
QoS and, 171–201
scheduling type, 178
SFID, 70, 109, 173
shadowing, 40
shaping and policing (QoS function), 60
shift index matrix, 304
SI bit, 123, 192, 196
single-hop networks, 35
SM, 321, 323
small-scale fading, 38
SNR, 44
spatial diversity of antennas, 43–44
SPID, 92, 216
sponsor channel, 238
sponsor node, 243
spreading codes, 55
SR retransmission policy, 65, 68, 86, 91,
    204
SS, 71–73
SS basic capability negotiation, 150–152
SSRTG, 79
SSTTG, 79
static provisioning, 59
STBC, 321
STC, 95, 321–324
STC subchannelization zone, 274
stop-and-wait retransmission policy, 65, 90,
    93, 144, 218, 367–368
STTD, 321
subchannelization, 95, 266–270
subframes, 76, 78
subheader
    fast-feedback allocation, 124
    fragmentation, 120
    grant management, 122
    MAC, 119, 383
    mesh, 123
    ordering of, 126
    packing, 121
sublayers, 69
switched beam adaptive antenna, 45

T

tail-biting method, 278
TDD
    basic concepts of, 50–52, 74, 157,
      161, 194
    framing, 76–78, 248–250
    WirelessMAN-SC, 247–250
TDM, 52, 78, 86, 157, 248
TDMA, 53, 78, 86, 247
TEK, 153, 226
temporal diversity of antennas, 43
throughput, 58
time relevance, 81–83, 161–162
time-frequency mapping, 290–294
TLV-encoded information, 136
tolerated jitter, 179
Tproc, 157, 162
traffic
    classification (QoS function), 59
    parameters, 173–181, 184
    priority, 176
    scheduling (QoS function), 60
transmission CS, 135
transmitter-initiated reset, 213
TSS&TP, 27, 359
TTG, 77, 156, 250
turbo code, 96
TUSC1 and TUSC2, 294

U

UBR, 183
UCD, 75, 78, 144, 163, 257
UGS, 97, 123, 178, 180, 183, 185, 187,
      192, 194–197, 346
UIUC, 75, 156, 163, 169, 252, 257, 335
UL burst, 78
UL channel encoding, 256–259
UL MAP, 76–78, 150
UL subframe, 78
unicast polling, 197–199
unlicensed spectrum, 30
unpacking, 97
unsolicited grant, 164

V

variable-length SDU and packing, 132
VCI, 99–101
vendor-specific QoS parameters, 178
VPI, 99–101

W

waiting-for-retransmission state, 209
waveform
    WirelessMAN-OFDM, 261–272
    WirelessMAN-OFDMA, 288
W-CDMA, 321
WiBro, 378
WiMAX Forum, 357–363
wireless QoS, 61–62
WirelessMAN-OFDM
    AAS, 319
    BW requests, 284
    channel encoding, 277–282
    conformance, 361
    control mechanisms, 282–286
    FEC, 278–281
    FFT size, 262–264
    fixed operation, 327–338
    frame structure, 272–277
    interleaving, 281
    mesh, 275–277
    mobile operation, 338–342
    modulation, 281
    overview, 54
    PMP network, 272
    power control, 285
    preambles, 270–272
    ranging, 282–284
    STC, 322
    subchannelization, 266–270
    waveform, 261–272
WirelessMAN-OFDMA
    AAS, 320
    BTC, 303
    CC, 303
    channel encoding, 302–305
    conformance, 361
    control mechanisms, 306–309
    CTC, 303

FEC, 302
FHDC, 323
frame structure, 300–302
interleaving, 305
LDPCC, 303–305
MIMO, 323
mobile operation, 343–356
modulation, 305
overview, 53, 287
permutations, 291–299
PMP network, 300
preambles, 301

STC, 323
time-frequency mapping, 290–294
waveform, 288
WirelessMAN-SC
conformance, 360–361
control mechanisms, 259
DL channel encoding, 252–256
frame structure, 248–252
overview, 93, 247
profiles, 358
UL channel encoding, 256–259
WirelessMAN-SCa, 93, 260

Additional Books in the
IEEE Standards Wireless Networks Series

This is the second edition of Bob O'Hara and Al Petrick's *IEEE 802.11 Handbook*. It has been referred to as the "LAN Bible". The new edition covers IEEE 802.11d, 802.11e, 802.11F, 802.11g, 802.11h, 802.11i, 802.11j, 802.11n, and more! The book is 365 pages, includes a Preface written by Stuart J. Kerry, Chairman of 802.11, a comprehensive list of abbreviations and acronyms, and handy index.

This handy pocket guide provides over 250 definitions for the jargon used in today's wireless industry. Written by Dr. James P.K. Gilb, Technical Editor of IEEE 802.15™, the dictionary is the perfect companion to the other handbooks in our Standards Wireless Networks Series and to the IEEE 802 wireless standards. Individuals who need to have a broad view of the wireless landscape will find that this book covers the most important topics and terms in today's market.*

Wireless Communications Standards: A Study of IEEE 802.11™, 802.15™, and 802.16™ is the only IEEE book of its kind that covers all of the current 802 wireless standards.*

*The authors of these books are also teaching courses based on IEEE wireless standards. For a deeper understanding of IEEE 802.16™, consider attending DoceoTech's instructor-led course called, "WiMAX and 802.16: Broadband Wireless Access." For more information, see http://www.doceotech.com.

http://standards.ieee.org/standardspress/

More Titles in the
IEEE Standards Wireless Networks Series

Wireless Multimedia: A Handbook to the IEEE 802.15.3™ Standard clarifies the IEEE 802.15.3 standard for individuals who are implementing compliant devices and shows how the standard can be used to develop wireless multimedia applications.*

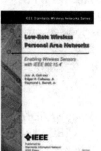

Low-Rate Wireless Personal Area Networks: Enabling Wireless Sensors with IEEE 802.15.4™ is an excellent companion to the standard for those interested in the field of "simple" wireless connectivity with a further focus on wireless sensors and actuators for the industry in general.

Please visit
http://standards.ieee.org/standardspress/

- Learn more about our products and authors
- Sign up to peer review future publications
- Submit a proposal and become an IEEE Author yourself!